Using Autodesk
Inventor™

Using Autodesk
Inventor™

Ron K. C. Cheng
Industrial Center
The Hong Kong Polytechnic University

autodesk
press

Australia • Canada • Mexico • Singapore • Spain • United Kingdom • United States

Using Autodesk Inventor™
Ron K. C. Cheng

Autodesk Press Staff

Business Unit Director:
Alar Elken

Executive Editor:
Sandy Clark

Acquisitions Editor:
James DeVoe

Developmental Editor:
John Fisher

Editorial Assistant:
Jasmine Hartman

Executive Marketing Manager:
Maura Theriault

Marketing Coordinator:
Paula Collins

Executive Production Manager:
Mary Ellen Black

Production Manager:
Larry Main

Production Editor:
Stacy Masucci

Art/Design Coordinator:
Mary Beth Vought

Cover Image:
Brucie Rosch

Library of Congress Cataloging-in-Publication Data
Cheng, Ron.
    Using Autodesk Inventor /
Ron K. C. Cheng.
        p. cm.
    ISBN 0-7668-2867-0
    1. Engineering graphics. 2. Engineering models–Data processing. 3. Autodesk Inventor I. Title.

    T353.C5195 2000
    620'.0042'02856693—dc21
                    00-045343

Notice To The Reader

The publisher and author do not warrant or guarantee any of the products described herein or perform any independent analysis in connection with any of the product information contained herein. The publisher and author do not assume and expressly disclaim any obligation to obtain and include information other than that provided to it by the manufacturer.

The reader is expressly warned to consider and adopt all safety precautions that might be indicated by the activities described herein and to avoid all potential hazards. By following the instructions contained herein, the reader willingly assumes all risks in connection with such instructions.

The publisher and author make no representations or warranties of any kind, including but not limited to, the warranties of fitness for particular purpose or merchantability, nor are any such representations implied with respect to the material set forth herein, and the publisher and author take no responsibility with respect to such material. The publisher and author shall not be liable for any special, consequential, or exemplary damages resulting, in whole or in part, from the readers' use of, or reliance upon, this material.

Trademarks

Autodesk, the Autodesk logo, and AutoCAD are registered trademarks of Autodesk, Inc., in the USA and other countries. Thomson Learning is a trademark used under license. Online Companion is a trademark and Autodesk Press is an imprint of Thomson Learning. Thomson Learning uses "Autodesk Press" with permission from Autodesk, Inc., for certain purposes. All other trademarks, and/or product names are used solely for identification and belong to their respective holders.

CONTENTS

PREFACE

Because computer and computer-aided design technology are rapid evolving, you can now use advanced 3D solid modeling systems throughout your design on your personal computer. Autodesk Inventor™ is mechanical engineering design software, an assembly-centric 3D solid modeling system for constructing 3D parametric solid parts, assemblies of solid parts, presentation of assemblies, and engineering drawings of solid parts, assemblies, and presentations. It uses four kinds of data files: you construct 3D solid parts in solid part files, assemblies of solid parts in assembly files, exploded view and animation of exploded view in presentation files, and engineering drawings of solid parts, assemblies, and presentations in drawing files.

Normally there are two stages in constructing a 3D solid part: analysis and synthesis. First, you think about how you will break the 3D object into simple 3D solid features and combine the features to form the 3D object. Then you think about how to construct the features and recombine them accordingly. Basically, there are two kinds of solid features: sketched and placed. Sketched solid features derive from sketches; you extrude, revolve, sweep, or loft the sketches to form sketched features. Placed solid features are pre-constructed; you select a placed solid feature from the menu and specify the location and parameters. To maintain a proper relative position between features, you need work features: work planes, work axes, and work points.

An assembly is a collection of parts that are put together properly to serve a purpose. In the computer application, an assembly file links to a set of solid part files. There are three approaches to constructing an assembly. In the first approach, you design and construct all the solid parts; then you start an assembly file and place the solid parts in the assembly. In the second approach, you start an assembly file; then you design and create the solid parts while you are working in the assembly environment, where you can perceive, visualize, and use the dimensions and features of the existing solid parts while designing new solid parts. The third approach is a hybrid approach; you design and construct some solid parts before constructing the assembly file. Then you put the solid parts together in the assembly and design other solid parts in the assembly environment. In the assembly, you can apply assembly constraints to align and mate the parts together. When an assembly is complete, you can output a bill of materials that can be linked to other databases. In order to explain how various components of an assembly are related to each other, you can construct presentation files that show the exploded view and animation of an assembly.

Apart from the general 3D solid modeling technology, Autodesk Inventor offers a number of advanced modeling technologies.

- **2D Layout Drawings** While modeling in 3D is a global trend in designing, proper deployment of 2D layout drawings is very useful in the initial design stage. You use 2D layout drawings to validate and evaluate a mechanism. You construct 3D solid parts based on the 2D layout drawings.

- **Adaptive Technology** By using adaptive technology, you can ensure that dimensions of corresponding parts in an assembly fit properly with each other by setting the dimensions of a feature of a solid part to adapt to the dimensions of the solid features of another solid part.

- **Spreadsheet Control** To control a set of dimensions in a solid part and across a number of solid parts, you can use a spreadsheet.

- **Design Catalogs** To reuse a solid feature in other designs, you can export a feature of a solid part to become an element of the design catalog and you import the element to other designs.

- **Workgroups** To take full advantage of the Internet and intranet in design, a team of designers can work together by using a common workgroup search path and library search path.

In general, constructing sheet metal components is similar to making other kinds of solid parts. However, there are some major differences. A sheet metal component is uniform is thickness throughout because a sheet metal component is constructed by the cutting and folding of sheet metal. When you fold sheet metal, you have to consider the bends and the seams. In thinking about the design of a 3D sheet metal component, you need to think about how to unfold the 3D object to a 2D flat pattern. To cope with the design of sheet metal, you use the special set of sheet metal application tools that take care of the bends, seams, and flat pattern.

In a modern digital factory, you transmit digital design data about the solid parts and assemblies of components to the downstream operating departments. However, there are occasions when 2D engineering drawings are necessary for the purpose of communication among the operators and designers. A drawing has two kinds of objects: drawing sheets and drawing views. You prepare a drawing sheet and the computer generates the orthographic drawing views on the drawing sheets. Then you add annotations and sketched geometry to complete the drawing.

ORGANIZATION OF THIS BOOK

This book contains six chapters, each chapter covering a major topic. Each chapter includes a summary and review questions. Exercises are included in Chapters 2 through 6, so you can practice the concepts learned in the chapter.

Chapter 1 outlines the concepts of parametric solid modeling, assembly modeling, and engineering drafting. It also provides a brief introduction to Inventor and familiarizes you with the design support system that helps you while you are designing.

Chapter 2 explains the key concepts of solid modeling. You will construct 3D solid parts using sketched solid features, placed solid features, and work features. You will also construct derived solid parts. Apart from learning how to construct solid parts, you will apply lighting to the environment, set material and color for the solid part, and export and import files.

Chapter 3 introduces the concepts of assembly modeling. You will construct assemblies of solid parts, exploded presentations of the assemblies, and a bill of materials for the assembly.

After learning the principles of solid modeling and assembly modeling in Chapters 2 and 3, you will learn various advanced modeling techniques in Chapter 4: 2D layout, adaptive technology, graphics slicing, design parameters, design elements, design notebook, relative motions in an assembly, and collaborative design.

Chapter 5 depicts the concepts of sheet metal modeling. You will construct 3D sheet metal parts and 2D flat patterns of 3D sheet metal parts, and you will convert 3D solid parts to 3D sheet metal parts.

Chapter 6 delineates the concepts of engineering drafting. You will output orthographic views from solid parts and orthographic and presentation views from assemblies of components.

Accompanying this book is a CD in which you will find all the data files for the tutorials in the book.

ACKNOWLEDGMENTS

This book would never have been realized without the contribution of many individuals.

A special thanks goes to the professionals who reviewed the manuscript in detail.

- Steven Brown, College of the Redwoods, Eureka, California
- John Clauson, Oakton Community College, Des Plaines, Illinois
- Scott Ertel, West Irondequoit High School, Rochester, New York

- David Rouch, Ohio Northern University, Ada, Ohio

- Mark Kurdi, Sheridan College, Brampton, Ontario, Canada

- Tom Singer, Sinclair Community College, Dayton, Ohio

Several people at Delmar Publishing Company also deserve special mention, particularly John Fisher, the developmental editor who worked closely with me on this book; Alar Elken, the publisher; Jasmine Hartman, the engineering editorial assistant; Stacy Masucci, the production editor; John Shanley of Phoenix Creative Graphics, the compositor; and Brucie Rosch, the cover designer.

Ron K. C. Cheng

CHAPTER 1

Introduction

OBJECTIVES

The aims of this chapter are to introduce the concept of parametric solid modeling, assembly modeling, and engineering drafting and to outline the key functions of Autodesk Inventor. The four kinds of Inventor data files are introduced, as are the Inventor user interface and the design support system. After studying this chapter, you should be able to

- Explain the principles of solid modeling, assembly modeling, and engineering drafting
- Describe the key functions of Autodesk Inventor
- List the four kinds of Inventor files
- Use the Inventor user interface and the design support system

OVERVIEW

There are three kinds of 3D models that can be created in a computer program: 3D wireframe model, 3D surface model, and 3D solid model. Of these, the 3D solid model is superior because it contains all the information regarding the edges, silhouettes, faces, and volume of the 3D object that it describes.

You use the parametric approach to design and develop products and systems. Based on a set of solid parts, you construct an assembly. In the assembly, you constrain the 3D parts to align or mate with each other in the 3D space. With a set of 3D parts and assemblies, you output 2D engineering drawings for conventional production methods and electronic data files for computerized manufacturing processes.

Autodesk Inventor is a 3D computer-aided design (CAD) application for making 3D parametric solid parts, assemblies and subassemblies of solid parts, presenting assemblies, creating engineering documents of 3D solids and assemblies and outputting electronic data files for downstream manufacturing operations. The application is user friendly and it provides a set of design support systems to help you design.

INTRODUCING 3D MODELS, ASSEMBLIES, AND ENGINEERING DRAFTING

Nowadays, computer-aided design tool is a prime means of communication and a medium for developing design ideas, products, and systems. Using the computer, you design the component parts and construct 3D models, assemble the 3D component parts as virtual assemblies, generate 2D projection drawings for making the parts and assemblies, and output the 3D models and assemblies in electronic format for down-stream computerized processes.

3D MODELS

To design and develop 3D objects in the computer, you use 3D models. Many years ago, when 3D was first introduced in computer applications, objects were represented as simple wireframe models. Over the years, computer-aided design applications have developed from simple electronic drawing boards to sophisticated 3D design and development tools.

The 3D wireframe model is the most primitive type of 3D object. It is a set of unassociated line and arc segments that are put together in the 3D space. The line and arc segments serve only to give the pattern of a 3D object; they are not related, one to another. As such, the model lacks any surface or volume information; it describes only the edges of the 3D object. Because of the scant information provided by the model, the use for a 3D wireframe model is limited.

The second type of 3D model, the surface model, is a set of surfaces that are put together in a 3D space to give the figure of a 3D object. When compared to a 3D wireframe model, a surface model has, in addition to edge data, information on the contour and silhouette of the surfaces. You use surface models to represent complicated free-form objects, which can be used in a computerized manufacturing system, or to generate photo-realistic rendering or animation.

With regard to information, a 3D solid model is superior to the other two kinds of models because it contains integrated mathematical data not only about the surfaces and edges but also about the volume of the object that the model describes. In addition to visualization and manufacturing, you use a solid model in more sophisticated design evaluation tasks.

3D Solids

There are many ways to construct a 3D solid in a computer. Among them, the most popular method is to construct 2D/3D wireframes and use the wireframes to construct 3D solid objects in four basic ways: by extruding, revolving, lofting, and sweeping.

To extrude, you construct a closed-loop 2D wireframe on a 3D plane and translate (extrude) the wire in a direction perpendicular to the 3D plane. The volume enclosed by the extrusion motion is the extruded solid.

To revolve, you construct a closed-loop 2D wireframe and a straight line on a 3D plane and revolve the closed-loop 2D wire about an axis defined by the straight line. The volume enclosed by the revolving motion is the revolved solid.

To loft, you construct a series of closed-loop 2D wireframes on a number of 3D planes and loft from one closed-loop 2D wireframe to the next closed-loop 2D wireframe. The volume enclosed by the lofting motion is the loft solid. In a loft solid, the cross section changes from the first closed-loop 2D wireframe to the next wireframe.

To sweep, you construct a closed-loop 2D wireframe on a 3D plane and an open-loop 3D wireframe and transit (sweep) the closed-loop 2D wireframe along the 3D wireframe. The volume enclosed by the sweeping motion is the sweep solid. In a sweep solid, the cross section of the sweep volume is constant along the sweeping path.

In addition to these four basic kinds of solids, there are other kinds of pre-constructed solids that you can use in a 3D model. You will learn about them in Chapter 2.

Parametric and Non-parametric Approaches

To construct the wireframes for making the solids, there are two major approaches: parametric and non-parametric.

The conventional way of constructing wireframes in the computer is to specify the location, orientation, and dimension of the lines accurately and explicitly. Once the lines are produced and the 3D solids are constructed from the wireframes, it is difficult to redefine the lines and modify the solids. This is commonly known as the non-parametric approach.

In the parametric approach, you start constructing the wireframes by making free-hand sketches. During the initial sketching stage, the geometry and size of the lines need not be accurate and precise. You concentrate on form and shape. To develop a proper geometric shape in accordance with your design intent, you specify the geometric relationships between the line segments of the sketch by assigning geometric constraints such as horizontal, vertical, parallel, etc. After developing the geometric shape, you specify the size by assigning parametric dimensions to the sketch. While geometric constraints and parametric dimensions are added to the sketch, the shape and size of the sketch change accordingly. In Chapter 2, you will learn the ways to use the parametric approach to construct wireframes for making 3D solids.

With the parametric wireframes, you construct parametric 3D solids. Not only can you change a sketch before making a solid from it, but you can also alter the parametric dimensions and geometric shape of the cross section of the solid feature at a later stage. In fact, you change the parameters of the wireframes and the solids flexibly any time you like.

ASSEMBLIES

A product or a system usually has more than one component part. After you construct the individual component parts in 3D solids, you put them together in an assembly to depict the entire product or system.

Simply stated, an assembly is a collection of component parts put together to form a useful whole. However, making an assembly in a computer application goes far beyond translating solid parts together in 3D space. You have to specify a positional relationship between component parts by assigning assembly constraints to the selected features of a solid part in relation to the selected features of another solid part. Features that you can select for placing assembly constraints are the vertices, edges, and faces of a solid part.

It is a normal engineering practice to translate (explode) the components of the assembly to illustrate how the parts are assembled together. Through the computer application, you can illustrate and animate the assembly explosion. In Chapter 3, you will learn how to construct and present an assembly.

ENGINEERING DRAFTING

Conventionally, 3D objects are represented in a 2D drawing through projection views. The most common kind of projection method is the orthographic projection method, in which a 3D object is projected onto a 2D plane to obtain a drawing view. In the traditional approach to constructing an engineering drawing, you think about how an object would look when viewed in a certain direction. Then you construct the 2D drawings in accordance with your perception. We all know that this process is very time consuming and can be inaccurate and incomplete. With 3D solid models, the creation of 2D projection views is automatic. You select an object and specify a projection direction. The computer generates 2D views from the 3D objects. In Chapter 6, you will learn how to output engineering drawings.

EXPORTING AND IMPORTING

Constructing 3D solids, assembling 3D solids, and then outputting engineering drawings from the 3D solids and assemblies are not the end in the process of design and development. In fact, they are a part in the iterative design and manufacturing cycle. You need to analyze the models and the assembly for amendment and modification, construct rapid prototypes for further evaluation, produce rapid tooling for small batch production, and mass manufacture the objects for general sale. To enable and facilitate these activities, it is essential to use the electronic data of the solids and assembly. Because these activities will require different kinds of computer applications that may be written in different programming languages using different file formats, it is necessary to export the 3D solids and assemblies to the file formats that are compatible with those applications. The file formats for exporting are listed later in Table 1–1.

To reuse existing designs from 3D solids that you construct by using other computer-aided design applications, you can open ACIS and STEP files and Mechanical Desktop solids and Mechanical Desktop assemblies. You can edit imported ACIS and STEP solids by manipulating their faces. (See Appendix A.) With Mechanical Desktop R4 or later releases installed in your computer, you can open Mechanical Desktop solids and assemblies and import all their parametric information to Inventor. Therefore, you can edit them in the same way as you edit Inventor solids and assemblies.

Along with geometric data about the 3D solids, you can incorporate additional information to your part file; you can insert various kinds of objects in your Inventor part files. For details, please refer to Appendix B.

AUTODESK INVENTOR FUNCTIONS

Autodesk Inventor is a 3D parametric solid modeling tool. It enables you to design and construct 3D solid parts, construct an assembly of solid parts, export solids and assemblies to various file formats, and produce engineering drawings of solid parts and assemblies.

CONSTRUCTING SOLID PARTS

To design and construct a solid part, you start by making rough sketches that reflect your design intent. While making the sketch, you concentrate on form and shape. Then you refine the sketch by specifying geometric constraints and parametric dimensions. Using the sketches, you extrude, revolve, loft, or sweep to make a 3D object. Figure 1–1 shows the 3D solid parts of a food grinder that you will construct in Chapter 2. Through constructing these solid parts, you will learn various 3D solid modeling methods.

Figure 1–1   *Parts of a food grinder*

In Chapter 4, you will learn more advanced solid modeling techniques. In particular, you will learn the use of 2D layout drawings to design an assembly of a working mechanism, the use of adaptive technology to design mating parts, and the use of design tools that enable collaborative design among several designers. Figure 1–2 shows the parts and assembly of an oscillator that you will construct in Chapter 4.

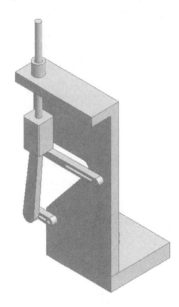

Figure 1–2 *Oscillator assembly*

Sheet metal parts are a special kind of solid part. You make a sheet metal component by cutting and folding a sheet of metal of uniform thickness. To meet the manufacturing requirement of providing rounded bends at the joints of faces and relieves at the bends, you need a special kind of solid modeling tool. Figure 1–3 shows a sheet metal component that you will construct in Chapter 5.

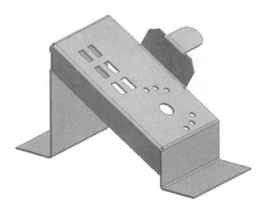

Figure 1–3 *Sheet metal component*

CONSTRUCTING ASSEMBLIES

There are three approaches to designing a device: bottom-up, top-down, and the hybrid approach. In the bottom-up approach, you construct all the solid parts of an assembly in individual part files and then link them together in an assembly file. In the top-down approach, you construct the part files while working in the assembly file. The hybrid approach is a combination of bottom-up and top-down approaches. No matter which approach you use, you place the component parts in a proper location and orientation relative to each other and maintain a proper alignment among the parts in an assembly by adding assembly constraints. Figure 1–4 shows the assembly of a food grinder that you will construct in Chapter 3.

Figure 1–4   *Assembly of the food grinder*

To illustrate how various parts of an assembly are put together, you construct an exploded view in a presentation file.

EXPORTING

To use the integrated data of a 3D solid model and an assembly of 3D solids in design analysis, rendering and visualization, or computerized manufacturing systems, you save the solid part and assembly to BMP, IGES, SAT, STEP, XGL, or ZGL format. To output a 3D solid part for rapid prototyping, you save the solid part to STL format. The file formats are described in Table 1–1.

Table 1-1   Export file formats

File Format	Description
BMP	Windows Bitmap (BMP) is a Windows image file format. The current screen display is saved to a 2D image file.
IGES	Initial Graphics Exchange Specification (IGES) format is an American standard established by the American National Standards Institute in 1979. Because it is a format for translating 3D wireframe lines and surfaces, an Inventor solid that you output in IGES format will exhibit as a surface model. Naturally, volume data will be lost.
SAT	Save As Text (SAT) is a file format of the ACIS object-oriented 3D geometric modeling engine by Spatial Technology Inc. It is a format for translating 3D lines, surfaces, and solids. ACIS supports two kinds of file formats: SAT and SAB. (SAB stands for Save As Binary.)
STEP	STandard for the Exchange of Product Model Data (STEP) is a product model data exchange standard that was initially developed by the International Organization for Standardization. Like SAT files, it is used for translating lines, surfaces, and solids.
STL	Stereolithography (STL) is a standard file format for use in most rapid prototyping machines. An STL file is a list of triangular surfaces that depict the 3D model. A 3D solid model saved in STL format is downgraded to a 3D model approximated by a set of triangular flat surfaces.
XGL	X Windows Graphics Library (XGL) is a standard file format that captures all the 3D information that can be rendered by the OpenGL rendering library.
ZGL	Compressed XG (ZGL) is a compressed XGL format that is about 10 times smaller than XGL files.

IMPORTING

There are two ways to import a solid part to Inventor:

- You can import a solid part that is saved in SAT or STEP format by opening the file. The SAT or STEP solid part becomes a base solid part in Inventor. Although there is no parametric information in the imported SAT or STEP solid, you can edit the model by manipulating the faces of the solid. (See Appendix A.)

- If you have installed Mechanical Desktop R4 in your computer and you have Mechanical Desktop parametric solids and assemblies, you can run Mechanical Desktop and then Inventor to open the DWG file. All the parametric information in the DWG file will be imported. The Mechanical Desktop parametric solid part becomes an Inventor parametric solid part file.

CONSTRUCTING ENGINEERING DRAWINGS

Engineering drafting is an engineering communication tool that depicts a 3D design in 2D engineering drawing views. You specify a solid part or an assembly and the computer application can automatically generate 2D orthographic views of 3D solid parts, sheet metal parts, and assemblies, flat patterns of sheet metal, and exploded views of assemblies. Figure 1–5 shows the engineering drawing of the food grinder that you will construct in Chapter 6.

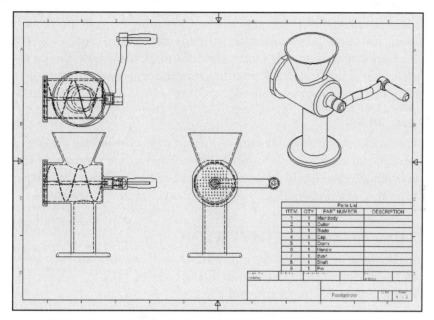

Figure 1–5   *Exploded engineering drawing of the food grinder*

INVENTOR FILE TYPES

Inventor uses four kinds of files:

- part files for constructing solid parts
- assembly files for assemblies of solid parts or sub-assemblies
- presentation files for exploded views of an assembly
- drawing files for engineering drawings

Text files are also used to organize the files into projects.

PART FILES

You construct a 3D solid part or a sheet metal part in a part file; the file extension is .ipt. A part file stores the definition of the parametric 3D solid part.

ASSEMBLY FILES

To construct an assembly or a sub-assembly, you use an assembly file; the file extension is .iam. An assembly file links to a set of parametric 3D solid parts and/or sub-assemblies of parametric 3D solid parts. It stores the information on how the component parts are assembled together. Information regarding the details of the parametric 3D solid parts is stored in the corresponding part files.

PRESENTATION FILES

To construct an exploded presentation of an assembly or animate the exploded presentation, you use a presentation file; the file extension is .ipn. A presentation file links to an assembly file. It stores the information on how the parts of the assembly are tweaked apart. Details regarding how the component parts are assembled are stored in the respective assembly file.

DRAWING FILES

To construct a 2D engineering drawing of a parametric 3D solid part, an assembly of 3D solid parts, and an exploded view of an assembly, you use a drawing file; the file extension is .idw. A drawing file links to a part file, an assembly file, or a presentation file. It stores the information about the 2D presentation of 3D objects.

STARTING AUTODESK INVENTOR

Now start Inventor by selecting Autodesk Inventor R3 icon from your desktop. In the What To Do panel of the QuickStart dialog box (see Figure 1–6), there are four icons: Getting Started, New, Open, and Projects.

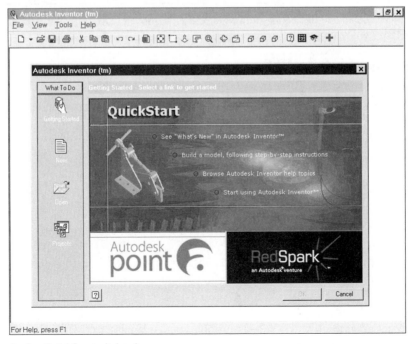

Figure 1–6 *QuickStart dialog box*

GETTING STARTED

By default, the Getting Started icon is selected, with four Getting Started topics:

- See "What's new" in Autodesk Inventor
- Build a model, following step-by-step instructions
- Browse Autodesk Inventor help topics
- Start using Autodesk Inventor

In addition to the four topics, there are two hyperlinks: Autodesk Point A and RedSpark.

- http://www.pointa.autodesk.com/
- http://www.redspark.com/

If your computer is already connected to the Internet, you can visit Autodesk Point A and RedSpark by selecting the hyperlinks.

See What's New

Now select See "What's new" in Autodesk Inventor. (See Figure 1–7.)

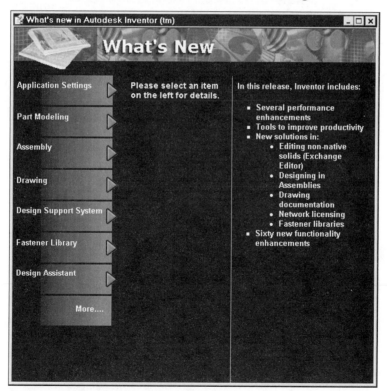

Figure 1–7 *What's New dialog box*

In the What's New dialog box, there are 11 topics: Application Settings, Part Modeling, Assembly, Drawing, Design Support System, Fastener Library, Design Assistant, Engineer's Notebook, DWG Support, Mechanical Desktop Support, and Others. If you are a novice or want to learn more about Autodesk Inventor Release 3, you should take some time to read these topics.

You can come back to this dialog box at a later stage while you work through the tutorials in this book by selecting Getting Started from the File menu.

Learn to Build a Model

To learn how to build a model, in the QuickStart dialog box, select Build a model, following step-by-step instructions from the Getting Started menu. This takes you to the Autodesk DesignProf dialog box (see Figure 1–8).

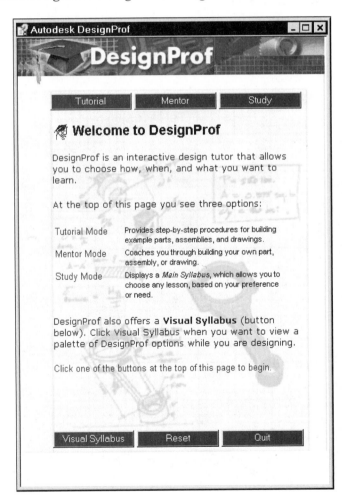

Figure 1–8 *DesignProf dialog box*

In the DesignProf dialog box, you work in tutorial, mentor, or study mode. Select Tutorial Mode. (See Figure 1–9.)

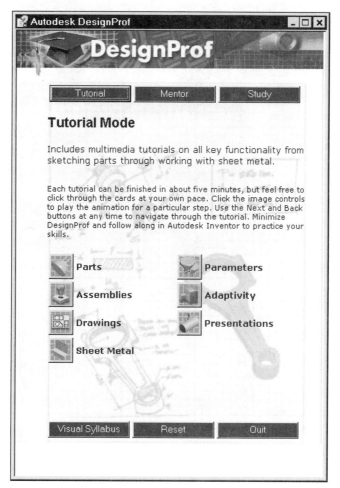

Figure 1–9   *Tutorial mode*

In tutorial mode, you learn about parts, assemblies, drawings, sheet metal, parameters, adaptivity, and presentation. Take a quick look at these tutorials and come back to them after you study this book. When you are done with the tutorials, select Quit and return to the QuickStart dialog box.

Browse for Help

To explore the help that is available, select the third topic, Browse Autodesk Inventor help topics, in the QuickStart dialog box. The Autodesk Inventor Help dialog box is displayed (see Figure 1–10). To access help later, select the Help Topic button on the Standard toolbar.

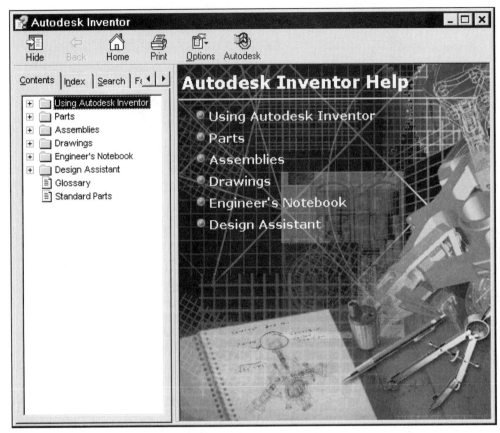

Figure 1–10 *Autodesk Inventor Help dialog box*

Start Using Inventor

To begin, select the fourth topic in the QuickStart dialog box, the Start Using Autodesk Inventor. It reminds you to select the New icon in the What To Do panel to start a new part file, assembly file, presentation file, or drawing file. (See Figure 1–11.)

In the Choose File panel, there are three tabs: Default, English, and Metric. In each of the tabs, there are four kinds of file template icons (ipt, iam, ipn, and idw). On the Default tab, the templates conform to a drafting system (English or Metric) that you configure when you install Inventor on the computer.

If you want to open an existing file, select Open in the What to Do panel. (See Figure 1–12.)

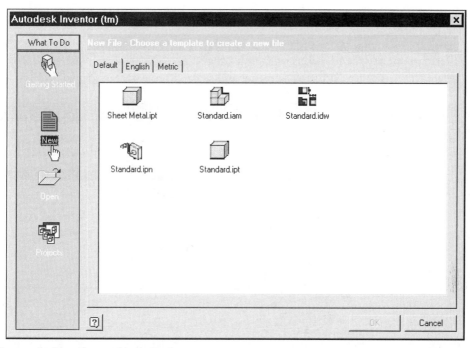

Figure 1–11 *Choose File dialog box*

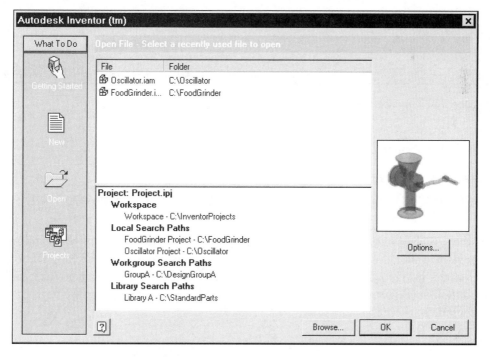

Figure 1–12 *Open File dialog box*

In the Open File panel, select a file to open.

Autodesk Inventor makes extensive use of "project files" to manage various files; they are text files that specify the locations of the files that make up a project. To select a project file, select Projects in the What to do panel and select a project. (See Figure 1–13.)

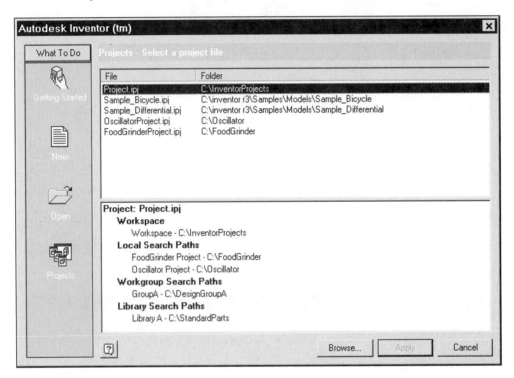

Figure 1–13 *Projects panel*

You will learn how to construct a project file in Chapter 2.

APPLICATION WINDOW

Now select the Cancel button. This brings you to the Inventor application window. (See Figure 1–14.)

In the application window, there are five major areas. At the top of the window, there is the standard Windows title bar which displays the name of the application. Below the title bar, is a set of pull-down menus, and below the menus is the Standard toolbar (Figure 1–15). Below the toolbar is a graphics area, and at the bottom of the window, there is a status bar.

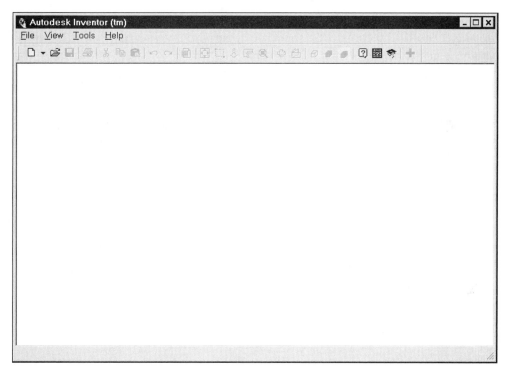

Figure 1–14 *Application window*

Figure 1–15 *Standard toolbar*

STARTING A NEW FILE

Having decided what to do (construct a part, assembly, presentation, or drawing file), you start a new file by selecting a template from the New dialog box. (See Figure 1–11.)

In the New dialog box, there are three tabs: Default, English, and Metric. In each tab, there are a number of template files. The English tab has English templates; the Metric tab has metric templates (including BSI, DIN, GB, ISO, and JIS standards); and the Default tab has templates configured for the default measurement system you selected when you installed Inventor.

Select the Default tab. You will find five template icons: Sheet Metal.ipt, Standard.iam, Standard.idw, Standard.ipn, and Standard.ipt.

There are two part file templates, Sheet Metal.ipt and Standard.ipt. Select the Sheet Metal.ipt template to construct a sheet metal part (you will learn about sheet metal parts in Chapter 5). Select the Standard.ipt template to construct a solid part (you will use this template in the Chapter 2.) Figure 1–16 shows the application window for constructing a solid part or a sheet metal part.

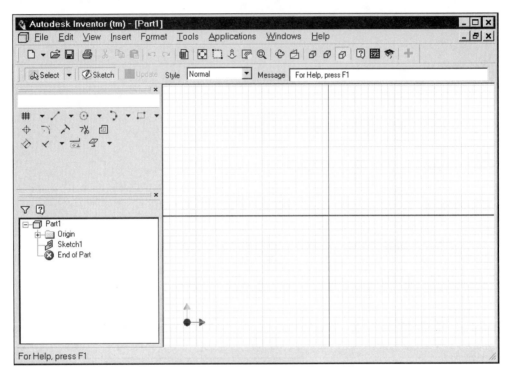

Figure 1–16 *Application window for a 3D solid part file*

Select the Standard.iam template to construct an assembly of parts (you will learn about assembly in Chapter 3). Figure 1–17 shows the application window for constructing an assembly.

Select the Standard.idw template to construct an engineering drawing (you will learn about engineering drawing in Chapter 6). Figure 1–18 shows the application window for constructing an engineering drawing.

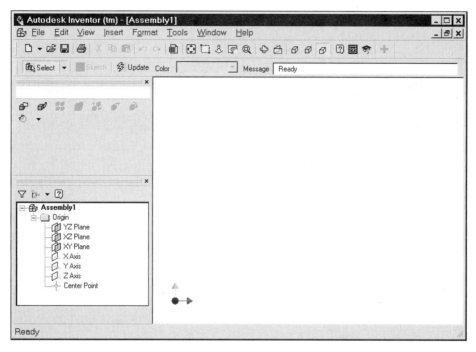

Figure 1–17 *Application window for an assembly file*

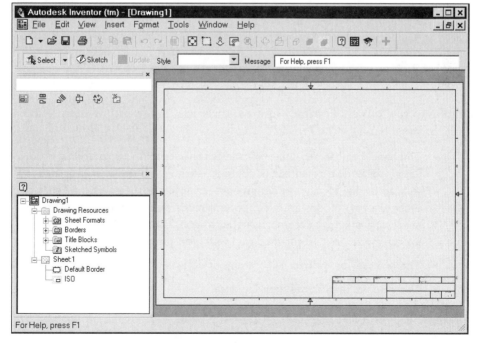

Figure 1–18 *Application window for a drawing file*

Select the Standard.ipn template to construct a presentation of assemblies (you will learn exploded presentation in Chapter 3). Figure 1–19 shows the application window for constructing a presentation of an assembly.

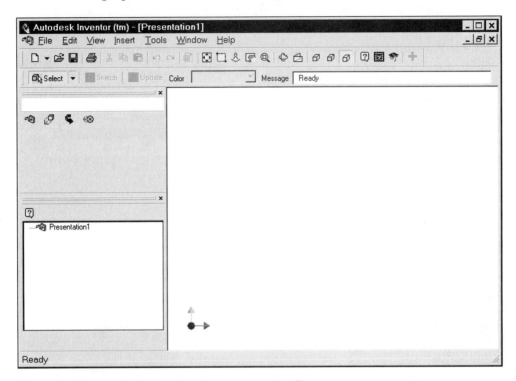

Figure 1–19 *Application window for a presentation file*

Common to the four application windows, there are three window areas: the panel bar, the browser bar, and the graphics area.

- The panel bar has a number of palettes that enable you to access various design tools. The panel bar is context sensitive, so the tools available will vary according to current design context. In each palette of the panel bar, there are two modes available: general mode and expert mode. In general mode, there is text accompanying each icon to depict the command. In expert mode, text is not displayed until you place the cursor on the icon.

- The browser bar shows a hierarchy of objects in the file.

- The graphics area is the working area.

To increase the graphics area, you can close the panel bar and the browser bar.

COMMAND SELECTION

There are several ways that you can select a command with the left mouse button.

- Select an item from the menu.
- Select an icon on the toolbar.
- Select an icon on the panel bar.
- Select an item from the shortcut menu.

SHORTCUT MENUS

Normally, your mouse has two buttons. You use the left button to select an object and use the right mouse button to activate a shortcut menu. Depending on the location of the mouse cursor and the kind of file you are working on, right-clicking the mouse will bring up different kinds of shortcut menus—they are context sensitive. You select an appropriate command from it. You will use the right-click button very often in the following chapters while you work through the tutorials.

DESIGN SUPPORT SYSTEM

Autodesk Inventor's design support system helps you use Inventor as a design tool. The design support system has seven components: Help Topics, Autodesk Online, What's New about Autodesk Inventor, Visual Syllabus, Design Professor, Design Doctor, and Sketch Doctor.

HELP TOPICS

The Autodesk Inventor Help dialog box, as shown in Figure 1–10, provides a comprehensive set of help topics on various aspects of Inventor. Apart from selecting Browse Autodesk Inventor help topics from the QuickStart dialog box shown in Figure 1–6, you can select the Help Topics button on the Standard toolbar. (See Figure 1–20.)

Figure 1–20 *Help button on the Standard toolbar*

AUTODESK ONLINE

At the top of the Help dialog box, there is a button that enables you to access Autodesk Online. (See Figure 1–21.) You can also select Autodesk Online from the Help menu.

Figure 1–21 *Autodesk Online dialog box*

In the Autodesk Online dialog box, there are six buttons: Autodesk Mechanical Solutions, Autodesk Inventor Home Page, Autodesk Inventor Support, Autodesk Inventor Update, Autodesk Point A, and RedSpark. Because these buttons bring you to the appropriate Web sites, you need to have your computer connected to the Internet before you use Autodesk Online.

WHAT'S NEW ABOUT AUTODESK INVENTOR

Access What's New about Autodesk Inventor from the Getting Started menu in the QuickStart dialog box shown in Figure 1–6 or by selecting What's New from the Help menu. Here you can discover all the new features of Inventor.

VISUAL SYLLABUS

The Visual Syllabus is a collection of lessons. To access these lessons, select the Visual Syllabus icon on the Standard toolbar. (See Figure 1–22.)

Figure 1–22 *Visual Syllabus button on the Standard toolbar*

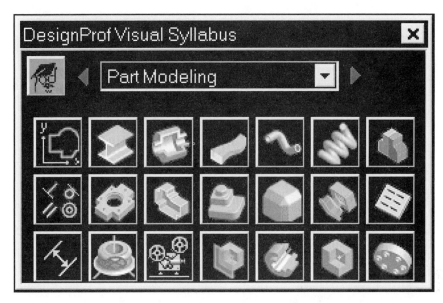

Figure 1–23 *DesignProf Visual Syllabus dialog box*

The Visual Syllabus lessons are organized in four categories: Part Modeling, Sheet Metal, Assembly Modeling, and Drawing. (See Figure 1–23.) You may find it helpful to spend time reading through the lessons before you start working on the remaining chapters of this book.

DESIGN PROFESSOR

As explained earlier, you can access DesignProf in the QuickStart dialog box, by clicking on the Getting Started icon and then selecting Build a model, following step-by-step instructions. In addition, you can select DesignProf (see Figure 1–24) on the Standard toolbar.

Figure 1–24 *DesignProf button on the Standard toolbar*

DESIGN DOCTOR

Unlike the other support tools, the Design Doctor is a context-sensitive diagnostic tool. It is available only when you encounter some error or problem. Figure 1–25 shows a problem encountered while a sketch is being extruded. Display the Design Doctor by selecting the Examine Profile Problem button. (See Figure 1–26.)

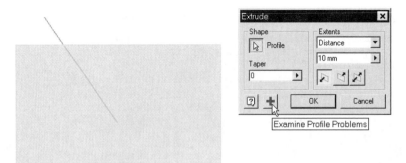

Figure 1–25 *Problem encountered*

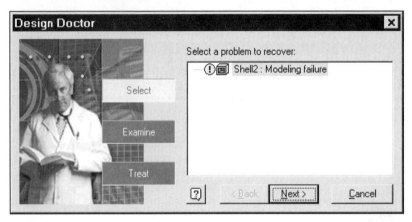

Figure 1–26 *Design Doctor dialog box*

SKETCH DOCTOR

Very similar to the Design Doctor, the Sketch Doctor is also a context-sensitive diagnostic tool. It is available when a problem is encountered in sketching. Figure 1–27 shows the Sketch Doctor dialog box after an attempt was made to extrude an open-loop curve.

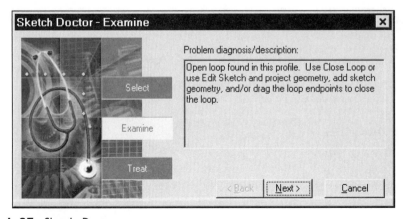

Figure 1–27 *Sketch Doctor*

SYSTEM SETTINGS

The way you construct solid parts, assemblies, presentations, and engineering drawings is affected by the system settings. You can modify system settings in the Options dialog box. To access the dialog box, select Options from the Tools menu. You can alter the settings through eleven tabs: General, Colors, Display, File Locations, Modeling Cross Section, Drawing, Notebook, Sketch, Part, Design Elements, and Adaptive. Only the General tab (See Figure 1–28) and the Colors tab are discussed here. Other tabs in the Options dialog box are discussed in later chapters.

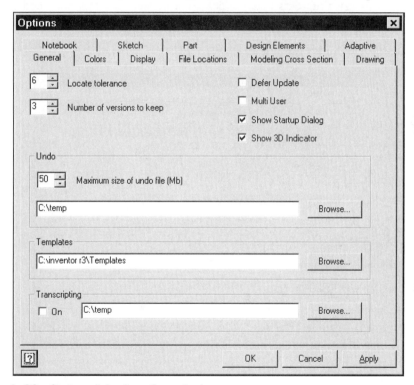

Figure 1–28 *Options dialog box: General tab*

In the General tab shown in Figure 1–28, there are nine settings:

- **Locate tolerance:** Sets the distance tolerance from which you select an object in terms of screen pixels. Set the tolerance to 6.

- **Number of versions to keep:** Sets the number of versions to keep as you save your files. Set the number of versions to 3.

- **Defer Update:** Enables updating the assembly as you modify the parts. Accept the default.

- **Multi User:** Enables safeguards when multiple users are editing the files. Accept the default. (You will learn more about this setting in Chapter 4.)

- **Show Startup Dialog:** Controls the Startup Dialog box displayed when you start Autodesk Inventor. Accept the default.

- **Show 3D Indicator:** Enables the display of the 3D indicator in the bottom left corner of the screen. Accept the default.

- **Undo:** Determines the maximum size and the location of the undo temporary file. Set the file size to 50.

- **Templates:** Specifies the location of the template files. Accept the default.

- **Transcripting:** Sets options for recording and replaying your modeling sequence and the location for storing the transcript file. Accept the default.

Select the Color tab. (See Figure 1–29.)

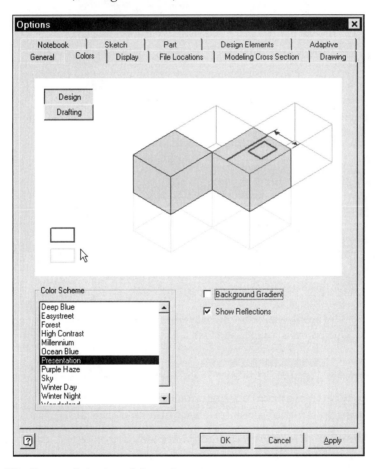

Figure 1–29 *Options dialog box: Colors tab*

On the Colors tab, select a color scheme and select the OK button.

SUMMARY

There are three kinds of 3D models in a computer: 3D wireframe, 3D surface, and 3D solid. Among them, the 3D solid model is superior because it provides integrated information about the vertices, edges, faces, and volume of the 3D object that it represents.

The most common method for constructing a 3D solid is to make sketches and extrude, revolve, loft, or sweep the sketches to form an extrude solid, revolve solid, loft solid, or sweep solid.

There are two major approaches to constructing 3D wires and 3D solids: parametric and non-parametric. With the non-parametric approach, the parameters of lines and the solids constructed from the lines are static and cannot be changed. With the parametric approach, you make free-hand sketches, assign geometric constraints, and specify parametric dimensions. Unlike a non-parametric solid, a parametric solid is flexible and you can modify it simply by specifying new parameters to the sketches and the solid.

With a set of 3D parametric solid models, you put them together in a virtual assembly by constraining the relative positions of their vertices, edges, and faces. To illustrate how the parts are assembled in an assembly, you construct an exploded view of the assembly.

To work with computerized manufacturing systems, you output the 3D parts and 3D assembly in an electronic format. To manufacture the parts in a more conventional way, you output 2D projection drawings from the 3D solids and assemblies.

Autodesk Inventor is a 3D parametric solid modeling tool. It enables you to construct parametric solid models, to compose an assembly from a set of solid parts or a set of assemblies of parts, to export solids and assemblies to standard file exchange formats, and to output 2D engineering drawings from the solid parts and assemblies.

Autodesk Inventor uses four kinds of files: you construct a 3D solid model in a part file (.ipt), you put components together to form an assembly in an assembly file (.iam), you use a presentation file to manage exploded views of an assembly (.ipn), and you output engineering drawings in a drawing file (.idw).

To assist you in your design work, Autodesk Inventor's design support system has seven components: Help Topics, Autodesk Online, What's New about Autodesk Inventor, Visual Syllabus, Design Professor, Design Doctor, and Sketch Doctor.

REVIEW QUESTIONS

1. Briefly explain the three kinds of 3D models in a computer application.

2. State the difference between the parametric and non-parametric approaches.

3. Give an outline of the design process, from making 3D models to outputting the parts and assembly.

4. What are the key functions of Autodesk Inventor?

5. What are the four kinds of Inventor files? What do you use them for?

6. Explore the use of the design support system and make a summary of the system.

CHAPTER 2

Solid Modeling

OBJECTIVES

The aims of this chapter are to delineate the key concepts of parametric solid modeling and to depict the ways to construct sketches and to build solids from sketches. Also explained are the ways to incorporate placed solid features by specifying the locations and parameters, to construct derived parts, to set light, material, color, and properties to a solid part. Finally, file export and compatibility are outlined. After studying this chapter, you should be able to

- Construct parametric sketches and build solid features from sketches
- Incorporate placed solids features to a 3D solid
- Use work features in constructing 3D solids
- Construct derived solid parts
- Set lighting, color styles, and material and assign properties to in a 3D solid part file
- Explain ways to export files and import files

OVERVIEW

The solid model of a 3D object is an integrated mathematical representation depicting the vertices, edges, faces, and volume of the object. Because any individual 3D object is unique in shape, it is impossible to derive a general mathematical expression that can represent all kinds of 3D objects. Over the years, various mathematical methods of representing 3D objects in the computer have been developed. Quite recently, the feature-based approach has become a commonly used 3D solid modeling method. Using this method, you decompose a complex 3D solid part into simple solid features that can be represented in the computer, construct the features accordingly, and combine the features while you construct them.

There are two main kinds of solid features, sketched and placed. Sketched solid features derive from sketches—you make free-hand sketches and let the computer construct the solid features from the sketches. Placed solid features are pre-constructed solid features that you select from a menu, specifying a location and parameters. To combine the solid features, you join, cut, and intersect. Autodesk Inventor has six kinds of sketched solid features (extruded, revolved, loft, sweep, coil, and split) and eight kinds of placed solid features (hole, shell, fillet, chamfer, rectangular pattern, circular pattern, mirror, and face draft).

To construct a sketch for making a sketched solid feature, you need a plane—the sketch plane. You can use the default XY plane, YZ plane, XZ plane, and any existing plane on a solid model as the sketch plane. If you want to position the sketch plane otherwise, you construct a work plane. In addition to work planes, you construct work axes and work points that help you establish geometric references.

Figure 2–1 shows the component parts for an assembly of a food grinder. You will construct some of the parts in this chapter. In the next chapter, you will construct the remaining parts and assemble the parts. Through making these parts, you will construct various kinds of sketched solid features, placed solid features, and work features.

Figure 2–1 *Parts of a food grinder*

PARAMETRIC FEATURE-BASED SOLID MODELING CONCEPTS

Autodesk Inventor is a parametric feature-based solid modeling application.

THE FEATURE-BASED APPROACH

Using the feature-based approach, you deduce a complex solid object to elements of simple solid features, construct the solid features, and combine the features to form the complex object. Autodesk Inventor has two kinds of solid features: sketched and placed.

Sketched solid features derive from sketches; you construct sketches and use the sketch to form a solid. There are six kinds of sketched solid features: extruded, revolved, loft, sweep, coil, and split. (See Figures 2–2 through 2–7.)

Placed solid features are pre-constructed solid features. You select a feature from the menu and specify its parameters. There are eight kinds of placed solid features: hole, shell, fillet, chamfer, rectangular pattern, circular pattern, mirror, and face draft features (see Figures 2–15 through 2–22).

In a feature-based solid modeling system, the construction of a 3D solid model involves two major tasks: analyzing and synthesizing. With a 3D object in mind, you study and analyze it critically to determine what kinds of solid features make up the object (sketched and placed). You think about how to construct the features and how to combine them together, by joining, cutting, or intersecting (see Figures 2–8 through 2–10). In thinking about the solid features, you need to consider how to establish the sketch planes (see Figure 2–11) for making the sketches and references for the placed features. You think about setting up geometric references by using work features—work planes, work axes, and work points (see Figures 2–12 through 2–14). After this careful thinking process, you start making the features and synthesizing them to make the 3D solid.

In order to better equip yourself to analyze and synthesize, you have to gain a good understanding of the various kinds of sketched solid features, Boolean operations, work features, and placed solid features.

THE PARAMETRIC APPROACH

In the early stage of your design, you have a shape of the sketched solid features in your mind, but you might not have decided on any precise dimensions. Using the parametric approach, you use the computer as an electronic sketching pad to record your design idea and you concentrate on forms and shapes rather than on dimensions. In the absence of exact dimensions, you construct a rough free-hand sketch. As the name implies, rough sketches are not precise at all—lengths are approximate. After constructing the sketch, you refine the geometry by specifying geometric constraints and modify the size by specifying dimensions. After sketching, you make a solid feature from the sketch or sketches.

SKETCHED SOLID FEATURES

Sketched solid features derive from sketches; extruded solid, revolved solid, loft solid, sweep solid, coil solid, and face split.

Extruded Feature

To make an extruded solid, you make a sketch and extrude the sketch in a direction perpendicular to the plane of the sketch. (See Figure 2–2.) You extrude the sketch in either direction or from mid-plane.

Figure 2–2 *Sketch and extruded features*

Revolved Feature

To make a revolved solid, you make a sketch and revolve it about an axis. (See Figure 2–3.)

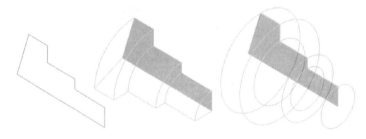

Figure 2–3 *Sketch and revolved features*

Loft Feature

To make a loft solid, you construct two or more sketches on a number of sketch planes and transit along the sketches. (See Figure 2–4.)

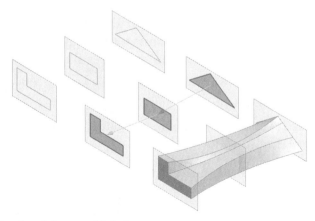

Figure 2–4 *Sketches, lofting, and loft feature*

Sweep Feature

To make a sweep solid, you construct two sketches. You use one sketch as the cross section and the other sketch as the path. Then you sweep the cross section along the path. (See Figure 2–5.)

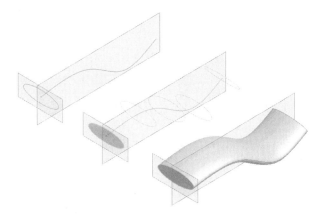

Figure 2–5 *Sketches, sweeping, and sweep feature*

Coil Feature

A coil solid is a special kind of sweep solid in which a sketch depicting the cross section is swept along a helical path. (See Figure 2–6.)

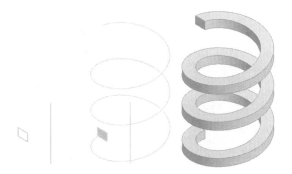

Figure 2–6 *Sketch, coiling, and coil feature*

Split Feature

A split feature is a special kind of sketched feature. You construct a sketch and split a face of a solid into two faces or split a solid into two solids and remove one of them. By splitting a face into two, you can apply face drafts in two directions. Note that a face split is a sketch feature and a face draft is a placed feature. Figure 2–7 shows a sketch and a face split into two by the sketch.

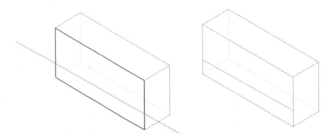

Figure 2–7 *Sketch and split feature*

BOOLEAN OPERATIONS

The first sketched solid feature you construct in a 3D solid part is the base solid feature. When you construct subsequent sketched solid features, you decide how to combine the new features to the existing feature—to join, cut, or intersect.

Join

To construct a solid that has the volume of two solids, you join the two together. Figure 2–8 shows an extruded solid joined to an extruded solid.

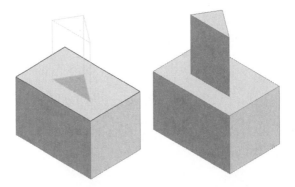

Figure 2–8 *Triangular solid joined to a rectangular solid*

Cut

To construct a solid with the volume of one solid removed from another, you subtract or cut the second solid from the first solid. Figure 2–9 shows an extruded solid cut from an extruded solid.

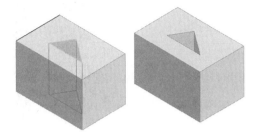

Figure 2–9 *Triangular solid cut from a rectangular solid*

Intersect

To construct a solid with a volume that contains the portion common to two solids, you intersect the solids. Figure 2–10 shows an extruded solid intersecting a revolved solid.

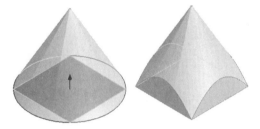

Figure 2–10 *Extruded solid intersecting a revolved solid*

SKETCH PLANES

To construct a sketched solid feature, you start with a sketch. To construct a sketch, you need a sketch plane, and there are three default planes where you specify a sketch plane: XY plane, XZ plane, and YZ plane. In addition, you can use any existing planes of a solid feature for making sketches. Figure 2–11 shows a sketch being constructed on the face of a solid.

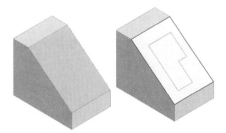

Figure 2–11 *Sketch plane established on the face of a solid*

WORK FEATURES

There are three kinds of work features: work planes, work axes, and work points.

Work Planes

In addition to the default planes and existing faces of a solid, you can construct work planes on which to set up sketch planes. There are nine ways to construct a work plane, as illustrated in Figure 2–12.

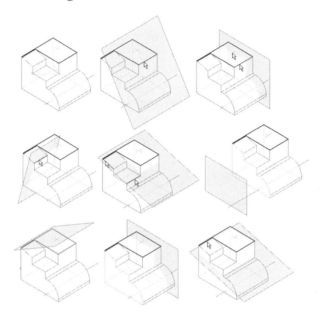

Figure 2–12 *Nine kinds of work planes*

The nine kinds of work planes shown in Figure 2–12 (from top left to lower right):

- Work plane passing through three points
- Work plane tangential to a face and through an edge
- Work plane normal to an axis and through a point
- Work plane normal to a face and through an edge
- Work plane passing through two coplanar edges
- Work plane offset from a face
- Work plane at angle to a face
- Work plane parallel to a plane and through a point
- Work plane tangential to a curve and parallel to a plane

Work Axes

To help in constructing a work plane and establishing a proper geometric relationship among existing objects, you construct work axes and work points. There are four ways to construct a work axis, as illustrated in Figure 2–13.

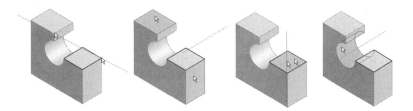

Figure 2–13   *Four kinds of work axes*

The four kinds of work axes shown in Figure 2–13 (from left to right):

- Work axis passing through two points
- Work axis passing through the intersection of two planes
- Work axis normal to a plane and through a point
- Work axis on the axis of a circular feature

Work Points

There are four ways to construct a work point, as illustrated in Figure 2–14.

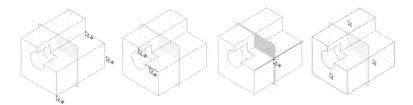

Figure 2–14   *Four kinds of work points*

The four kinds of work points as shown in Figure 2–14 (from left to right):

- Work point on selected points (endpoints, midpoints, and intersecting points) on the solid part
- Work point at the intersection of two edges
- Work point at the intersection of a work plane and an edge
- Work point at the intersection of three planes

PLACED SOLID FEATURES

Placed solid features are pre-constructed common engineering features; you can simply specify their type and parameters and place them on an existing solid model. They are hole, shell, fillet, chamfer, rectangular pattern, circular pattern, mirror, and face draft features.

Hole Feature

To construct a hole feature, you specify type, size, and location. Figure 2–15 shows a hole feature (a set of three holes) placed on a solid.

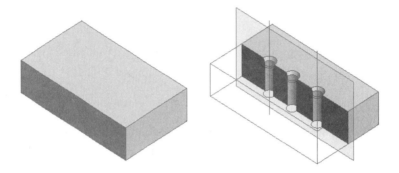

Figure 2–15 *A rectangular solid and a cutaway view of the hole feature placed on the solid*

Shell Feature

A shell feature makes a solid object hollow; you state the thickness of the shell. Figure 2–16 shows a shell feature placed on a loft solid.

Figure 2–16 *Loft solid made hollow*

Fillet Feature

A fillet feature rounds off the edges of a solid; you select edges and specify fillet radii. Figure 2–17 shows fillet features placed on a solid.

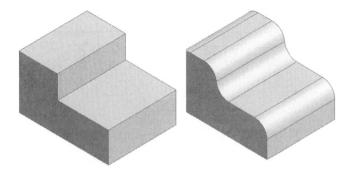

Figure 2–17 *Fillet features placed on a solid*

Chamfer Feature

A chamfer feature bevels the edges of a solid; you select edges and specify bevel distances or the bevel angle. Figure 2–18 shows chamfer features placed on a solid.

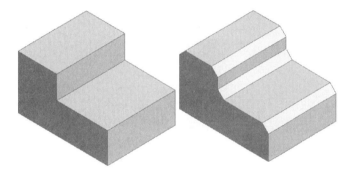

Figure 2–18 *Chamfer features placed on a solid*

Rectangular Pattern

To copy a solid feature in a rectangular array, you select a feature and specify the directions, distances, and the number of repetitions. (See Figure 2–19.)

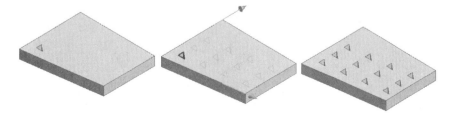

Figure 2–19 *Rectangular pattern feature*

Circular Pattern

To copy a solid feature in a circular array, you select a feature and specify an axis, angular distance, and number of repetitions. (See Figure 2–20.)

Figure 2–20 *Circular pattern feature*

Mirror Feature

To construct a mirror copy of a solid feature, you select a feature and specify a mirror plane. (See Figure 2–21.)

Figure 2–21 *Mirror feature placed*

Face Draft Feature

To taper the faces of a solid, you place a face draft; you select an edge or a split line and specify a draft angle. (See Figure 2–22.)

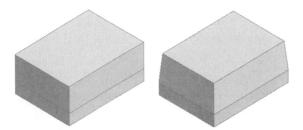

Figure 2–22 *Face draft feature placed along a split line*

By now, you should have a general picture of the three kinds of features (sketched solid feature, work features, and placed solid features) that you use in making a solid model.

CONSTRUCTING SKETCHED SOLID FEATURES

Now you will construct various kinds of sketched solid features. To reiterate, there are six kinds of sketched solid features: extruded, revolved, loft, sweep, coil, and split. To construct a sketched solid feature, you construct sketches.

SKETCHING

Sketches need not be accurate and precise from the outset. You concentrate on the form and shape of the sketch in accordance with your design intent. You can use lines, splines, circles, ellipses, arcs, rectangles, and points to construct sketch objects. To modify a sketch, you can trim, extend, fillet, and offset.

Geometric Constraints

After you are satisfied with the general form and shape of your sketch, you refine it by specifying the geometric relationships among the elements of the sketch. For example, you might want to set a line to be perpendicular to another line and set an arc to be concentric with a circle. To specify these geometric relationships, you assign geometric constraints to the sketch. There are ten kinds of geometric constraints: perpendicular, parallel, tangent, coincident, concentric, collinear, horizontal, vertical, equal, and fix, as described in Table 2–1:

Table 2–1   Ten kinds of geometric constraints

Geometric constraint	Description
Perpendicular	Sets two lines to be perpendicular to each other.
Parallel	Sets two lines to be parallel to each other.
Tangent	Sets a line, an arc, or a circle to be tangential to an arc or circle.
Coincident	Sets a point, the end point of a line, or the center of a circle or an arc to be coincident with another point, the end point of another line, or the center of another circle or arc.
Concentric	Sets the center of an arc or a circle to be coincident with the center of an arc or a circle.
Collinear	Sets two lines to be collinear.
Horizontal	Sets a line to be horizontal or a point, the end point of a line, the center of a circle or an arc to lie horizontally with another point, the end point of another line, or the center of another circle or arc.
Vertical	Sets a line to be vertical or a point, the end point of a line, the center of a circle or an arc to lie vertically with another point, the end point of another line, or the center of another circle or arc.
Equal	Sets two lines to have equal length or two arcs or circles to have equal radius.
Fix	Fixes the length of a line, the radius of a circle or an arc, or the location of a point, the end point of a line, and the center of a circle or an arc.

While you make your sketch, Inventor may assign some of these constraints automatically to your sketch in accordance with the way you construct your sketch. For example, if you construct a line that is nearly horizontal, Inventor will assign a horizontal constraint to it. If you do not want a geometric relationship that Inventor assigned to your sketch or want to delete a relationship that you assigned earlier, you can display the geometric constraint symbols, right-click, and delete the constraints.

Parametric Dimensions

To refine the sketch, you add parametric dimensions (Autodesk Inventor dimensions are parametric dimensions). A parametric dimension differs from conventional dimensions in that it not only reports the size of the objects that it describes, but it also drives the size of the related objects. For example, if you construct a line that is 12.5 units long, when you add a parametric dimension to the line, the exact length is displayed. You can accept the length or you can enter a new dimension. If you enter a length of 14 units, the length of the line will change to 14 units.

You can modify geometric constraints and dimensions any time during the design process. You can even change their parameters after you construct a solid from them.

MODIFYING SYSTEM SETTINGS

Before you construct a sketch and make a sketched solid from it, you can modify the related system settings.

Sketch

1. Select Options from the Tools menu and select the Sketch tab. (See Figure 2–23.)

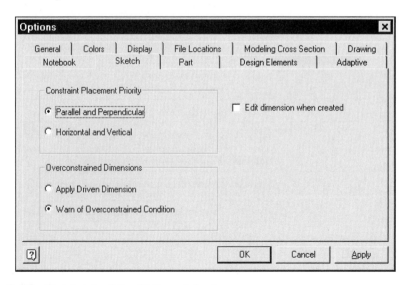

Figure 2–23 *Sketch tab of the Options dialog box*

The Sketch tab has three areas: Constraint Placement Priority, Overconstrained Dimension, and Edit Dimension When Created.

- **Constraint Placement Priority** When you construct lines in a sketch, select the kind of constraint (Parallel and Perpendicular or Horizontal and Vertical) to have higher priority over the other kinds of constraint.

- **Overconstrained Dimension** When you over-dimension a sketch, you can select Apply Driven Dimension to have it become a driven dimension or Warn Of Overconstrained Condition to receive a warning message.

- **Edit dimension when created** If you select this check box, when you add dimensions to a sketch, you will be prompted to edit the dimension instead of having default dimensions applied.

2. Select Parallel and Perpendicular, select Warn Of Overconstrained Condition, and deselect the Edit Dimension When Created check box.

Part

3. Now select the Part tab of the Options dialog box to modify the settings related to the construction of solid parts. (See Figure 2–24.)

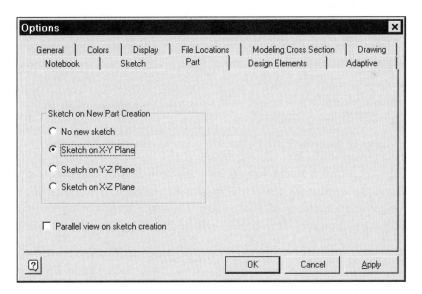

Figure 2–24 *Part tab of the Options dialog box*

The Part tab has two areas: Sketch On New Part Creation, with four options, and Parallel View On Sketch Creation. They concern the sketch plane and the display of a new sketch.

- **No New Sketch**   If selected, there will be no default sketch on new part creation. You have to set up a sketch plane.

- **Sketch On X-Y Plane**   If selected, a sketch on the X-Y Plane is set up on new part creation.

- **Sketch On Y-Z Plane**   If selected, a sketch on the Y-Z Plane is set up on new part creation.

- **Sketch On X-Z Plane**   If selected, a sketch on the X-Z Plane is set up on new part creation.

- **Parallel View On Sketch Creation**   If this check box is selected, whenever a new sketch plane is set up, the display will be changed to the top view of the new sketch plane.

4. Select the Sketch On X-Y Plane option and deselect the Parallel View On Sketch Creation check box. This way, a sketch is constructed on the X-Y plane of the new part file and the display will not change when a new sketch plane is set up.

File Locations

Autodesk Inventor manages files in four sets of working/search directories:

- **Workspace**   Your working directory, the default location where you save your files.

- **Local Search Paths**   Locations in your computer where Autodesk Inventor searches for files to open.

- **Group Search Paths**   Locations in your computer or in a computer on the network that you and your teammates in a design group will access when opening a file.

- **Library Search Paths**   Locations to store the library parts.

Search paths are useful in managing an assembly, which is a collection of component parts and linked part files. When you open an assembly file, Inventor will search first the library path, then the workspace path, then the local paths, and finally the workgroup paths. (You will learn how to construct an assembly in the next chapter.)

There are two ways to set these working/search directories: You can add the directories in the File Locations tab of the Options dialog box manually and you can use a project path file to specify the file locations.

5. Now create the following folders in your computer.

- C:\InventorProjects

- C:\DesignGroupA

- C:\FoodGrinder

- C:\Oscillator

- C:\StandardParts

Specify File Locations

6. Because you cannot modify the search paths if any Inventor file is opened, close all files before you proceed.

7. Now select the File Locations tab of the Options dialog box in Inventor. (See Figure 2–25.)

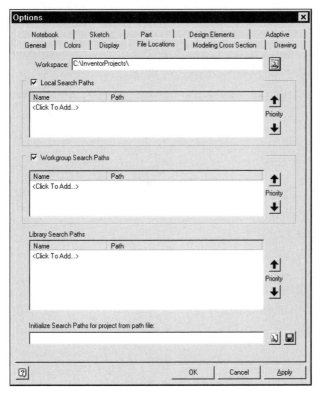

Figure 2–25 *File Locations tab of the Options dialog box*

In the File Locations tab, there are five fields: Workspace, Local Search Paths, Workgroup Search Paths, Library Search Paths, and Initialize Search Paths For Projects From Path File.

8. Select and clear the file displayed in the Initialize Search Paths For Project From Path File box. Then select the Workspace field and specify a workspace (C:\InventorProjects).

9. To assign a local search path, select the Local Search Paths check box. Then select Click to Add in the Local Search Paths box. (See Figure 2–26.)

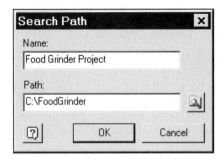

Figure 2–26 *Search Path dialog box*

10. Select a folder in the Search Path dialog box, type the path name, and select the OK button. A path is set.

11. Similarly, set the Workgroup Search Paths and Library Search Paths to C:\DesignGroupA and C:\StandardParts, respectively.

Use Project Path File

Another way to set the search paths is to set up a project file to depict the default locations for workspace, local search paths, workgroup search paths, and library search paths. You construct your own path files using a text editor. The file extension is .ipj and the format of the project path file is as follows:

```
[Included Path File]
PathFile=(path/filepath.txt)
[Workspace]
Workspace=(path)
[Local Search Paths]
LocalPath1=(path1)
LocalPathn=(pathn)
[Workgroup Search Paths]
WorkgroupPath1=(path1)
WorkgroupPathn=(pathn)
[Library Search Paths]
LibPath1=(path1)
LibPathn=(pathn)
```

12. Now use the text editor to construct a project file (file name: Project.ipj) with the following content and save it in the directory C:\InventorProjects.

```
[Included Path File]
PathFile=C:\InventorProjects\project.ipj
[Workspace]
```

```
Workspace=C:\InventorProjects
[Local Search Paths]
FoodGrinder Project=C:\FoodGrinder
Oscillator Project=C:\Oscillator
[Workgroup Search Paths]
GroupA=C:\DesignGroupA
[Library Search Paths]
Library A=C:\StandardParts
```

To reiterate, you must not have open any Inventor files while you specify a project file.

13. Select the Initialize Search Paths For Project From Path File box on the File Locations tab of the Options dialog box in Inventor.

14. Then select the project path file (Project.ipj). The working/search directories are set (see Figure 2–27).

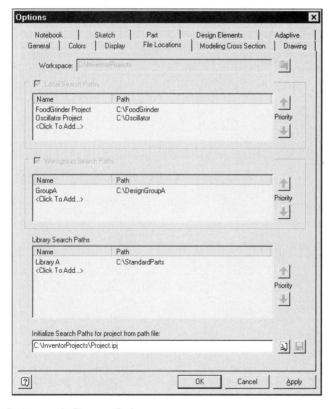

Figure 2–27 *Project path file specified*

15. Select the OK button to close the Options dialog box.

STARTING A PART FILE

1. Now start a new part file by selecting New from the File menu.

2. In the New dialog box, select the Default tab.

3. Select the Standard.ipt template and select the OK button.

After you start a new part drawing, you have, in addition to the Standard toolbar, the following objects: Command Bar toolbar, panel bar, browser bar, and graphics window. Figure 2–28 shows the graphics screen.

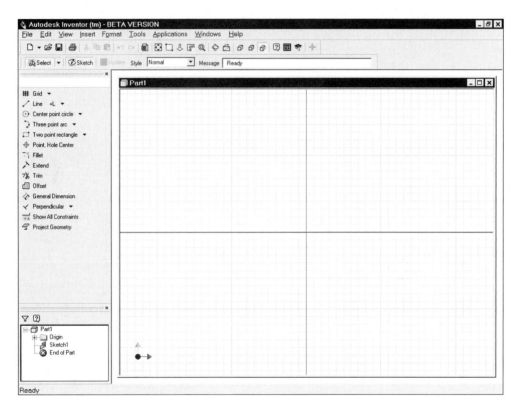

Figure 2–28 *New part file started*

Command Bar Toolbar

The Command Bar toolbar has five areas. (See Figure 2–29.)

Figure 2–29 *Command Bar toolbar*

From the left, the first area is a Select list box. Here you determine the priority of selection. When you select an object in your graphics area and there is more than one object at the location of your cursor, selection priority settings will determine the kind of object to be selected. For example, when you place your cursor on a solid part and you set the priority to Face, you will select a face. Selection priorities available depend on the kinds of file you open, as shown in Table 2–2.

Table 2–2 Selection priority settings available for each file type

File	Selection priority
Part File	Feature, Face, Sketch
Assembly File	Component, Leaf Part, Feature, Face, Sketch
Presentation File	Component, Leaf Part
Drawing File	Edge, Feature, Part

When the cursor is in the graphics area, you can set selection priority by holding down the SHIFT key and right-clicking to bring up the selection priority shortcut menu. Figure 2–30 shows the priority shortcut menus for part files, assembly files, presentation files, and drawing files, respectively.

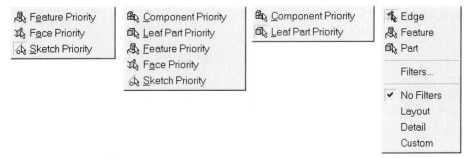

Figure 2–30 *Selection priority shortcut menus*

The second area is a Sketch button for opening a new sketch. You will use this button to start a new sketch for making sketched solid features.

The third area is an Update button. You will use this button to update changes that you make to the solid part.

The fourth area is a list box where you set the style of sketch lines or the color of a solid. Sketch line styles choices are Normal or Construction. Construction lines assist in making sketches; they are not used in defining the cross sections or paths for making sketched solid features.

In the fifth area of the Command Bar toolbar, you get feedback messages from the system.

Panel Bar

Tools are available in the panel bar or the Sketch toolbar. The panel bar has a number of palettes that enable you to access various design tools in accordance with current design context. In a new part file, the sketch panel is displayed initially. It provides tools for you to construct sketches. To access other panels, right-click to display the shortcut menu. (See Figure 2–31.)

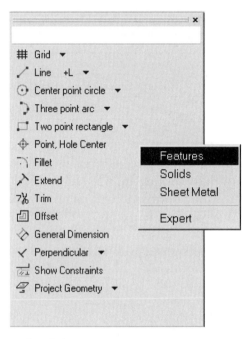

Figure 2–31 *Sketch panel and shortcut menu*

In the shortcut menu shown in Figure 2–31, there are four items: Features, Solids, Sheet Metal, and Expert. When you select Features, Solids, or Sheet Metal, the panel bar changes to display the appropriate tools. When you select Expert, the annotations that explain the meaning of the icons in the panel will disappear. Thus, the panel will occupy a smaller space on your screen.

Select Features in the shortcut menu to display the Features panel, which provides tools for you to construct sketched solid features, placed solid features, and work features. Figure 2–32 shows the Features panel. Note that some of the icons in your screen are grayed out in a new part file. For example, the Hole icon is grayed out because you cannot construct a hole without having a solid part in your file.

Place the cursor on the panel and right-click. Then select Solids from the shortcut menu. The Solids panel is now displayed. (See Figure 2–33.)

Figure 2–32 *Features panel*

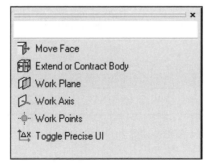

Figure 2–33 *Solids panel*

The Solids panel provides tools to manipulate imported non-parametric solid parts. You will learn how to manipulate an imported solid in Appendix A of this book. Now right-click and select Sheet Metal from the shortcut menu. The Sheet Metal panel is displayed. (See Figure 2–34.)

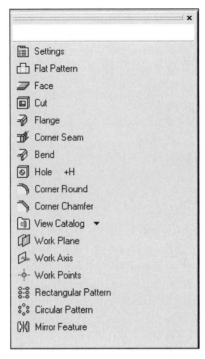

Figure 2–34 *Sheet Metal panel*

The Sheet Metal panel provides tools for constructing sheet metal parts. You will learn how to construct sheet metal parts in Chapter 5.

Toolbars

To display the Sketch toolbar, select View ➤ Toolbars ➤ Sketch. Figure 2–35 shows the Sketch toolbar.

Figure 2–35 *Sketch toolbar*

To access the Features toolbar, select View ➤ Toolbars ➤ Features. Figure 2–36 shows the Features toolbar.

Figure 2–36 *Features toolbar*

The other toolbars that you can display are Solids, Precise Input, Collaboration, and Sheet Metal. (See Figure 2–37.)

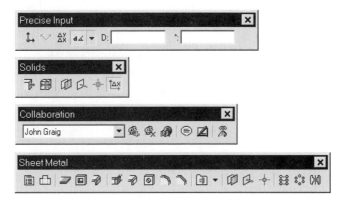

Figure 2–37 *Solids, Precise Input, Collaboration, and Sheet Metal toolbars*

As I've have explained, sketches need not be precise at all because you should concentrate on form and shape and refine a sketch by adding geometric constraints and dimensions later. However, you may prefer to enter precise coordinates while making a sketch. To enter precisely, you use the Precise Input toolbar.

The Collaboration toolbar concerns working collaboratively among your teammates. You will learn about it in Chapter 4.

Sketching Tools

In the Sketch panel and Sketch toolbar shown Figures 2–31 and 2–35, there are 14 button areas. Table 2–3 describes the choices.

Table 2–3   Sketch toolbar and panel options

Option	Description
Grid/ Edit Sketch Coordinates	Grid enables you to manipulate grid settings. Edit Sketch Coordinates enables you to move the coordinate system to selected geometry of the solid part. You will find it useful when using the Precise Input tool.
Line/ Spline	Line enables you to construct a line. Spline enables you to construct a spline.
Center Point Circle/ Tangent Circle/Ellipse	Center Point Circle enables you to construct a circle by specifying the center and a point on the circumference of the circle. Tangent Circle enables you to construct a circle by selecting three coplanar lines or edges. Ellipse enables you to construct an ellipse.

Option	Description
Three Point Arc/Tangent Arc/ Center Point Arc	Three Point Arc enables you to construct an arc by specifying the end points and a point on the arc. Tangent Arc enables you to construct an arc that is tangential to a selected line or arc. Center Point Arc enables you to construct an arc by first specifying the center and then the end points.
Two Point Rectangle Three Point Rectangle	Two Point Rectangle enables you to construct a rectangle with horizontal and vertical sides by specifying the diagonal points. Three Point Rectangle enables you to construct a rectangle by specifying two points define an edge and a diagonal point.
Point, Hole Center	Point, Hole Center enables you to construct a point or a hole center.
Fillet	Fillet enables you to construct a rounded corner at the intersection of two non-parallel lines.
Extend	Extend enables you to extend a line or an arc to meet another line or arc.
Trim	Trim enables you to trim away a portion of a line or arc.
Offset	Offset enables you to construct an offset geometric from a selected geometry.
General Dimension	General Dimension enables you to construct a parametric dimension or a driven dimension.
Perpendicular/Parallel/ Tangent/Coincident/ Concentric Collinear/ Horizontal/Vertical/ Equal/ Fix	Perpendicular constrains two selected lines to be perpendicular to each other. Parallel constrains two selected lines to be parallel to each other. Tangent constrains a line and an arc or two arcs to be tangential to each other. Concentric constrains two circles/arcs to be concentric to each other. Collinear constrains two lines to be collinear to each other. Horizontal constraints a line to be horizontal. Vertical constrains a line to be vertical. Equal constrains two lines to be equal in length or two circles/arcs to be equal in radius. Fix constrains an end point to be fixed.
Show Constraints	Show constraints enables you to display and delete the constraint applied to a line, arc, or ellipse.
Project Geometry/ Project Cut Edges	Project Geometry enables you to construct sketch objects by projecting selected geometry to the current sketch plane. Project Cut Edges enables you to construct sketch objects by projecting the intersecting edges between the solid part and the current sketch plane.

Features Construction Tools

The Features panel and Features toolbar (Figures 2–32 and 2–36) have 19 button areas. Table 2–4 describes the choices.

Table 2–4   Features toolbar and panel options

Option	Description
Extrude	Extrude enables you to construct an extruded feature from a sketch.
Revolve	Revolve enables you to construct a revolved feature from a sketch.
Hole	Hole enables you to place a hole feature.
Shell	Shell enables you to place a shell feature.
Loft	Loft enables you to construct a loft feature from sketches.
Sweep	Sweep enables you to construct a sweep feature from sketches.
Coil	Coil enables you to construct a coil feature from sketches.
Fillet	Fillet enables you to place a fillet feature.
Chamfer	Chamfer enables you to place a chamfer feature.
Face Draft	Face Draft enables you to place a face draft feature.
Split	Split enables you to split a face into two faces or split a solid into two solids and remove one of them.
View Catalog/ Create Design Element/ Insert Design Element	View Catalog displays the design catalog. Create Design Element enables you to construct a design element. Insert Design Element enables you to insert a design element. (You will learn about design elements in Chapter 4.)
Derived Part	Derived Part enables you to construct a derived solid part.
Rectangular Pattern	Rectangular Pattern enables you to place a rectangular pattern of features.
Circular Pattern	Circular Pattern enables you to place a circular pattern of features.
Mirror	Mirror enables you to place a mirror feature.
Work Plane	Work Plane enables you to construct a work plane.
Work Axis	Work Axis enables you to construct a work axis.
Work Point	Work Point enables you to construct a work point.

In this chapter you will learn how to use the sketching tools and most of the feature construction tools.

Shortcut Keys

Apart from using the appropriate toolbars, you can use the shortcut keys shown in Table 2–5.

Table 2–5 Shortcut keys

Shortcut Key	Function
S KEY	Sketch
L KEY	Sketch line
D KEY	Sketch/Drawing dimension
E KEY	Extrude
R KEY	Revolve
H KEY	Hole
F1 KEY	Help
F2 KEY	Pan
F3 KEY	Zoom
F4 KEY	Rotate
F5 KEY	Previous View

Browser Bar

Below the panel bar is the browser bar, which shows the hierarchy of objects in a file. Figure 2–38 shows the browser bar of a part file. Sketches and features that you construct will display as objects in the hierarchy. If you want to edit an object, select the object in the browser bar, right-click, and select Edit. If you want to remove an object, select the object in the browser bar, right-click, and select Delete.

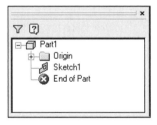

Figure 2–38 *Browser bar*

When browsing the items in the browser bar, you can use the HOME, END, PGUP, and PGDN keys on the keyboard. To expand the hierarchy, you right-click and select Expand All. To collapse the hierarchy, you right-click and select Collapse All.

Graphics Window

The graphics window is your main working area. There are grid points and axis lines. (See Figure 2–39.) Initially, you have an origin that consists of the YZ, XZ, and XY planes, X, Y, and Z axes, and the center point. Optionally, you can have a default sketch plane.

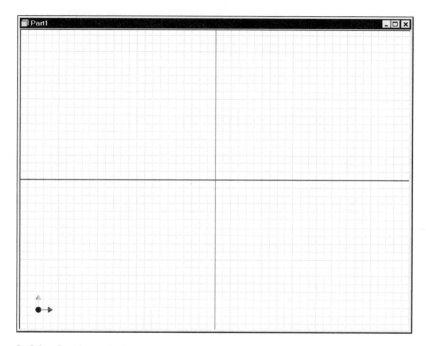

Figure 2–39 *Graphics window*

At the lower left corner of the graphics window, there is a 3D indicator. If the indicator is not displayed, select Tools ➤ Options, and then select the General tab of the Options dialog box. Select the Show 3D Indicator check box, and then select the OK button to exit.

SETTING UNITS OF MEASUREMENT

Before you construct a solid part, you set the units of measurement.

1. Select Format ➤ Dimensions to display the Properties dialog box. (See Figure 2–40.)

Here you specify length, time, angle, and mass units and determine linear and angular dimension display precision. Note that display precision concerns the number of decimal places displayed on the screen and does not affect the precision of the data stored in your file.

2. Now set the Length Units to millimeter, Angle Units to degree, Linear Dim Display Precision to 3 decimal places, Time Units to second, Mass Units to kilogram, and Angular Dim Display Precision to 2 decimal places.

Because you will assemble the solid parts that you will construct in this chapter in an assembly, it is necessary that you use the same units of measurement for all the parts.

3. Select the OK button to close the Properties dialog box.

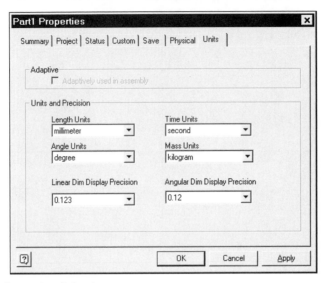

Figure 2–40 *Properties dialog box*

EXTRUDING A SOLID

Now you will construct a solid feature for making the blade of the food grinder (see Figure 2–41). It is an extruded solid; you will construct a sketch and extrude the sketch to form a solid.

Figure 2–41 *Extruded solid*

Sketch

Now construct a circle on the default sketch plane. When you make the circle, exact size is unimportant because you will later modify it by adding parametric dimensions.

1. Select Center point circle from the Sketch toolbar (or the Sketch panel).

2. Then select a center point and a point on the circumference of the circle to describe a circle. (See Figure 2–42.)

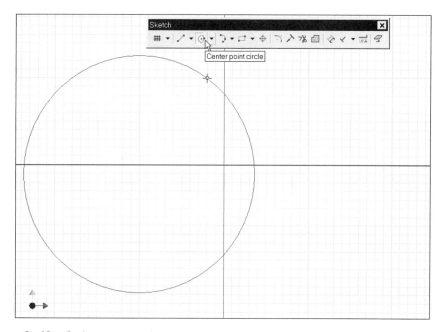

Figure 2–42 *Circle constructed*

A circle of unknown size is constructed. To set the size of the circle, you will add a parametric dimension. A parametric dimension serves two purposes: It reports the size and it controls the size of the geometry. General Dimension on the toolbar and panel enables you to construct a parametric dimension.

3. Select General Dimension from the Sketch toolbar or panel.

4. Select the circle.

5. Select a point to specify the location of the dimension. (See Figure 2–43.) Note that the value shown here will not be the same as that shown in your sketch.

The initial dimension shown in the Edit Dimension dialog box tells you the exact size of the circle.

6. Now select the dimension in order to display the Edit Dimension dialog box.

7. Set the dimension value to 80 units and select the Checkmark button in the dialog box to confirm the change.

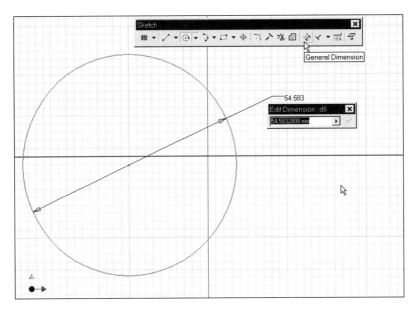

Figure 2–43 *Dimension added to the sketch*

After you set a new value, the circle changes in size and it becomes a circle with a diameter of 80 units.

8. To adjust the display to show the entire sketch in your screen, select Zoom All from the Standard toolbar. (See Figure 2–44.)

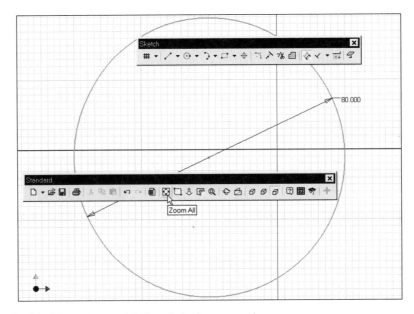

Figure 2–44 *Dimension modified and display zoomed*

Having constructed a circle with a diameter of 80 units, you will now construct a concentric circle by offsetting.

9. Select Offset from the Sketch toolbar or panel.

10. Select the circle and right-click to display the shortcut menu. (See Figure 2–45.)

11. In the shortcut menu, Loop Select and Constrain Offset should already be selected by default. Make sure that the check marks appear adjacent to them. This way, you select the entire loop by selecting anywhere in the loop and you carry the constraints applied to the selected loop to the offset loop.

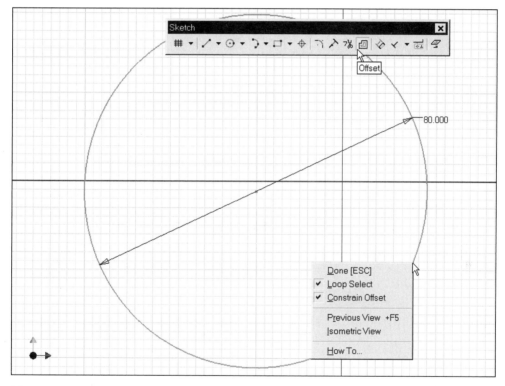

Figure 2–45 *Right-click shortcut menu*

12. Now select a point outside the shortcut menu and then select the circle.

13. Select a point inside the existing circle to construct an offset circle as shown in Figure 2–46. Again, we do not know the exact size of the offset circle at this stage. However, the size is unimportant because you will next add a general dimension to properly set its size.

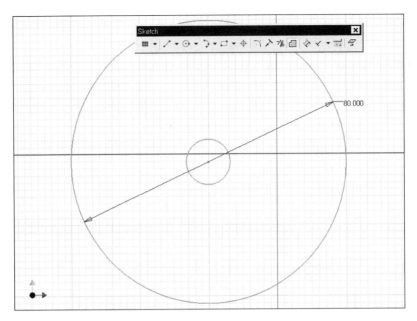

Figure 2–46 *Offset circle constructed*

14. Select General Dimension from the Sketch toolbar or panel and add a dimension (6 units) to the offset circle as shown in Figure 2–47.

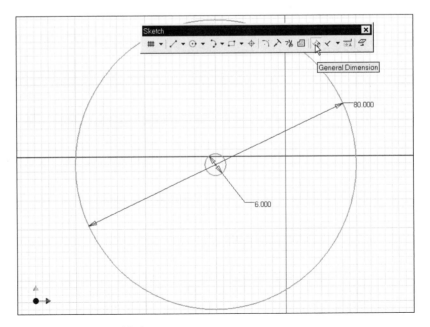

Figure 2–47 *Dimension added*

Now you will construct three line segments.

15. Select Line from the Sketch toolbar or panel.

16. Select a point inside the large circle and near the upper quadrant.

17. Move the cursor slowly downward on the screen in a nearly vertical direction. You will find a vertical constraint symbol (looks like the letter "I") displaying near the cursor position.

18. While the symbol is still displaying, select a point to specify the end point of the line. Now you have a vertical line segment. (See Figure 2–48.)

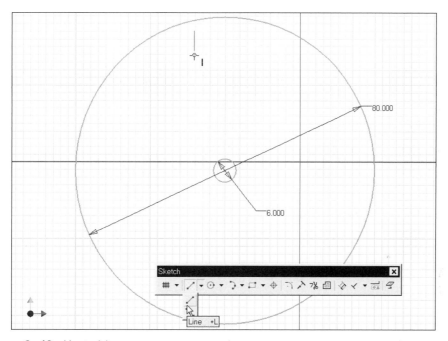

Figure 2–48 *Vertical line segment constructed*

19. Move the cursor slowly to the right in a nearly horizontal direction. You will find a perpendicular constraint symbol (looks like an inverted letter "T"). It denotes that the line segment is perpendicular to the last line segment.

20. Select a point. Now you have a second line that is perpendicular to the first line. (See Figure 2–49.)

21. Move the cursor slowly upward to the position as shown in Figure 2–50. You will find a dashed horizontal line and a parallel constraint symbol (looks like "//"). If you select a point now, you will have a parallel line and one of the end points of this line will have the same Y-coordinate value as the other vertical line. However, do not select a point now.

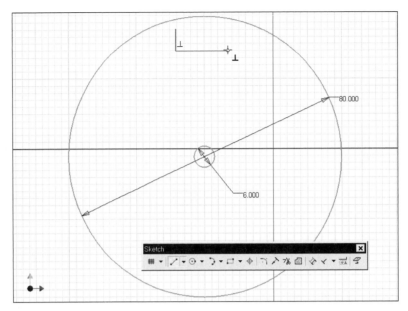

Figure 2–49 *Horizontal line segment constructed*

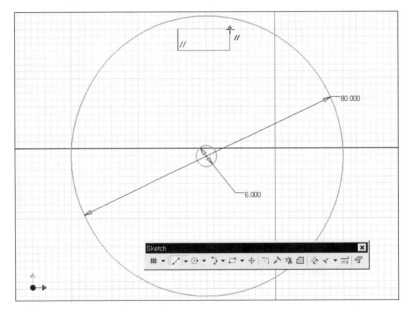

Figure 2–50 *Dashed horizontal line and parallel constraint symbol*

22. Now move the cursor further upward until it touches the circle.

23. Select the circle when you find a small dot and a coincident constraint symbol (looks like ".∩/"). They denote that the end point of the line will be constrained to the circumference of the selected circle. (See Figure 2–51.)

24. The upper end point of the left vertical line needs to be extended. Select Extend from the Sketch toolbar or panel and then select the upper end of this vertical line to extend the line to meet the circle. (See Figure 2–52.) The line is extended.

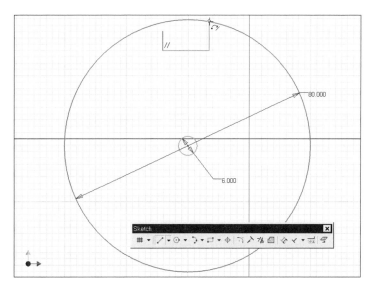

Figure 2–51 *End point of a line constrained to the circumference of a circle*

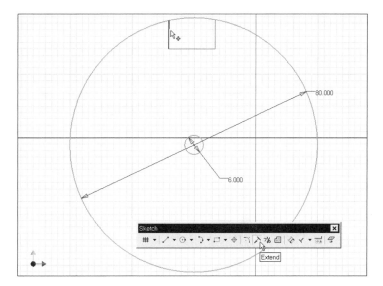

Figure 2–52 *Line extended*

25. A portion of the circle needs to be trimmed away. Select Trim from the Sketch toolbar or panel.

26. Select the circle as shown in Figure 2–53. The dashed portion of the circle will be trimmed.

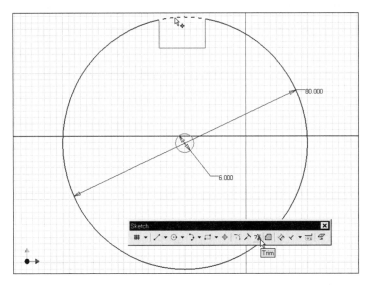

Figure 2–53 *Trimmed portion dashed*

27. The length of the two vertical lines should be equal. Select Equal from the Sketch toolbar or panel (see Figure 2–54) and then select the two vertical lines.

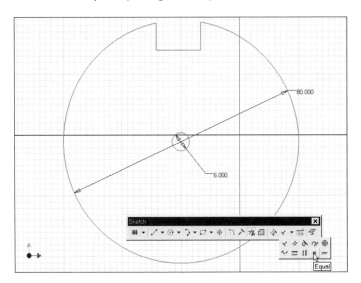

Figure 2–54 *Lines constrained to be equal in length*

Note that the "=" sign might not be showing in your toolbar and that all choices of constraints will be available as "fly-out" when you choose this constraint.

28. To complete the sketch, select General Dimension from the Sketch toolbar or panel.

29. Add dimensions (4 units for the horizontal line and 2 units for the vertical lines) to the sketch, as shown in Figure 2–55.

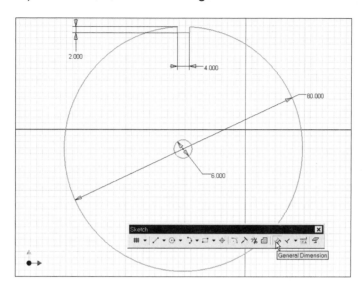

Figure 2–55 *Sketch completed*

30. The sketch is complete. Deselect Sketch on the Command Bar toolbar to exit sketch mode. (See Figure 2–56.)

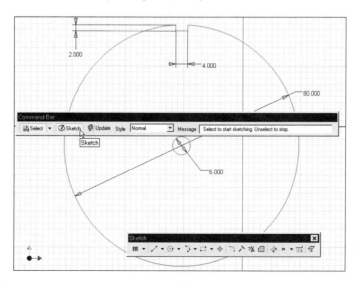

Figure 2–56 *Deselecting Sketch on the Command Bar toolbar*

Save Options

31. Now save your file. Select Save from the File menu.

32. Select FoodGrinder Project in the Locations panel and set the file name to Blade.ipt to save the file in the FoodGrinder directory. (See Figure 2–57.)

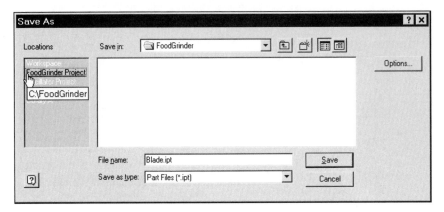

Figure 2–57 *Save As dialog box*

33. Before you select the OK button of the Save As dialog box, select the Options button. (See Figure 2–58.)

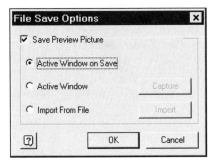

Figure 2–58 *File Save Options dialog box*

There are three ways to save a preview picture: Active Window On Save, Active Window, and Import From File.

- **Active Window On Save** The preview picture will be the most updated graphics window.

- **Active Window** You use this option in conjunction with the Capture button to specify a preview picture to be saved in the file. The preview picture will be the graphics window that you capture.

- **Import From File**   You use this option in conjunction with the Import button to specify a 120 × 120 pixel image as the preview picture. The preview picture will be the image file that you specify.

34. In the File Save Options dialog box, select Save Preview Picture and Active Window On Save. Then select the OK button.

35. On returning to the Save As dialog box, select the OK button. The file is saved.

Extrude the Solid

Now you will construct an extruded solid from the sketch. An extruded solid is a sketched solid feature. Construct it by extruding a sketch in a direction perpendicular to the plane of the sketch. First you construct a sketch, and then you select the areas where you want to extrude. After that, you select a direction for extrusion. You can extrude a set of 2D sketches in three ways: extrude in one direction, extrude in the other direction, and extrude in both directions.

36. Right-click and select Isometric View from the shortcut menu to set the display to an isometric view.

37. Select Extrude from the Features toolbar or panel. (See Figure 2–59.)

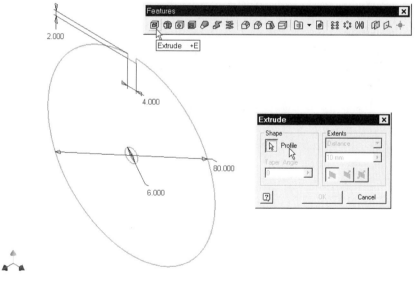

Figure 2–59   *Extrude dialog box*

38. In the Extrude dialog box, select the Profile button, if it is not already selected.

39. Select a point between the inner loop and the outer loop of the sketch such that the area highlighted in Figure 2–60 is selected.

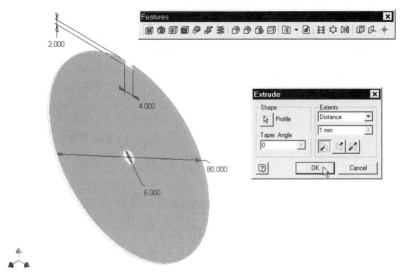

Figure 2–60 *Area to extrude selected*

40. Set the distance of extrusion to 1 mm and select the OK button. An extruded solid is constructed.

Set Display Mode

41. There are three kinds of display for a solid: Shaded Display, Hidden Edge Display, and Wireframe Display. To set the display to wireframe, select Wireframe Display from the Standard toolbar. (See Figure 2–61.)

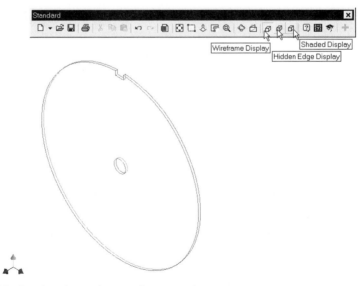

Figure 2–61 *Display changed to wireframe mode*

The extruded solid is complete.

Design Doctor and Sketch Doctor

If there is any problem in your sketch and you try to extrude it to a solid, you may find a dialog box similar to that shown in Figure 1–26 of Chapter 1 or the Sketch Doctor shown in Figure 1–27 of Chapter 1. Following the instructions delineated in the Design Doctor or Sketch Doctor dialog boxes will help you overcome the problem.

Edit the Object

42. Autodesk Inventor is a parametric solid modeling system. You can edit the parameters of the objects that you construct by selecting the object in the browser, right-clicking, and selecting Edit Feature or Edit Sketch. (See Figure 2–62.)

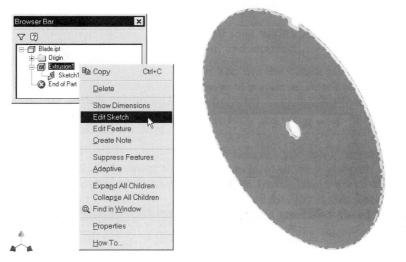

Figure 2–62 *Editing the feature*

If you select Edit Sketch, the sketch will be displayed (similar to Figure 2–56). You can modify the sketch by using the Sketch toolbar or panel. If you select Edit Feature, the Extrude dialog box will be displayed (similar to Figure 2–60). You can modify the feature by changing the extruding parameters.

43. After you are satisfied with the changes that you make, select Update from the Command Bar toolbar to update the solid.

44. Now save and close the file.

REVOLVING A SOLID

Now you will construct a revolved solid for making the handle of the food grinder. (See Figure 2–63.)

Figure 2–63 *Revolved solid*

Sketch

A revolved solid is a sketched solid feature. You construct a sketch and revolve the sketch about an axis. The axis can be a line of the sketch, an existing edge, or a work axis. In making this revolved solid, you will construct a sketch consisting of a spline and three lines. You will use the sketch as the profile and revolve the profile about the line.

1. Start a new part file and use Standard.ipt on the Default tab as the template file. On the default sketch plane, you find grids and axes displayed. They indicate the approximate screen size and the location of the origin. Now you will turn them off.

2. Select Grid from the Sketch toolbar or panel. In the Grid Settings dialog box, deselect the Grid and Axes buttons. (See Figure 2–64.)

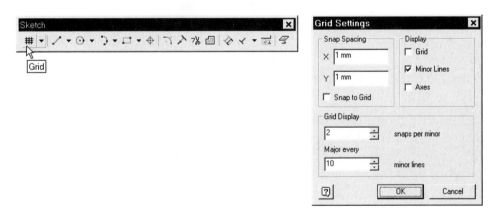

Figure 2–64 *Grid and axes displays being turned off*

Now you will construct a set of points. Using the points, you will construct a spline.

3. Select Point, Hole Center from the Sketch toolbar or panel and construct five points as shown in Figure 2–65.

4. If you want to control the exact location of the points while you are sketching, you can use the Precise Input dialog box. To display the Precise Input dialog box, select View ➤ Toolbars ➤ Precise Input.

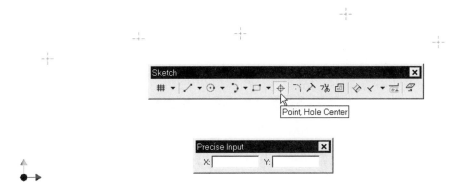

Figure 2–65 *Points constructed and Precise Input dialog box displayed*

5. To specify the exact location of the points, select General Dimension from the Sketch toolbar or panel and add dimensions as shown in Figure 2–66.

6. After adding the dimensions, select Zoom All from the Standard toolbar to set the display.

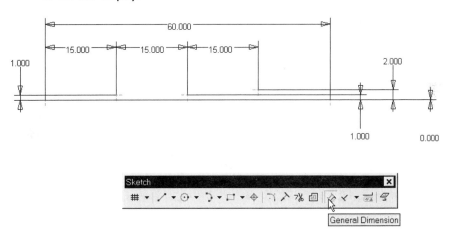

Figure 2–66 *Dimensions added to control the locations of the points*

Now you will construct a spline to pass through the five points that you constructed.

7. Select Spline from the Sketch toolbar or panel and select a point as shown in Figure 2–67.

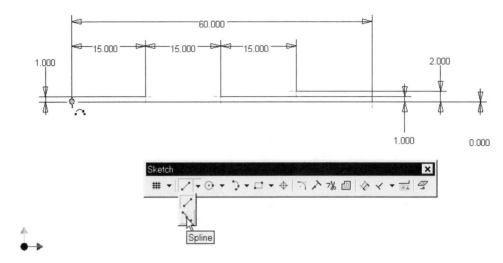

Figure 2–67 *Spline being constructed and a point selected*

8. Select the remaining points as shown in Figure 2–68.

9. Right-click to display a shortcut menu and select Create. A spline is constructed.

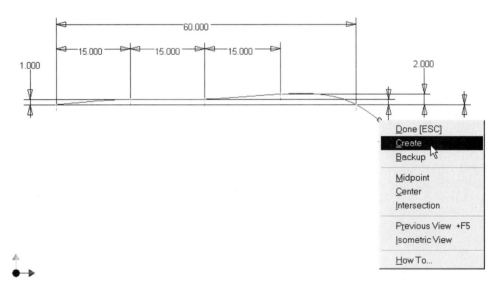

Figure 2–68 *Points selected and shortcut menu displayed*

Now you will construct three line segments.

10. Select Line from the Sketch toolbar or panel and select the left end point of the spline as shown in Figure 2–69.

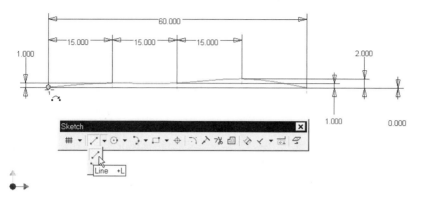

Figure 2–69 *End point of the spline selected*

Precise Input

11. Use the Precise Input toolbar to enter a precise coordinate to construct the line segments. Select View ➤ Toolbars ➤ Precise Input to view the toolbar.

12. Select Specify A Point Using Distance From The Origin And Angle From X-axis button on the Precise Input toolbar.

13. Type 8 in the D: box and 270 in the ° box (see Figure 2–70) and press ENTER. A line segment that is 8 units in length, in the direction of 270 degrees from the last selected point, is constructed.

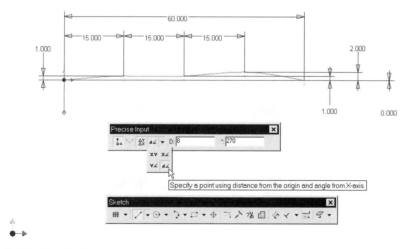

Figure 2–70 *Precise line segment constructed*

14. Now type 60 in the D: box and 0 in the ° box and press ENTER. A line segment that is 60 units in length, in the direction of 0 degrees from the last point, is constructed. (See Figure 2–71.)

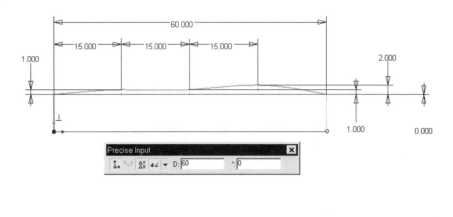

Figure 2–71 *Second precise line segment constructed*

15. Select the right end point of the spline as shown in Figure 2–72 to construct the third line segment.

16. Close the Precise Input toolbar.

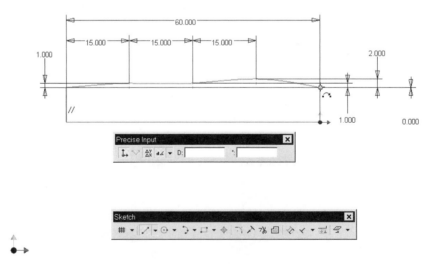

Figure 2–72 *Third line segment constructed*

17. To find out the geometric constraints that are applied to the horizontal line, select Show Constraints from the Sketch toolbar or panel and select the horizontal line. (See Figure 2–73.) The geometric constraints that are applied to the selected line are displayed.

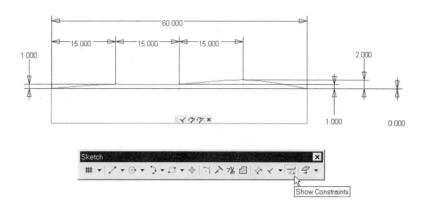

Figure 2–73 *Constraint symbols displayed on the horizontal line*

18. To complete the sketch, select General Dimension from the Sketch toolbar or panel and add a dimension as shown in Figure 2–74.

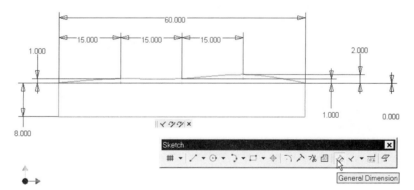

Figure 2–74 *Dimension added*

19. The sketch for the handle of the food grinder is complete. Deselect Sketch on the Command Bar toolbar to exit sketch mode.

20. Save the file (file name: Handle.ipt).

21. Right-click and select Isometric to set the display to an isometric view.

Revolve

Now you will revolve the sketch to form a revolved solid.

22. Select Revolve from the Features toolbar or panel. (See Figure 2–75.) To make a revolved solid, you specify a profile and an axis. Because there is only one closed loop in the sketch, it is selected automatically as the profile.

78

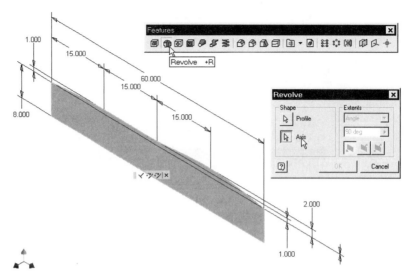

Figure 2–75 *Profile selected automatically*

23. To specify the axis, select the Axis button (if not already selected) and select the lower edge of the sketch. (See Figure 2–76.)

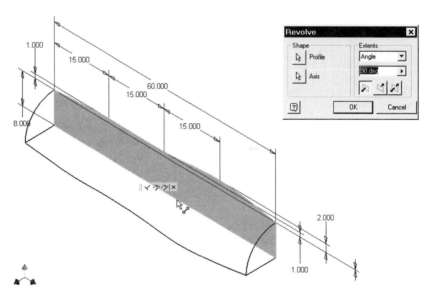

Figure 2–76 *Lower edge of the 2D sketch selected as the centerline*

24. The default angle of revolution is 90 degrees. To have a complete revolution, set Extents to Full in the Revolve dialog box. (See Figure 2–77.) Then select the OK button.

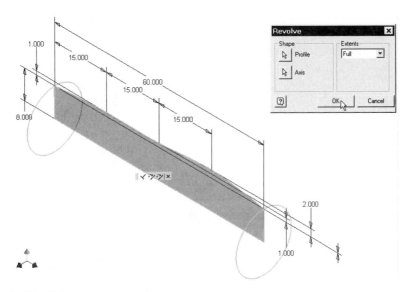

Figure 2–77 *Full extents selected*

25. The revolved solid is complete. Save and close your file.

CONSTRUCTING A SWEEP SOLID

Now with a new part file, construct a sweep solid to make the crank of the food grinder. (See Figure 2–78.)

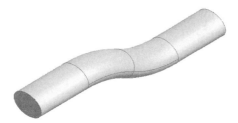

Figure 2–78 *Sweep solid*

Profile Sketch

A sweep solid is a sketched solid feature. To make a sweep solid, you construct two sketches, using one sketch as the cross section and the other sketch as the path. The cross section profile has to be a closed loop and the path can be an open loop or a closed loop. In making this sweep solid, you will construct an ellipse as the profile and a set of lines and arcs as the path.

1. Start a new part file.

2. To locate the center and major and minor axes end points of the ellipse, construct three points. Select Point, Hole Center from the Sketch toolbar or panel to construct three points as shown in Figure 2–79.

Figure 2–79 *Three points constructed*

3. Before using the points to construct the ellipse, select General Dimension from the Sketch toolbar or panel to add four dimensions as shown in Figure 2–80. (The vertical distance is 6 units and the horizontal distance is 9 units.)

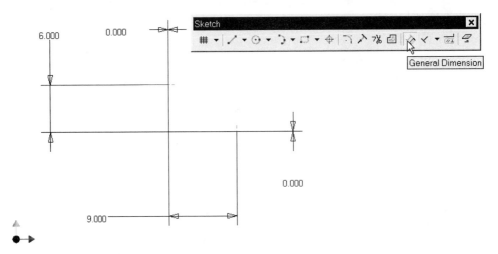

Figure 2–80 *Dimensions added*

4. Select Ellipse from the Sketch toolbar or panel.

5. Select the lower left point to establish the center of the ellipse. (See Figure 2–81.)

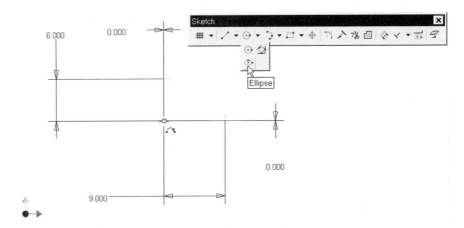

Figure 2–81 *Center point of the ellipse established*

6. Select the other two points to define the major and minor axes of the ellipse. (See Figure 2–82.) A sketch for the profile of the sweep solid is complete.

7. Deselect Sketch on the Command Bar toolbar to exit sketch mode.

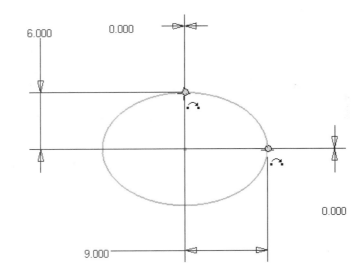

Figure 2–82 *Ellipse constructed*

8. Right-click and select Isometric View from the shortcut menu to set the display to an isometric view. (See Figure 2–83.)

9. To obtain the desired view, you can select Zoom All, Zoom Window, Zoom, and Pan from the Standard toolbar.

82

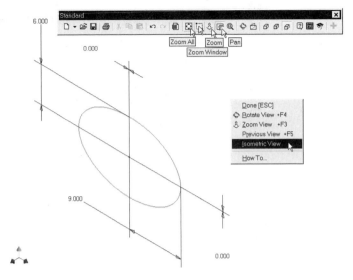

Figure 2–83 *Display set to an isometric view*

Work Plane

Now you will construct a sketch for making the path of the sweep solid. To make this sketch, you need to construct a work plane that is parallel to the YZ plane and through the center of the ellipse. Using the work plane, you will set up a sketch plane.

10. Select Work Plane from the Features toolbar or panel. Then select the center point of the ellipse as shown in Figure 2–84.

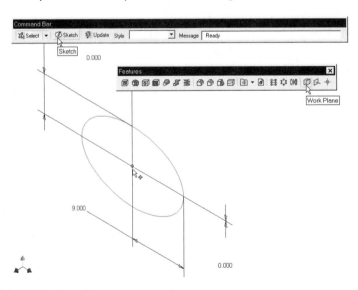

Figure 2–84 *Work plane being constructed*

11. Expand the browser Origin object to display the three default planes.

12. Then select YZ Plane in the browser bar. A work plane parallel to the YZ plane and through the center of the ellipse is constructed. (See Figure 2–85.)

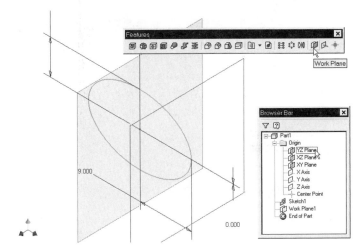

Figure 2–85 *Work plane constructed*

Sketch Path

13. Now set up a sketch on the work plane that you constructed. Select Sketch from the Command Bar toolbar and select the new work plane by clicking on an edge. (See Figure 2–86.)

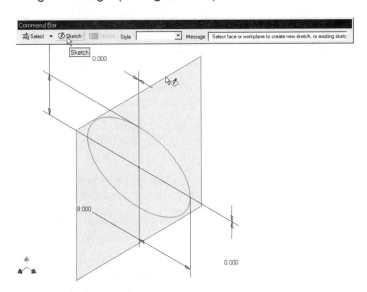

Figure 2–86 *Sketch plane being set up on the new work plane*

14. Construct a sketch on the new sketch plane. Select Line from the Sketch toolbar or panel to construct two horizontal line segments and select Tangent Arc from the Sketch toolbar or panel to construct two arcs as shown in Figure 2–87.

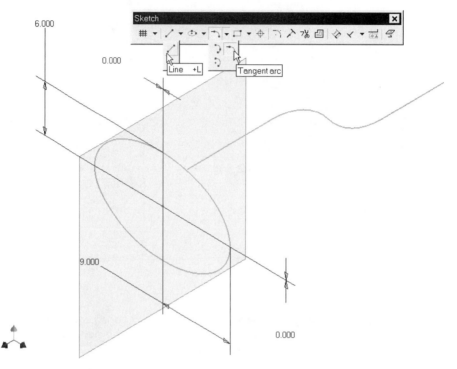

Figure 2–87 *Line and arc segments constructed*

Now apply geometric constraints to the sketch.

15. Select Horizontal from the Sketch toolbar or panel and select the horizontal lines to apply horizontal constraint to them.

16. Select Equal from the Sketch toolbar or panel, select the lines to apply equal length constraint to the lines, and select the arcs to apply equal radius to the arcs.

17. Select Tangent from the Sketch toolbar or panel and the lines and arcs to apply tangential constraint to them. (See Figure 2–88.)

To constrain this sketch with reference to the first sketch, you will project geometry of the first sketch to the current sketch plane.

18. Select Project Geometry from the Sketch toolbar or panel and select the ellipse. (See Figure 2–89.) By using Project Geometry from the Sketch toolbar or panel, you can construct geometry from edges and silhouettes of objects not residing on the current sketch plane.

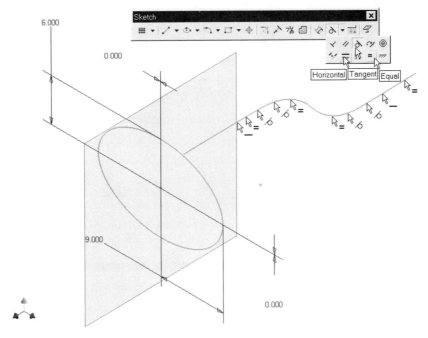

Figure 2–88 *Constraints applied*

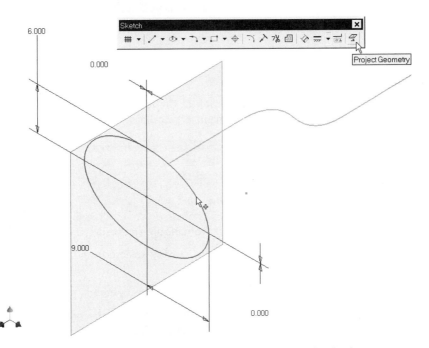

Figure 2–89 *Selected geometry being projected onto the current sketch plane*

19. To properly define the path, select General Dimension from the Sketch toolbar or panel to apply dimensions to the sketch as shown in Figure 2–90. Note that you should apply a coincident geometric constraint to the left end point of the current sketch and the center of the ellipse to make them coincident.

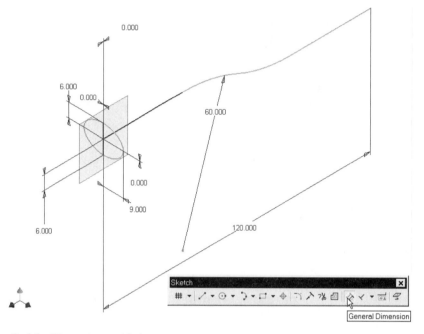

Figure 2–90 *Dimensions added*

20. The sketches for making the crank of the food grinder are complete. Deselect Sketch on the Command Bar toolbar.

21. Save your file (file name: Crank.ipt).

Sweep

Now you will construct a sweep solid.

22. Select Sweep from the Features toolbar or panel. (See Figure 2–91. Note that the closed loop sketch is automatically selected.)

23. Then select the Path button (if not already selected) and select the other sketch as the sweeping path.

24. Select the OK button. The sweep solid is constructed.

25. To hide the work plane, select the Work Plane in the browser bar, right-click, and deselect Visibility. (See Figure 2–92.)

26. The sweep feature is complete. Save and close your file.

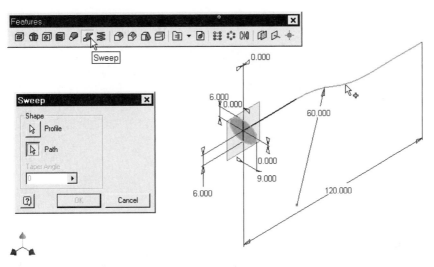

Figure 2–91   *Path being selected*

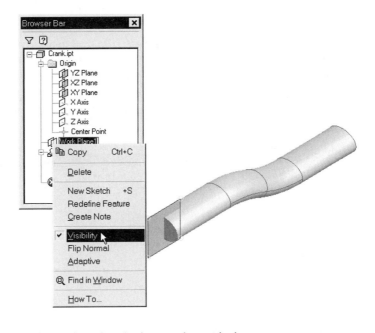

Figure 2–92   *Work plane selected in the browser bar, with shortcut menu*

CONSTRUCTING A LOFT SOLID

Now, with a new part file, you will construct a loft solid for making the funnel of the food grinder. (See Figure 2–93.)

88

Figure 2–93 *Work plane invisible*

First Sketch

A loft solid is a sketched solid feature. You construct a set of sketches on different sketch planes and loft along the sketch. In making this loft solid, you will construct two sketches. You will construct an ellipse and a circle on a sketch and project the circle to the second sketch.

1. Start a new part file.

2. Now you will construct an ellipse. Select Point, Hole Center from the Sketch toolbar or panel to construct three points for locating the center, major axis, and minor axis of the ellipse.

3. Select General Dimension from the Sketch toolbar or panel to place four dimensions to the points as shown in Figure 2–94. (The vertical and horizontal distances are 18 and 12 units respectively.)

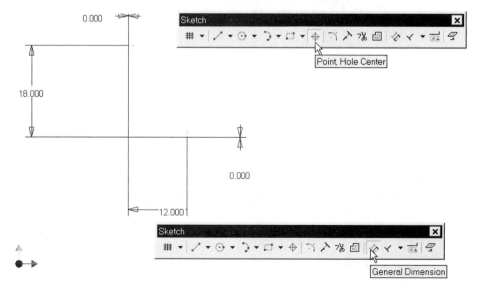

Figure 2–94 *Three points constructed and dimensions added*

4. Using the three points as center and axis end points, select Ellipse to construct an ellipse.

5. Select Center Point Circle to construct a circle with the center coincident with the center of the ellipse.

6. Select General Dimension to add a dimension to the circle (110 units in diameter). (See Figure 2–95.)

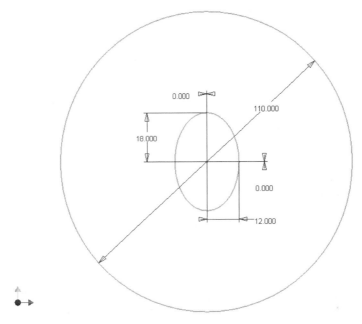

Figure 2–95 *Ellipse and circle constructed*

7. The first sketch is complete. Deselect Sketch on the Command Bar toolbar to exit sketch mode.

Rotate 3D View

Now you will learn how to manipulate the 3D view.

8. Select Rotate from the Standard toolbar. Then select a point on the screen, hold down the left mouse button, and drag the mouse. You rotate the display view in 3D. (See Figure 2–96.)

9. To rotate the object in 3D, click on or inside the circle. To rotate around the center of the circle, click outside the circle. To rotate about an axis, click on the lines.

10. While setting the 3D view, you can set the display to one of the common views. Press the space bar. Then select one of the arrows. (See Figure 2–97)

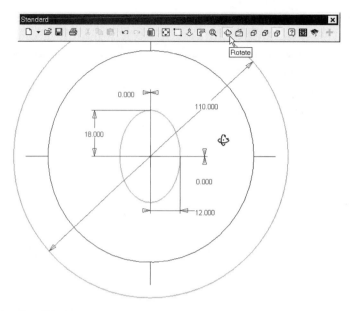

Figure 2–96 *Rotating 3D view*

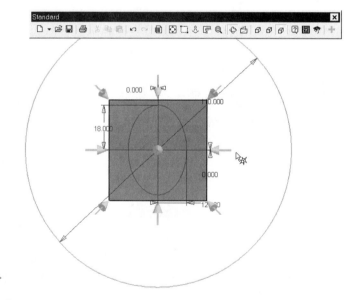

Figure 2–97 *Common views*

11. To return to 3D rotation, press the space bar again.

12. To set to isometric view, right-click to display the shortcut menu and select Isometric View. (See Figure 2–98.)

13. Right-click and select Done to exit 3D view.

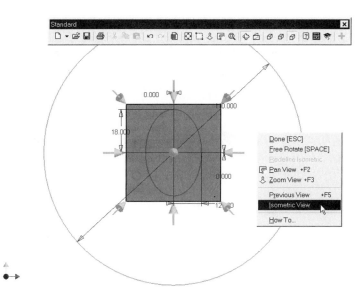

Figure 2–98 *Shortcut menu*

Work Plane

Now you will construct a work plane that is parallel to the XY plane and at a distance.

14. Select XY Plane in the browser bar, right-click to display the shortcut menu, and select Visibility to make the XY plane visible. (See Figure 2–99.)

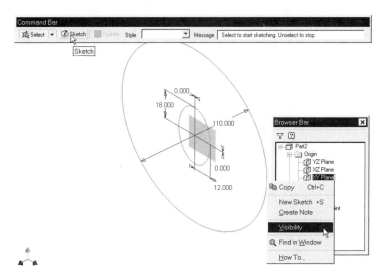

Figure 2–99 *XY plane made visible*

15. Select Work Plane from the Features toolbar or panel.

16. Then select the XY plane in the screen, hold down the left mouse button, and drag the mouse to a new position. (See Figure 2–100.) A new work plane that is offset from the selected XY plane is constructed.

17. In the Edit Dimension dialog box, type 100 to set the offset value to 100 units. Now the new work plane is offset 100 units from the XY plane. (See Figure 2–101.)

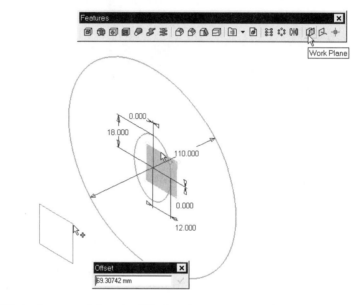

Figure 2–100 *Work plane parallel to the XY plane being constructed*

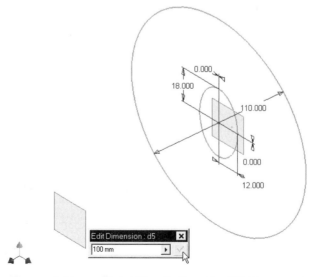

Figure 2–101 *New work plane offset 100 units from the XY plane*

Sketch planes and work planes are unlimited in size. You can construct a sketch anywhere on the sketch plane. However, if the display size of a sketch plane or work plane is too small, you may find it difficult to visualize the exact location of a sketch. Therefore, you can modify the display size.

18. Select a corner of the work plane. Then hold down the left mouse button and drag to a new position. The display size of the work plane is modified. (See Figure 2–102.)

19. Enlarge the XY plane as shown in Figure 2–103. To reiterate, the display size of the work planes has no effect on the solid model constructed on them.

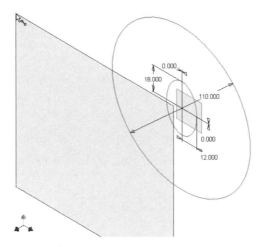

Figure 2–102   *Work plane enlarged*

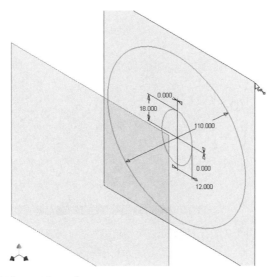

Figure 2–103   *XY plane enlarged*

Second Sketch

Now set up a new sketch plane on the new work plane.

    20. Select Sketch from the Command Bar toolbar and select the work plane as shown in Figure 2–104 to set up a new sketch plane.

On the new sketch plane, you will construct a circle by projecting a circle from the first sketch.

    21. Select Project Geometry from the Sketch toolbar or panel and select the circle on the first sketch. (See Figure 2–105.)

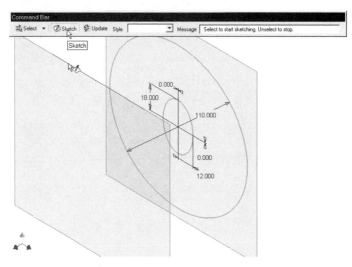

Figure 2–104   *New sketch plane set up*

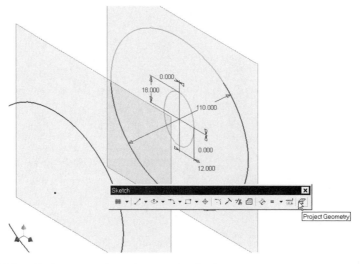

Figure 2–105   *Circle projected on the new sketch plane*

22. The sketches for a feature of the main body of the food grinder are complete. Deselect Sketch on the Command Bar toolbar to exit sketch mode.

23. Save your file (file name: Mainbody.ipt).

Loft

Now you will construct a loft solid.

24. Select Loft from the Features toolbar or panel. (See Figure 2–106.)

25. Select the Sections area of the Loft dialog box. Then select the XY plane and the ellipse as shown in Figure 2–107. Be careful to avoid selecting the circle.

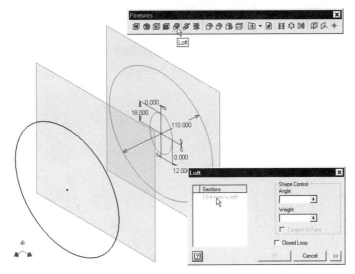

Figure 2–106   *Loft dialog box activated*

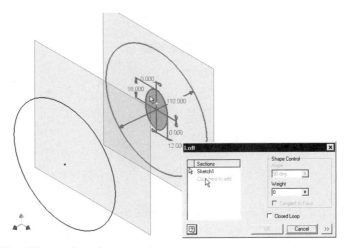

Figure 2–107   *Ellipse selected as a section*

26. Select the Sections area to add another section. Then select the circle highlighted in Figure 2–108.

27. Now select the OK button. The loft solid is complete. (See Figure 2–109.)

28. To hide the work plane, select it, right-click, and deselect Visibility.

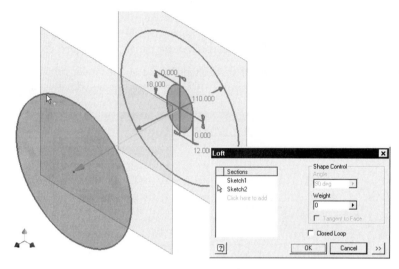

Figure 2–108   *Second section selected*

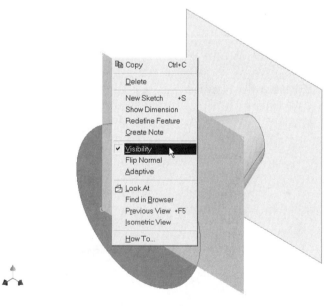

Figure 2–109   *Loft solid constructed and work plane being hidden*

29. The loft solid is complete. Save and close your file.

CONSTRUCTING A COIL SOLID

Now with a new part file, you will construct a coil solid to make the cutter of the food grinder. (See Figure 2–110.)

Figure 2–110   *Coil solid*

Sketch

A coil solid is a special kind of sweep solid feature in which the path is a helix. To make a coil solid, you construct a 2D sketch to depict the cross section of the coil, specify an axis, and specify the parameters of the helix. The axis can be a line, an edge, or a work axis.

1. Start a new part file.
2. First you will construct a rectangle that depicts the cross section of the coil. Select Two Point Rectangle from the Sketch toolbar or panel.
3. Select two points on the screen to construct a rectangle as shown in See Figure 2–111.

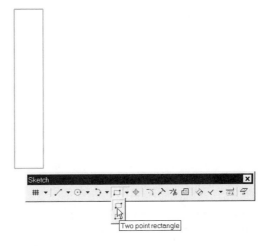

Figure 2–111   *Rectangle constructed*

4. Select Line from the Sketch toolbar or panel to construct a horizontal line as shown in Figure 2–112. You will use it as the axis of the coil.

5. Select General Dimension from the Sketch toolbar or panel and add three dimensions (1, 1, and 37 units) to the sketch as shown in Figure 2–113.

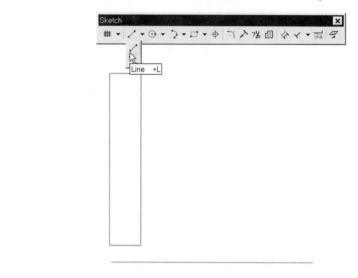

Figure 2–112 *Horizontal line constructed*

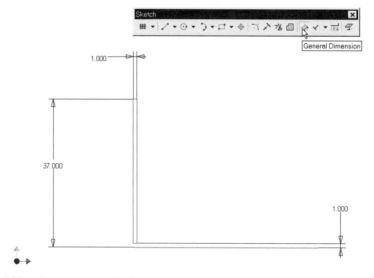

Figure 2–113 *Dimensions added*

6. The sketch for the cutter of the food grinder is complete. Deselect Sketch on the Command Bar toolbar to exit sketch mode.

7. Save your file (file name: Cutter.ipt).

Coil

Now you will construct a coil solid.

8. Set the display to an isometric view.

9. Select Coil from the Features toolbar or panel. (See Figure 2–114.)

10. Select the rectangle as the profile and select the line as the axis.

11. Now set the parameters of the coil. Select the Coil Size tab of the Coil dialog box.

12. Set Type to Revolution and Height.

13. Set the height to 117 units and the number of revolution to 2 as shown in Figure 2–115 and select the OK button. (See Figure 2–110.)

14. The coil solid is complete. Save and close your file.

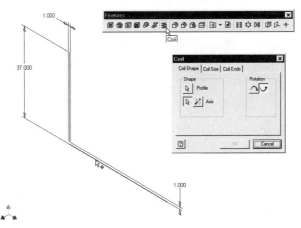

Figure 2–114 *Coil profile selected and coil axis selected*

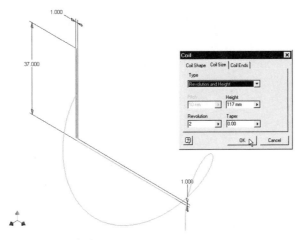

Figure 2–115 *Coil Size specified*

SPLITTING A SOLID

There are two kinds of split features: face split and part split—they are both sketched features. You construct a sketch and use the sketch to split a face into two faces or a solid part into two solids and remove one of them. Now you will make a revolved solid and construct a sketch to split the solid. Figure 2–116 shows a solid split and Figure 2–117 shows a face split.

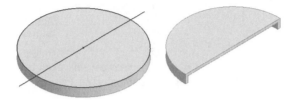

Figure 2–116   *Solid split and a portion removed*

Figure 2–117   *Face of a solid split*

Two Solid Parts

With a new part file, you will construct a revolved solid. Then you will save the solid in two files. After that, you will split the solid in one file and split a face of the solid in another file.

1. Start a new part file.

2. Now construct a sketch as shown in Figure 2–118.

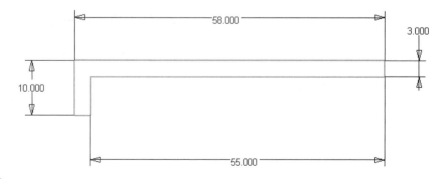

Figure 2–118   *Sketch constructed*

3. Deselect Sketch on the Command Bar toolbar to exit sketch mode.

4. Set the display to an isometric view.

5. Select Revolve from the Features toolbar or panel to revolve the sketch as shown in Figure 2–119.

6. Select the OK button. The revolved solid is complete.

7. Save the file (file name: Split1.ipt).

8. Select Save Copy As from the File menu to save to another file, named: Split2.ipt.

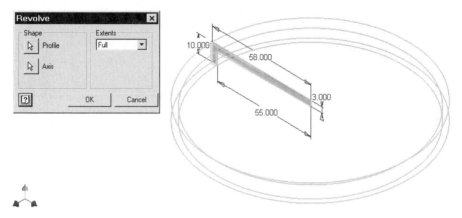

Figure 2–119   *Sketch being revolved*

Split Part

Now you will construct a sketch and use it to split the part and remove a portion of it.

9. Select Sketch from the Command Bar toolbar and select the top face highlighted in Figure 2–120 to set up a new sketch plane.

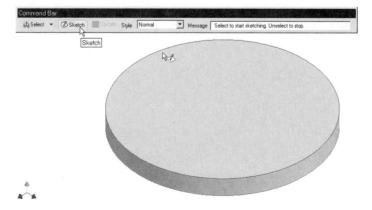

Figure 2–120   *Sketch set up*

10. Select Line from the Sketch toolbar or panel to construct a line.

11. Select Vertical from the Sketch toolbar or panel and select the line to apply a vertical constraint.

12. Select Coincident from the Sketch toolbar or panel and select the line and the center point indicated in Figure 2–121 to set the line to pass through the center point.

13. The sketch is complete. Deselect Sketch on the Command Bar toolbar to exit sketch mode.

14. Select Split from the Features toolbar or panel. (See Figure 2–122.)

15. Select the Split Part button and select the line.

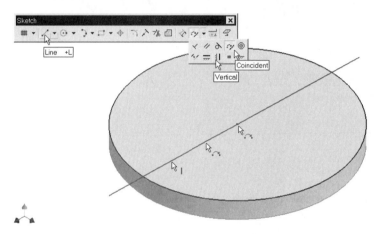

Figure 2–121   *Line constructed*

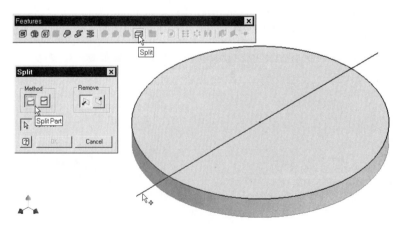

Figure 2–122   *Part being split*

16. Select the Remove button. (See Figure 2–123.) Then select the OK button.

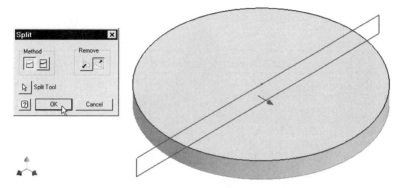

Figure 2–123   *Side to remove selected*

The part is split. (See Figure 2–124.)

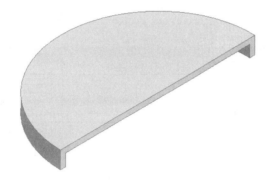

Figure 2–124   *Part split*

17. Save and close the file.

Split Face

Now, with the file Split2.ipt, you will construct a sketch and use the sketch to split a face of the solid.

1. Open the file Split2.ipt.

2. Select XY Plane in the browser bar, right-click, and select Visibility to make the plane visible. (See Figure 2–125.)

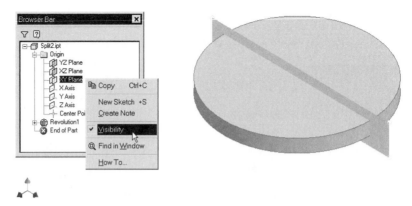

Figure 2–125 *XY plane being made visible*

3. Select Sketch on the Command Bar toolbar and select the XY plane to construct a new sketch plane.

4. Select Look At from the Standard toolbar to set the display to the top view of the current sketch plane. (See Figure 2–126.)

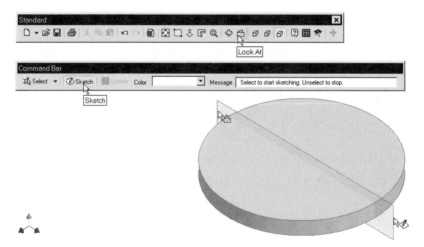

Figure 2–126 *Sketch plane constructed*

5. Select Line from the Sketch toolbar or panel and construct a line.

6. Select General Dimension from the Sketch toolbar or panel to add a dimension. (See Figure 2–127.) The sketch is complete.

7. Deselect Sketch on the Command Bar toolbar to exit sketch mode.

8. Now you will use the sketch to split the face of the solid. Select Split from the Features toolbar or panel.

9. Select the Split Face button in the Split dialog box.

10. Select the line as the split line and select the cylindrical face of the solid as the face to split. (See Figure 2–128.)

Figure 2–127 *Sketch constructed*

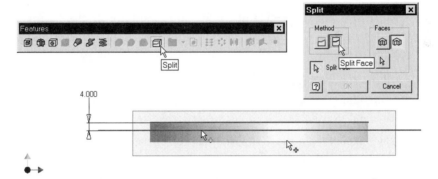

Figure 2–128 *Face being split*

11. The face is split. Right-click and select Isometric to set the display to an isometric view. (See Figure 2–129.)

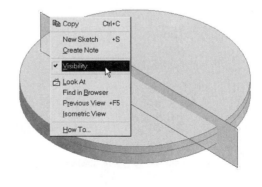

Figure 2–129 *Isometric view*

12. The face split is complete. Save and close your file.

To reiterate, you split a solid part or a face of the solid part. When you split a solid part, a portion of the solid is removed. When you split a face of a solid, you can apply face draft in two directions. Face draft is a placed feature; you will learn how to place a face draft feature later in this chapter.

BOOLEAN OPERATIONS

In making a solid part that has two or more sketched solid features, you construct them one by one and combine them by joining, cutting, or intersecting.

CONSTRUCTING WITH JOIN

Joining a sketched solid feature to a solid part produces a solid that has the volume enclosing the new solid feature and the existing solid part.

You will work on two part files—Mainbody.ipt and Crank.ipt. In the Mainbody part file, you will construct a revolved solid and join it to the solid part. In the Crank part file, you will construct two revolved solids and join them to the part.

Main Body of the Food Grinder

1. Now open the file Mainbody.ipt. You will construct a sketch on the XY plane and use the sketch to construct a revolved solid.

2. Select Sketch from the Command Bar toolbar and select the XY plane to start a new sketch.

3. Select Look At from the Standard toolbar and select the XY plane to set the display to the top view of the selected plane. (See Figure 2–130.)

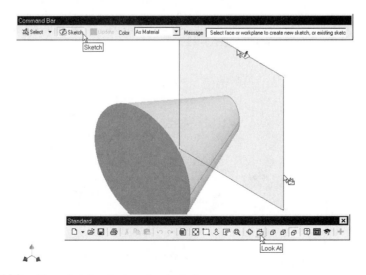

Figure 2–130 *New sketch set up and selected*

4. A sketch plane is set up and the display is set to the new sketch plane. Select Wireframe Display from the Standard toolbar to set the display to wireframe mode. (See Figure 2–131.)

5. Now you will construct a sketch on the sketch plane. Select Line from the Sketch toolbar or panel to construct three horizontal lines and three vertical lines as shown in Figure 2–132.

6. To ensure that the lines are either horizontal or vertical, select Vertical and Horizontal from the Sketch toolbar or panel to apply the necessary constraints to the lines. However, if these lines are drawn properly, they are already constrained and attempting to add Vertical and Horizontal constraints will overconstrain the sketch.

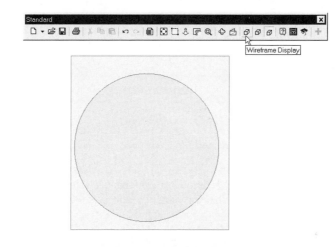

Figure 2–131 *Display set to the top view of the sketch plane*

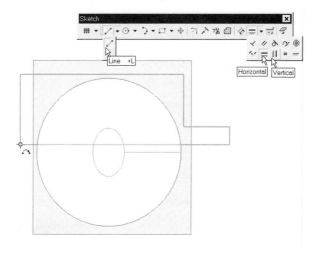

Figure 2–132 *Display set to wireframe mode and line segments constructed*

7. To set up a reference between the current sketch and the circular edge of the solid part, select Project Geometry from the Sketch toolbar or panel and select the circular edge of the loft solid to project it to the current sketch plane. (See Figure 2–133.)

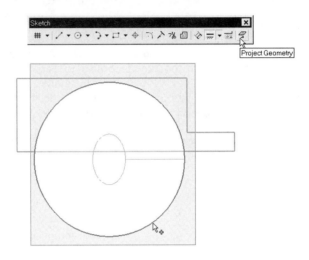

Figure 2–133 *Circular edge selected for projection*

8. Using the projected geometry as reference, select General Dimension from the Sketch toolbar or panel to add two dimensions as shown in Figure 2–134.

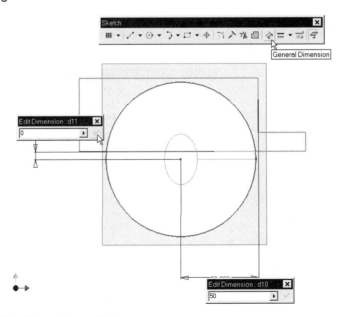

Figure 2–134 *Dimensions added*

9. Select General Dimension from the Sketch toolbar or panel and complete the dimensions (12, 40, 120, and 142 units) as shown in Figure 2–135.

10. The sketch is complete. Right-click to display the shortcut menu and select Isometric View to set the display to an isometric view.

11. Select Revolve from the Features toolbar or panel to construct a revolved solid.

12. Select the Profile button and select the sketch as shown in Figure 2–136.

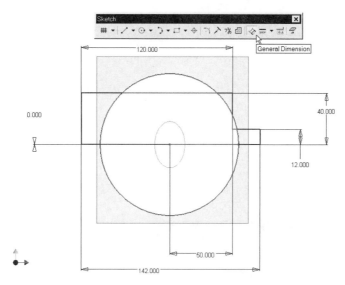

Figure 2–135 *Dimensioning completed*

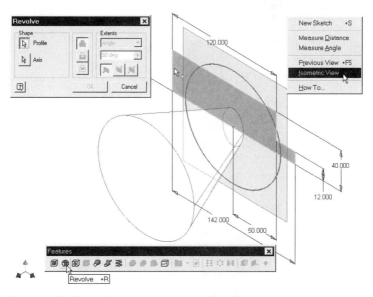

Figure 2–136 *Profile for making the revolved solid selected*

13. Select the Axis button and select the lower edge of the sketch as the centerline. (See Figure 2–137.)

14. Set Extents to Full. Then select the Join and the OK buttons. (See Figure 2–138.)

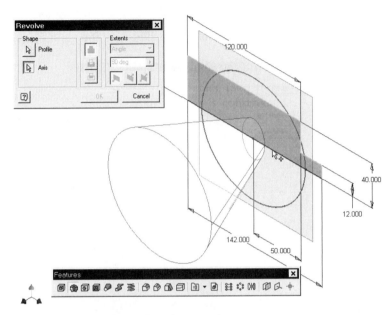

Figure 2–137 *Lower edge selected as the centerline*

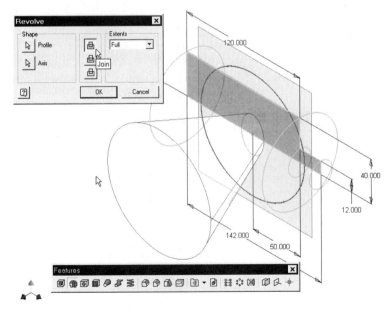

Figure 2–138 *Extents selected and Join button selected*

15. A revolved solid feature is constructed and joined to the loft solid. Now select the sketch plane and right-click. Deselect Visibility to hide the sketch plane. (See Figure 2–139.)

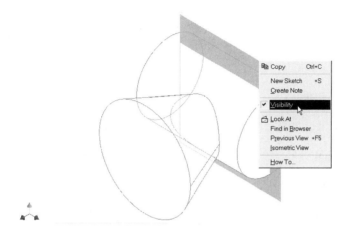

Figure 2–139 *Sketch plane selected and shortcut menu displayed*

The sketch plane is hidden. (See Figure 2–140.)

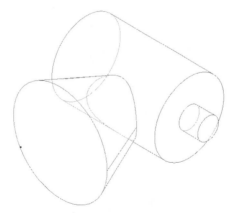

Figure 2–140 *Sketch plane hidden*

16. The join operation is complete. Save and close the file.

Crank of the Food Grinder

Now you will work on the crank of the food grinder.

1. Open the file Crank.ipt.

2. Select Sketch from the Command Bar toolbar and select the vertical face indicated in Figure 2–141 to set up a sketch plane.

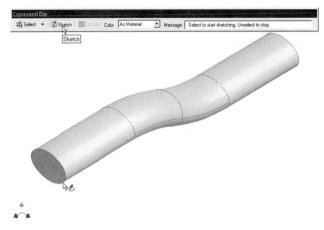

Figure 2–141   *Vertical face selected*

3. Select Two Point Rectangle and construct a rectangle as shown in Figure 2–142.

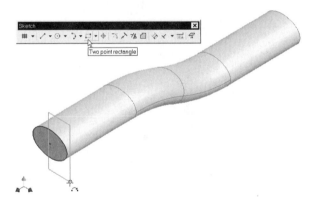

Figure 2–142   *Rectangle constructed*

4. Select General Dimension from the Sketch toolbar or panel and add two dimensions (0 and 9 units). (See Figure 2–143.)

5. Add two more dimensions (12 and 18 units) as shown in Figure 2–144.

6. The sketch is complete. Deselect Sketch on the Command Bar toolbar to turn off sketch mode.

7. Now select Revolve from the Features toolbar or panel.

8. Select the rectangle as the profile and select left vertical edge of the sketch as the axis.

9. Select the Join button, set Extents to Full, and select the OK button. (See Figure 2–145.) A revolved solid is constructed and joined to the solid part.

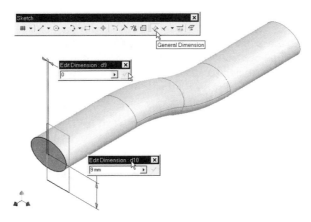

Figure 2–143 *Two dimensions constructed*

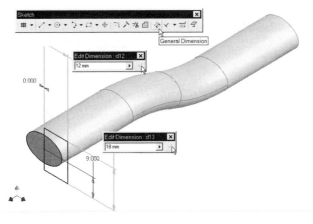

Figure 2–144 *Sketch dimensioned*

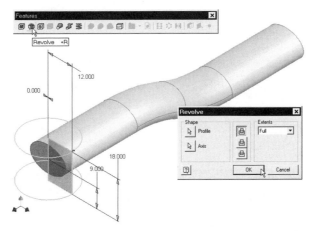

Figure 2–145 *Sketch being revolved and joined to the sweep solid*

Now you will construct another revolved solid.

10. Select Sketch from the Command Bar toolbar and select the other vertical face of the sweep solid to set up another sketch plane. (See Figure 2–146.) Note that it is easier to select this face if you switch the display to hidden edge or wireframe display.

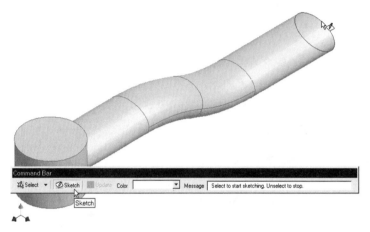

Figure 2–146 *Sketch plane set*

11. Select Wireframe Display from the Standard toolbar to set the display to wireframe mode.

12. Select Two Point Rectangle from the Sketch toolbar or panel to construct a rectangle as shown in Figure 2–147.

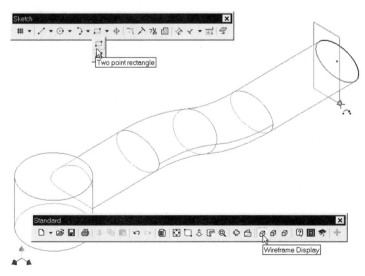

Figure 2–147 *Rectangle constructed and dimensions added*

13. Select General Dimension from the Sketch toolbar or panel and add four dimensions (0, 9, 12, and 18 units) to the sketch as shown in Figure 2–148.

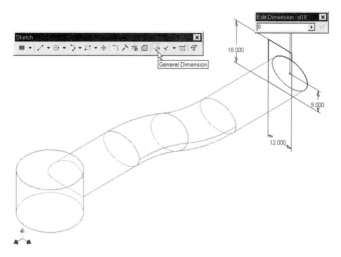

Figure 2–148   *Sketch being dimensioned*

14. The sketch is complete. Turn off sketch mode and then select Revolve from the Features toolbar or panel to revolve the sketch to form a revolved solid.

15. Select the sketch as profile, the right vertical edge as the axis, select the Join button, set Extents to Full, and select the OK button. (See Figure 2–149.)

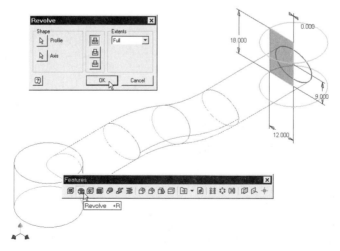

Figure 2–149   *Sketch being revolved*

16. The join operation is complete. Save your file.

CONSTRUCTING WITH CUT

Cutting a sketched solid feature from a solid part produces a solid that has the volume of the original solid part without that of the new solid feature.

Crank of the Food Grinder

Continue to work on the file Crank.ipt. If you have closed the file, open it again. Now you will construct two extruded solids and combine them to the solid part by cutting.

1. Select Sketch from the Command Bar toolbar and select a face indicated in Figure 2–150 to set up a new sketch plane.

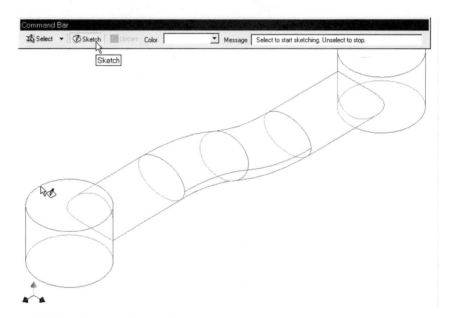

Figure 2–150   *New sketch plane being set up*

2. Select Offset from the Sketch toolbar or panel.

3. Select the circular outer edge.

4. Select a point inside the circular edge to construct an offset circle. (See Figure 2–151.)

5. Select Line from the Sketch toolbar or panel and construct a line as shown in Figure 2–152.

6. Select General Dimension from the Sketch toolbar or panel and add dimensions as shown in Figure 2–153.

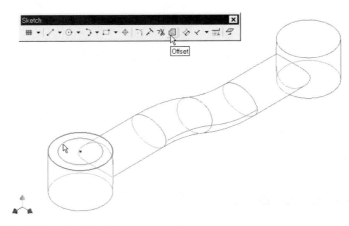

Figure 2–151 *Offset circle being constructed*

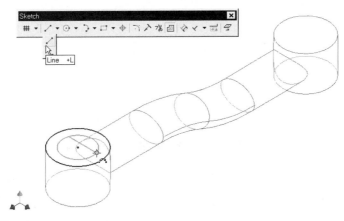

Figure 2–152 *Line constructed*

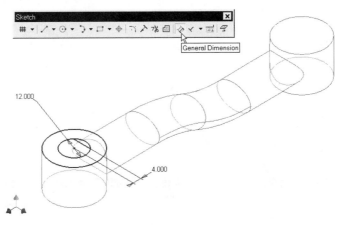

Figure 2–153 *Dimensions added*

7. Turn off sketch mode and then select Extrude from the Features toolbar or panel.

8. Select a point near the center of the sketch such that the area highlighted in Figure 2–154 is selected.

9. Select the Cut button, set Extents to All, and select the OK button.

10. The extruded solid feature is cut from the solid part. Now select Sketch from the Command Bar toolbar and select a face as shown in Figure 2–155.

11. Select Offset from the Sketch toolbar or panel and select the circular edge to construct an offset circle.

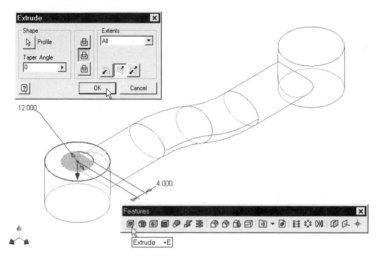

Figure 2–154   *Sketch extruded and solid cut*

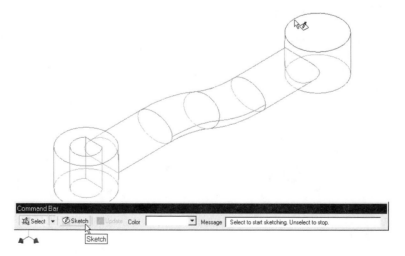

Figure 2–155   *Sketch plane constructed*

12. Select General Dimension from the Sketch toolbar or panel to construct a dimension.

13. Turn off sketch mode and select Extrude to extrude the sketch to cut through the solid. (See Figure 2–156.)

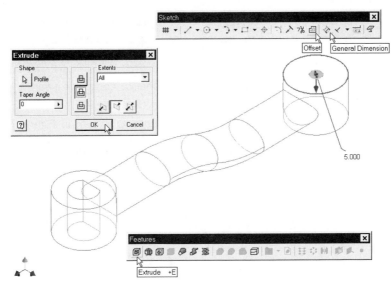

Figure 2–156 *Offset circle constructed, dimensioned, and being extruded*

The extruded feature is cut from the solid part. (See Figure 2–157.)

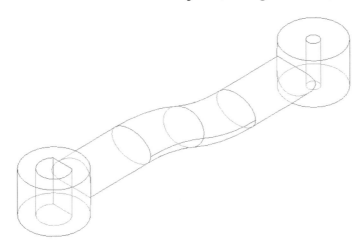

Figure 2–157 *Sketch extruded, and the solid cut*

14. The cut operation is complete. Save and close your file.

CONSTRUCTING WITH INTERSECT

Intersecting a sketched solid feature with a solid part produces a solid that has the volume common to the existing solid part and the new sketched solid feature.

Cap of the Food Grinder

You will construct the cap of the food grinder. You will construct a revolved solid and then construct an extruded solid to intersect with the revolved solid.

1. Start a new part file. You will construct a sketch for making a revolved solid.

2. Select Line from the Sketch toolbar or panel to construct a series of vertical and horizontal lines.

3. Select General Dimension to add dimensions as shown in Figure 2–158.

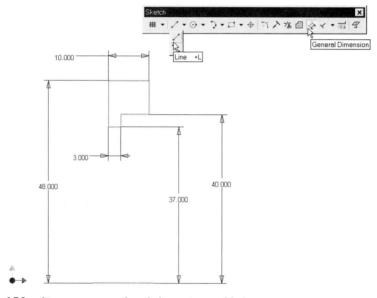

Figure 2–158 *Lines constructed and dimensions added*

4. The sketch is complete. Set the display to an isometric view.

5. Turn off sketch mode and then select Revolve from the Features toolbar or panel.

6. Select the closed loop of the sketch as the profile, and select the lower horizontal line as the axis.

7. Set Extents to Full and select the OK button. (See Figure 2–159.)

A revolved solid is complete.

8. Now you will construct an extruded solid. Select Sketch from the Sketch toolbar or panel and select the circular edge as shown in Figure 2–160 to set up a new sketch plane.

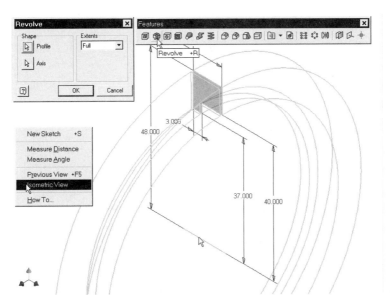

Figure 2–159   *Sketch being revolved*

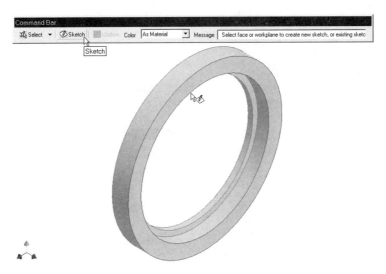

Figure 2–160   *New sketch plane constructed*

9. Select Look At from the Standard toolbar and select the circular edge to set the display to the top view of the selected face.

10. Select Two Point Rectangle to construct a rectangle on the new sketch plane.

11. Select General Dimension to add dimensions as shown in Figure 2–161.

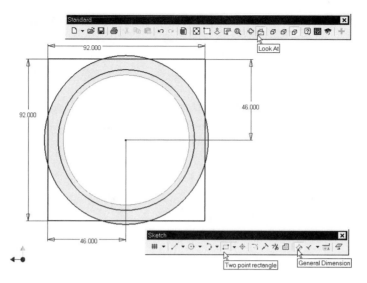

Figure 2–161 *Sketch constructed and dimensions added*

12. Set the display to an isometric view.

13. Turn off sketch mode and then select Extrude from the Features toolbar or panel.

14. To select the areas highlighted in Figure 2–162, you need to select both the rectangle and the interior of the circle.

15. After selecting the profile to extrude, select the Intersect button, set Extents to All, and select the OK button.

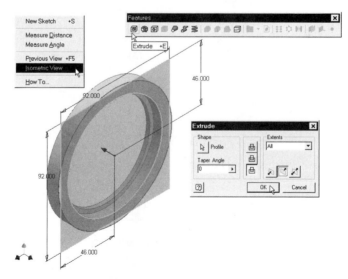

Figure 2–162 *Sketch being extruded*

The extruded solid is intersected with the revolved solid. (See Figure 2–163.)

Figure 2–163   *Features intersected*

16. The intersect operation is complete. Save and close your file (file name: Cap.ipt).

PLACED SOLID FEATURES

Because placed solid features are pre-constructed features that you select and specify, you do not need to make any sketch to construct a placed feature. You specify the type and the appropriate parameters. There are eight kinds of placed solid features: hole, shell, fillet, chamfer, rectangular pattern, circular pattern, mirror, and face draft.

PLACING A HOLE FEATURE

A hole feature is a placed solid feature. You set up a sketch plane, construct a center point location, and specify the hole parameters.

Handle of the Food Grinder

1. Open the file Handle.ipt. You will place a hole in the handle of the food grinder.

2. Select Wireframe Display from the Standard toolbar.

3. Select Sketch from the Command Bar toolbar and select the vertical face highlighted in Figure 2–164 to set up a new sketch plane.

4. To place a hole, you need to construct a center point. Because there is already a center point on the sketch plane, you can place the hole on it directly. Select Hole from the Features toolbar or panel.

5. Select the center point as shown in Figure 2–165. (If there is no existing center point, you construct a center point by selecting Hole, Hole Center from the Sketch toolbar or panel and specifying a location.)

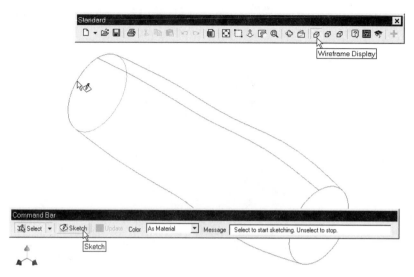

Figure 2–164 *New sketch plane being constructed*

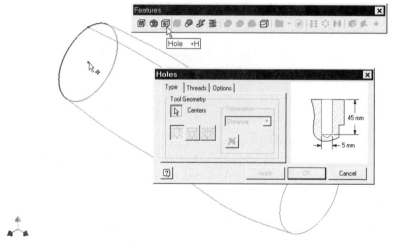

Figure 2–165 *Hole center selected*

6. Set hole size (diameter = 5 units and depth = 45 units) and select tool geometry and hole termination method as shown in Figure 2–166.

7. Select the Options tab and set the drill point angle as shown in Figure 2–167.

8. Select the OK button. A hole feature is placed in the solid part. (See Figure 2–168.)

9. The hole feature is complete. Save and close your file.

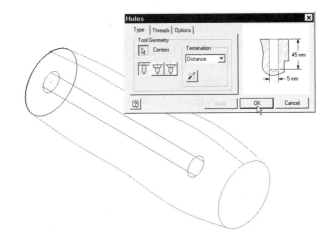

Figure 2–166 *Hole size, tool geometry, and termination method selected*

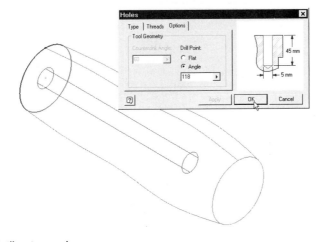

Figure 2–167 *Drill point angle set*

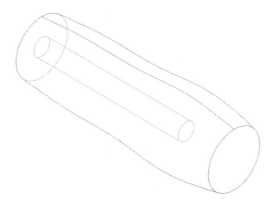

Figure 2–168 *Hole feature placed*

PLACING A SHELL FEATURE

A shell feature is a placed solid feature; it makes a solid hollow. While constructing a shell feature, you can remove some of the faces of the solid.

Main Body of the Food Grinder

1. Open the file Mainbody.ipt. You will place a shell feature on the main body of the food grinder. While placing the shell feature, you will remove two faces.

2. Select Rotate from the Standard toolbar and rotate the display as shown in Figure 2–169. Then right-click and select Done.

3. Select Shell from the Features toolbar or panel.

4. Set the Direction to Inside and Thickness to 3 units.

5. Select the Remove Faces button and select the faces highlighted in Figure 2–170.

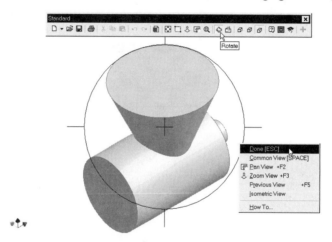

Figure 2–169   *Display set to wireframe mode*

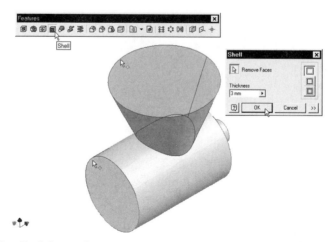

Figure 2–170   *Shell feature being placed*

6. Select the OK button. The solid is made hollow and the selected faces are removed. (See Figure 2–171.)

7. The shell feature is complete. Save your file.

Figure 2–171   *Shell feature placed*

PLACING FILLET FEATURES

A fillet feature is a placed feature; it causes an edge to be rounded. You select an edge and specify a fillet radius. Now you will place a fillet feature on the main body, the crank, and the handle of the food grinder.

Main Body of the Food Grinder

1. Open the file MainBody.ipt if you have closed it.

2. Select Rotate from the Standard toolbar and rotate the display as shown in Figure 2–172. Then right-click and select Done.

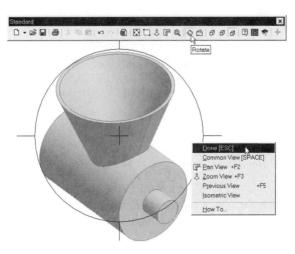

Figure 2–172   *Display rotated*

128

3. Select Fillet from the Features toolbar or panel.

4. Set the fillet radius to 4 units.

5. Select three edges highlighted in Figure 2–173.

6. Select the OK button. The edges are filleted. (See Figure 2–174.)

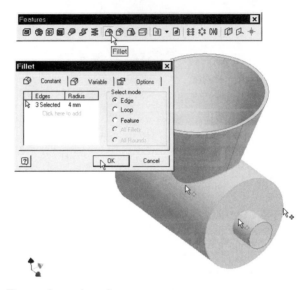

Figure 2–173   *Three edges selected*

Figure 2–174   *Three edges filleted*

7. The fillet feature is complete. Save and close your file.

Crank of the Food Grinder

1. Open the file Crank.ipt.

2. Select Fillet from the Features toolbar or panel and select six edges as shown in Figure 2–175.

3. Set the fillet radius to 2 units and then select the OK button. The edges are filleted. (See Figure 2–176.)

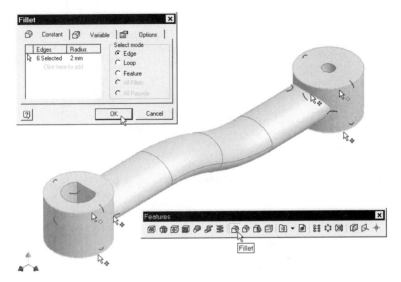

Figure 2–175 *Six edges selected*

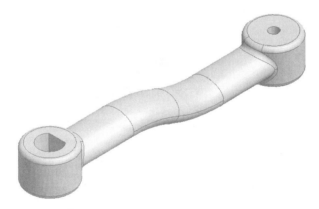

Figure 2–176 *Six edges filleted*

4. The crank of the food grinder is complete. Save and close your file.

Handle of the Food Grinder

1. Now open the file Handle.ipt.

2. Select Fillet from the Features toolbar or panel.

3. Set the fillet radius to 3 units and select two edges as shown in Figure 2–177.

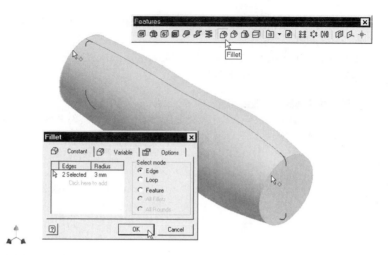

Figure 2–177   *Fillet radius set and edges selected*

4. Select the OK button. (See Figure 2–178.)

Figure 2–178   *Edges filleted*

5. The crank of the food grinder is complete. Save and close your file.

Engineering Drawing Output

Now you have completed two component parts of the food grinder, the crank and handle. If you wish to learn how to produce engineering drawings from these solid parts, you may proceed to Chapter 6. Now continue to work on this chapter to learn the other methods for constructing placed solid features.

PLACING A CHAMFER FEATURE

A chamfer feature is a placed feature; it causes an edge to be beveled. You specify an edge and state the size of the bevel. You will place a chamfer feature for an edge of the main body of the food grinder.

Main Body of the Food Grinder

1. Open the file Mainbody.ipt.

2. Select Chamfer from the Features toolbar or panel.

3. Set the Distance to 2 units and select an edge as shown in Figure 2–179.

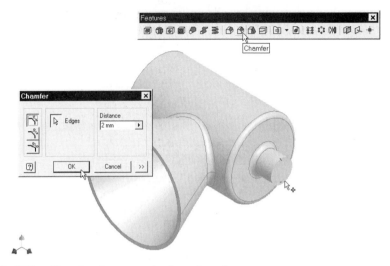

Figure 2–179 *Chamfer size set and edge selected*

4. Select the OK button. (See Figure 2–180.)

Figure 2–180 *Edge chamfered*

5. The chamfer feature is complete. Save and close your file.

PLACING A RECTANGULAR PATTERN FEATURE

Rectangular pattern features are placed features. Selected features are repeated in a rectangular pattern.

Blade of the Food Grinder

1. Open the file Blade.ipt. You will construct three extruded solids and make three rectangular patterns of the extruded features on the blade of the food grinder.

2. Now make a sketch and construct an extruded solid. Select Sketch from the Command Bar toolbar and select a face to construct a sketch plane.

3. Select Center Point Circle from the Sketch toolbar or panel and construct a circle.

4. Select General Dimension from the Sketch toolbar or panel and add three dimensions (Horizontal distance = 24 units, vertical distance = 24 units, and diameter = 2 units). (See Figure 2–181.)

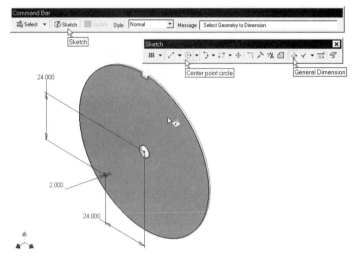

Figure 2–181 *Sketch plane constructed, circle constructed, and dimensions added*

5. Turn off sketch mode and select Extrude from the Features toolbar or panel.

6. Select the circle to extrude it so that it cuts through the solid. If you find it difficult to select the circle, move the cursor over the circle, right-click to display the shortcut menu, and choose Select Other to cycle through the possible selections. An extruded solid is constructed. (See Figure 2–182.)

7. Make another extruded solid. Select Sketch from the Command Bar toolbar and select the face highlighted in Figure 2–183 to set up a sketch plane.

8. Construct a circle and add dimensions (horizontal distance = 35 units, vertical distance = 6 units, and diameter = 2 units) as shown in Figure 2–183.

9. Select Extrude from the Features toolbar or panel and extrude the circle to cut through the solid.

10. Now construct the third circle (horizontal distance = 15 units, vertical distance = 30 units, and diameter = 2 units) and extrude it through the solid as shown in Figure 2–184.

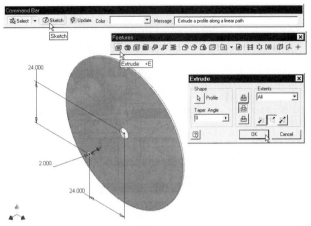

Figure 2–182   *Circle extruded to cut through the solid*

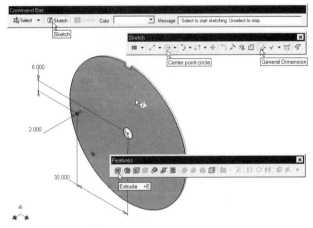

Figure 2–183   *Circle constructed*

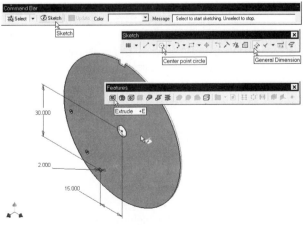

Figure 2–184   *Third circle constructed and being extruded*

Now you will construct a rectangular pattern of an extruded solid. A rectangular pattern needs one or two straight edges to define the directions of the pattern. Here you will use the horizontal edge of the solid to define a direction.

11. Select Rectangular Pattern from the Features toolbar or panel and select the highlighted extruded cut feature.

12. Select the Direction 1 button and select the highlighted edge.

13. Set Count to 6 and Spacing to 6 units and select the OK button. (See Figure 2–185.) A rectangular pattern is constructed.

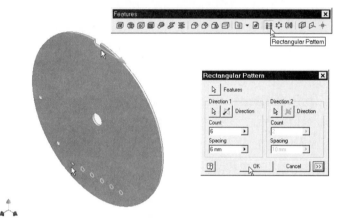

Figure 2–185 *Extruded hole being patterned*

Now you will make the second rectangular pattern feature. In making this rectangular pattern, you will define two directions.

14. Select Rectangular Pattern from the Features toolbar or panel again.

15. Select the extruded cut feature indicated in Figure 2–186.

16. Select the horizontal edge highlighted in Figure 2–186 and set Direction 1 Count to 9 and Spacing to 6 units.

17. Now select the Direction 2 button and select the vertical edge highlighted in Figure 2–187. (Select the Direction button if the direction arrow is not pointing upward.)

18. Then set the Direction 2 Count to 3 and Spacing to 6 units.

19. Select the OK button. The selected extruded cut feature will repeat in a rectangular pattern of three rows and nine columns. (See Figure 2–187.) The second rectangular pattern feature is constructed.

20. To construct the third rectangular pattern feature, select the Rectangular Pattern button of the Features toolbar or panel again.

21. Select the extruded cut feature highlighted in Figure 2–188.

22. Specify five columns at 6 units spacing and three rows at 6 units spacing.

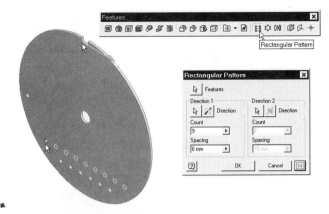

Figure 2–186   *Column count and spacing set*

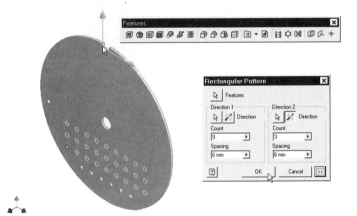

Figure 2–187   *Row count and spacing set*

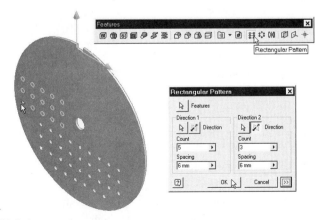

Figure 2–188   *Hole being patterned*

23. Select the OK button. (See Figure 2–189.)

24. The rectangular patterns are complete. Save and close your file.

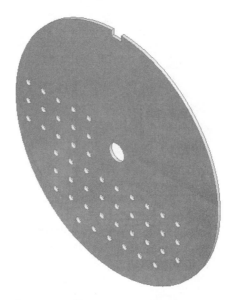

Figure 2–189   *Rectangular pattern features constructed*

PLACING A CIRCULAR PATTERN FEATURE

Circular pattern features are placed features. Features are repeated in a circular pattern about an axis.

Cap of the Food Grinder

You will construct two extruded features and make two circular patterns of the extruded features on the cap of the food grinder. You will set up a sketch plane on a face of the solid, construct a sketch, and extrude the sketch to cut the solid.

1. Open the file Cap.ipt.

2. Now construct an offset circle. Select Sketch from the Command Bar toolbar and select a vertical face to set up a new sketch plane as shown in Figure 2–190.

3. Select Offset from the Sketch toolbar or panel and select the circular edge to construct an offset circle.

4. Select Line from the Sketch toolbar or panel and construct two vertical lines as shown in Figure 2–191.

5. Select General Dimension from the Sketch toolbar or panel and add three dimensions (15 units, 30 units, and 84 units) as shown in Figure 2–192.

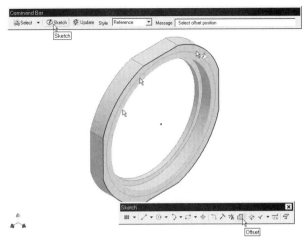

Figure 2–190 *Offset circle constructed on the new sketch plane*

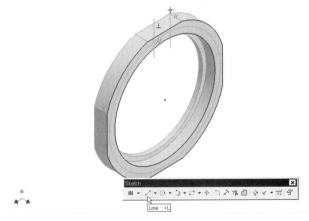

Figure 2–191 *Vertical lines constructed*

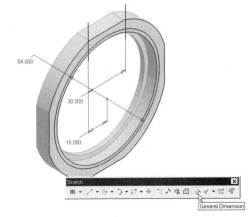

Figure 2–192 *Dimensions added*

6. Select Trim from the Sketch toolbar or panel and select the locations highlighted in Figure 2–193 to trim the sketch.

7. The sketch is complete. Now extrude the sketch to cut the solid. Select Extrude from the Features toolbar or panel and select the area highlighted in Figure 2–194 as the profile for extrusion.

8. Set Extents to 7 units and select the Cut button.

9. Select the OK button. The sketch is extruded to cut the solid.

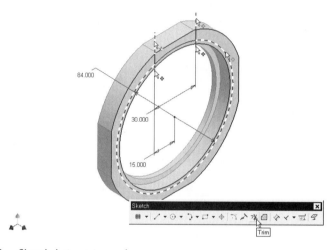

Figure 2–193   *Sketch being trimmed*

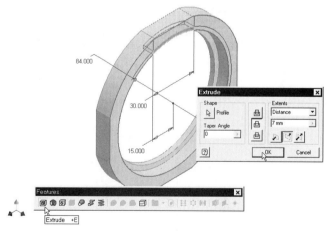

Figure 2–194   *Extruded solid being constructed*

Now you will construct a circular pattern of the extruded cut feature. To make a circular pattern, you need a straight edge, a circular edge, or an axis to define the rotation axis. Here you will use a circular edge.

10. Select Circular Pattern from the Features toolbar or panel and select the extrude cut feature and the circular edge highlighted in Figure 2–195 as Features and Rotation Axis, respectively.

11. Select the >> button to expand the Circular Pattern dialog box and set Creation Method to identical and Positioning Method to Incremental.

12. Set Placement Count to 4 and Angle to 90.

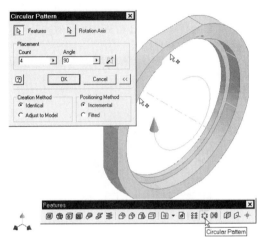

Figure 2–195 *Placement count and angle set*

13. Select the OK button. A circular pattern feature is constructed. (See Figure 2–196.)

14. Now select Rotate from the Standard toolbar to rotate the display.

Figure 2–196 *Circular pattern features placed*

15. Rotate the display as shown in Figure 2–197. Then right-click and select Done.

Now you will construct another sketch and use the sketch to construct an extruded solid to cut the solid.

16. To make a sketch, select Sketch from the Command Bar toolbar and select a circular edge to set up a new sketch plane. (See Figure 2–197.)

17. To display a wireframe so that edges on the other side can be selected, select Wireframe Display from the Standard toolbar.

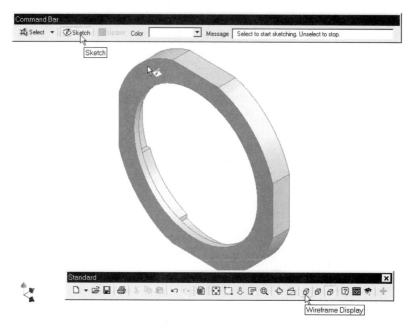

Figure 2–197 *Display rotated and new sketch plane constructed*

18. Select Project Geometry from the Sketch toolbar or panel.

19. Select two straight edges and a circular edge as shown in Figure 2–198 to project them onto the current sketch plane.

20. Select Center point arc from the Sketch toolbar or panel.

21. Select the center point of the circular solid.

22. Select the end points of the projected lines as shown in Figure 2–199. An arc is constructed.

23. Select Extrude from the Features toolbar or panel.

24. Select the area highlighted in Figure 2–200 as the profile.

25. Extrude the profile a distance of 7 units to cut the solid.

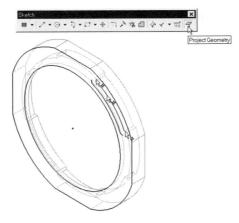

Figure 2–198 *Lines and arc projected*

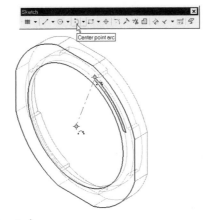

Figure 2–199 *Arc being constructed*

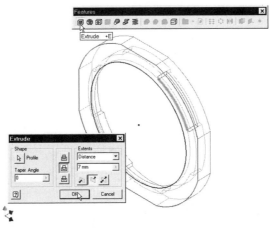

Figure 2–200 *Extruded cut feature constructed*

The sketch is extruded to cut the solid.

26. Now construct a circular pattern of this extruded cut feature. Select Circular Pattern from the Features toolbar or panel.

27. Select the highlighted feature.

28. Set Placement Count to 4 and Incremental Angle to 90. (See Figure 2–201.)

29. Select the OK button. The circular pattern features are placed. (See Figure 2–202.)

30. The cap of the food grinder is complete. Save and close your file.

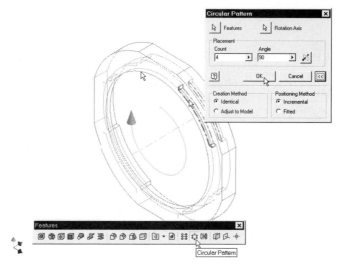

Figure 2–201   *Circular pattern features being constructed*

Figure 2–202   *Circular pattern features constructed*

PLACING A MIRROR FEATURE

The mirror feature is a placed solid feature. To construct a mirror feature, you need a mirror plane, which can be an existing plane, sketch plane, or work plane.

Blade of the Food Grinder

1. Open the file Blade.ipt.

This solid part has three rectangular pattern features. You will construct three mirror features of these pattern features. To make the mirror features, you will construct work planes, and to locate the work planes, you will construct a work axis.

2. Select Work Axis from the Features toolbar or panel and select the inner circular edge as highlighted in Figure 2–203 to construct a work axis.

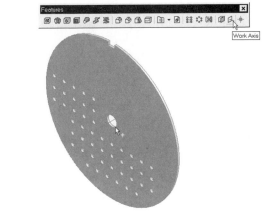

Figure 2–203   *Work axis constructed*

A work axis is constructed. Now you will construct a work plane that is parallel to the YZ plane and passes through the work axis.

3. Select Work Plane from the Features toolbar or panel.

4. Select YZ Plane in the browser bar. (See Figure 2–204.)

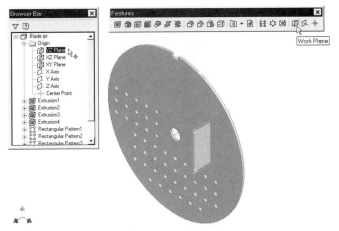

Figure 2–204   *YZ plane selected*

5. Select the work axis to set the new work plane to pass through the axis. (See Figure 2–205.)

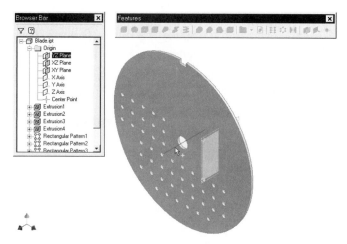

Figure 2–205   *Work axis selected*

6. In the Angle dialog box, type 0 to set the new work plane to parallel to the YZ plane. (See Figure 2–206.)

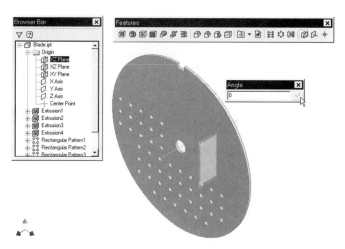

Figure 2–206   *Angle between the new work plane and the relative plane set*

A work plane parallel to the YZ plane and passing through the work axis is constructed. Now construct another work plane that is parallel to the XZ plane and passes through the work axis.

7. Select Work Plane and select the XZ plane. (See Figure 2–207.)

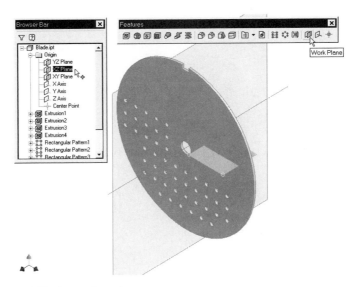

Figure 2–207 *XZ plane selected*

8. Select the work axis and set the angle to 0. A work plane parallel to the XZ plane and passing through the work axis is constructed. (See Figure 2–208.)

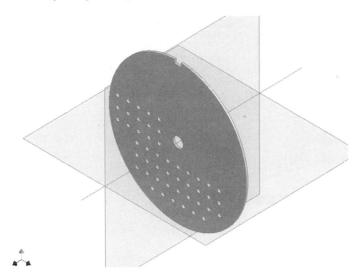

Figure 2–208 *Work plane constructed*

9. Now you have two work planes. You will use them as mirror planes. Select Mirror Feature from the Features toolbar or panel.

10. Select the two pattern features highlighted in Figure 2–209 so that the resulting mirror features resemble Figure 2–210.

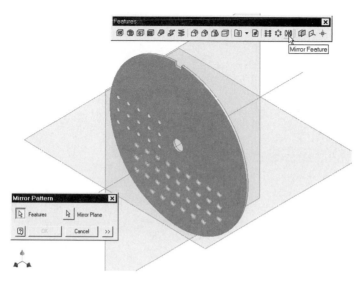

Figure 2–209   *Pattern features selected*

11. Select the Mirror Plane button and select the horizontal work plane. (See Figure 2–210.)

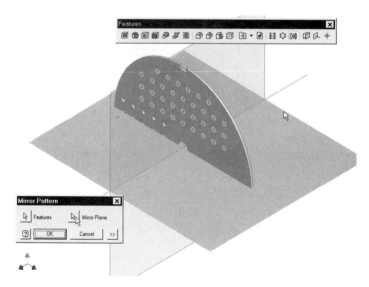

Figure 2–210   *Horizontal work plane selected*

12. Two sets of pattern features are mirrored. Now select Mirror Feature from the Features toolbar or panel again.

13. Then select the features and vertical plane highlighted in Figure 2–211.

14. Select the OK button. The mirror features are complete. (See Figure 2–212.)

15. Now select the work planes and work axis one by one and, for each, right-click and deselect Visibility.

The work planes and the work axis are hidden. (See Figure 2–213.)

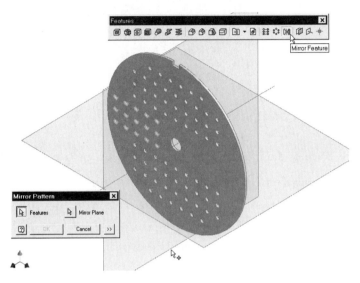

Figure 2–211   *Pattern features and vertical work plane selected*

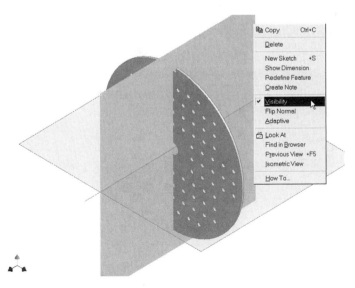

Figure 2–212   *Vertical work plane selected*

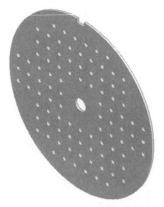

Figure 2–213 *Mirror features completed*

16. The blade of the food grinder is complete. Save and close your file.

PLACING FACE DRAFT FEATURES

A face draft feature applies a taper along the selected faces of a solid. You specify edges or split lines and state the taper angle.

1. Open the file Split2.ipt. You will place two face draft features, one along an edge and one along a split line.

2. Select Face Draft from the Features toolbar or panel.

3. Then select the split line and the face indicated in Figure 2–214 and set the draft angle to 10 degrees.

4. Select the OK button. A face draft is placed along the split line on the selected face.

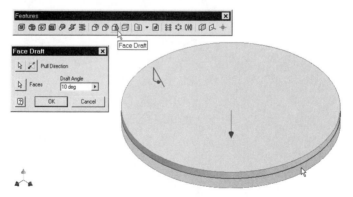

Figure 2–214 *Face draft place along the split line and on the face*

5. Now you will rotate the display. Select Rotate from the Standard toolbar and rotate the display as shown in Figure 2–215.

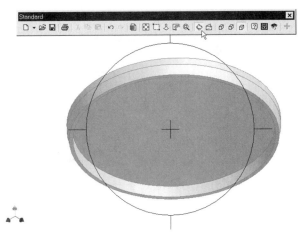

Figure 2–215   *Display rotated*

6. To place another face draft, select Face Draft from the Features toolbar or panel.

7. Select the face and edge indicated in Figure 2–216.

8. The face draft features are placed. Save and close your file.

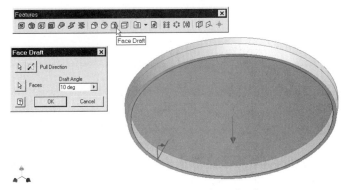

Figure 2–216   *Face draft placed along an edge and on the face*

CONSTRUCTING COMPLEX SOLID PARTS

In the preceding sections, you learned how to construct sketches, make sketched features and placed features, and use work features. Now you will have more practice on solid modeling and you will complete the cutter and the main body of the food grinder.

CUTTER OF THE FOOD GRINDER

1. Open the file Cutter.ipt. You will join a revolved feature, cut an extruded feature, and place a chamfer feature to complete the solid. The completed part is shown in Figure 2–226.

Revolve Join

2. Select XY Plane in the browser bar, right-click, and select Visibility to make the plane visible.

3. Select and drag the corners of the plane to enlarge its display size and reposition it as shown in Figure 2–217.

4. Select Look At from the Standard toolbar and select the plane to display the top view of the selected plane.

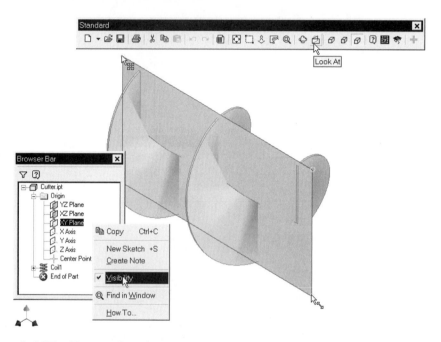

Figure 2–217 *Plane made visible, enlarged, and repositioned*

5. Select Sketch from the Command Bar toolbar and select the XY plane to set up a new sketch plane.

6. Select Two Point Rectangle from the Sketch toolbar or panel and construct a rectangle. (See Figure 2–218.)

7. Select General Dimension from the Sketch toolbar or panel and add dimensions as shown in Figure 2–219.

8. Right-click and select Isometric View to set the display to an isometric view.

9. Select Revolve from the Features toolbar or panel and select the rectangle as the profile and the lower edge of the rectangle as the centerline.

10. Set Extents to Full, select the Join button, and select the OK button. (See Figure 2–220.)

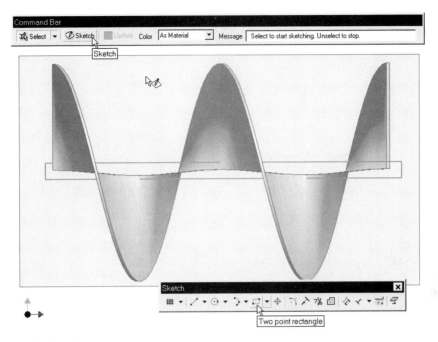

Figure 2–218 *Display set*

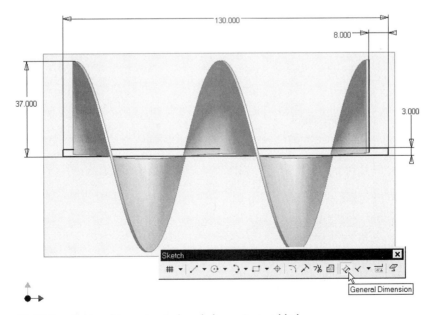

Figure 2–219 *Rectangle constructed and dimensions added*

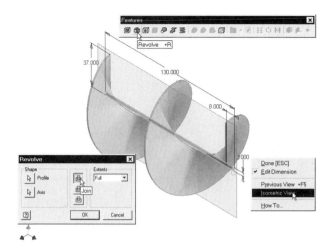

Figure 2–220 *Rectangle being revolved*

Extrude Cut

11. Select XY Plane in the browser bar, right-click, and deselect Visibility to hide the plane.

12. Select Sketch from the Command Bar toolbar and select the vertical end face as shown in Figure 2–221 to set up a new sketch plane.

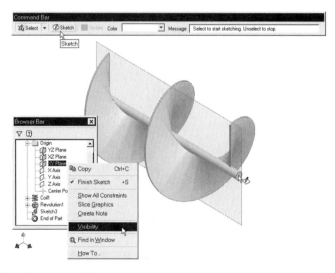

Figure 2–221 *Vertical end face selected*

13. Select Two Point Rectangle from the Sketch toolbar or panel to construct a rectangle on the sketch plane. (See Figure 2–222.)

14. Select General Dimension from the Sketch toolbar or panel and add dimensions as shown in Figure 2–223.

15. Select Extrude from the Features toolbar or panel.

16. Select the rectangle as the profile.

17. Set Extents to Distance of 7 units and select the Cut and OK buttons. (See Figure 2–224.)

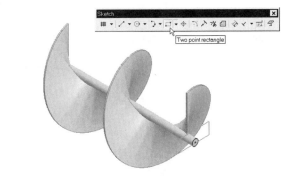

Figure 2–222   *Rectangle constructed*

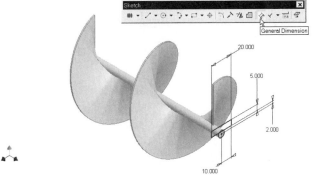

Figure 2–223   *Dimensions added*

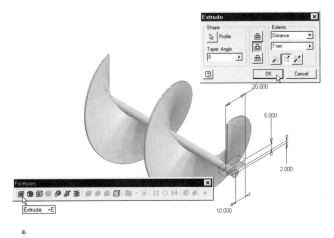

Figure 2–224   *Rectangle being extruded*

Chamfer

18. Now select Chamfer from the Features toolbar or panel.

19. Set Distance to I unit.

20. Select the edge as shown in Figure 2–225.

21. Select the OK button to chamfer the edge. (See Figure 2–226.)

22. The cutter of the food grinder is complete. Save and close your file.

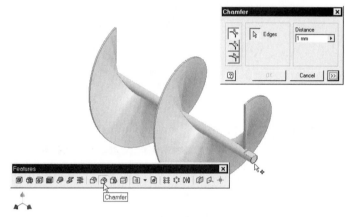

Figure 2–225   *Edge selected*

Figure 2–226   *Cutter completed*

MAIN BODY

1. Open the file Mainbody.ipt.

You will add a number of sketched and placed features to the solid. The completed solid part is shown in Figure 2–245.

Hole

2. Select Sketch from the Command Bar toolbar.

3. Select the vertical face as shown in Figure 2–227 to set up a new sketch plane.

4. Select Hole from the Features toolbar or panel to construct a hole on the sketch plane.

5. Set Termination to Through All and diameter to 12 units and select the OK button.

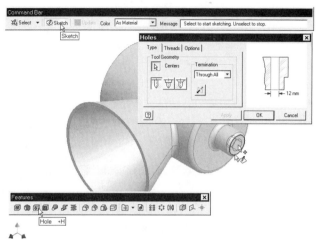

Figure 2–227   *Hole feature being placed*

Extrude Join

6. Select Wireframe Display from the Standard toolbar to set the display to wireframe mode.

7. Select Sketch from the Command Bar toolbar and select the vertical face highlighted in Figure 2–228 to set up a sketch plane.

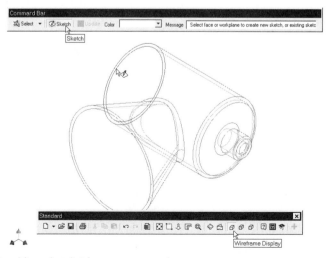

Figure 2–228   *New sketch plane constructed*

8. Select Offset from the Sketch toolbar or panel to construct an offset circle and select Line to construct two vertical lines. (See Figure 2–229.)

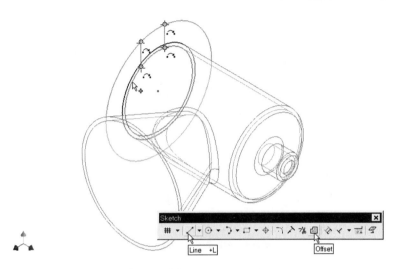

Figure 2–229 *Offset circle and vertical lines constructed*

9. Select General Dimension and add dimensions as shown in Figure 2–230.
10. Select Extrude from the Features toolbar or panel.
11. Select the area highlighted in Figure 2–231.
12. Set Extents to Distance of 3 units and select the Join button and the OK button.

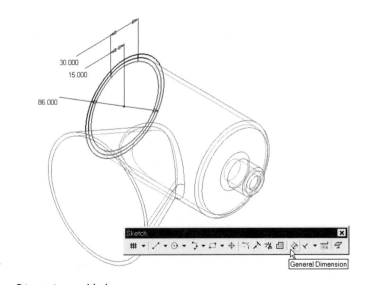

Figure 2–230 *Dimensions added*

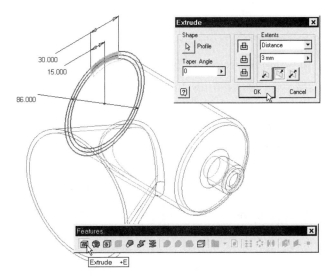

Figure 2–231   *Selected area being extruded*

Circular Pattern

13. Select Circular Pattern from the Features toolbar or panel.

14. Select the Features button and select the extruded feature highlighted in Figure 2–232.

15. Select the Rotation Axis button and select a circular edge to construct a circular pattern.

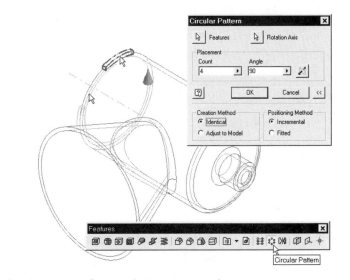

Figure 2–232   *Circular pattern features being constructed*

Extrude Join

16. Select XY Plane in the browser bar.

17. Right-click and select Visibility to set it visible.

18. Select and drag the corners of the XY plane to adjust its display size and location as shown in Figure 2–233.

19. Select Work Plane from the Features toolbar or panel and select the XY plane.

20. Then hold down the mouse and drag to a new location.

21. Set the offset distance to −140 units. (See Figure 2–234.) A work plane offset from the selected XY plane is constructed.

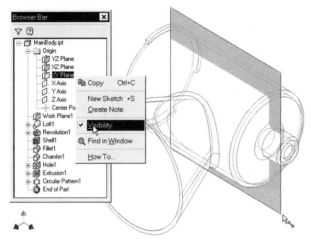

Figure 2–233 *XY plane made visible and adjusted*

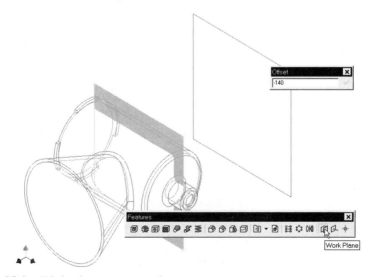

Figure 2–234 *Work plane constructed*

22. Select Sketch from the Command Bar toolbar and select the new work plane to set up a new sketch plane.

23. Select Project Geometry from the Sketch toolbar or panel and select the circular edge highlighted in Figure 2–235 to project the geometry on the current sketch plane as reference geometry.

24. Select Center Point Circle from the Sketch toolbar or panel and construct two concentric circles.

25. Select General Dimension to add dimensions as shown in Figure 2–236.

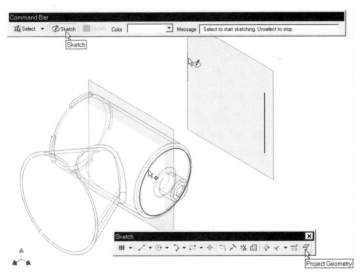

Figure 2–235   *Sketch plane constructed and geometry projected*

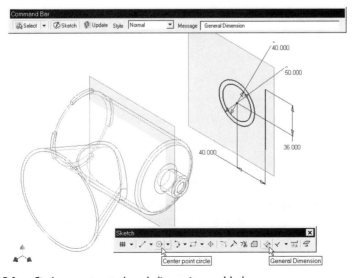

Figure 2–236   *Circles constructed and dimensions added*

26. Select Extrude from the Features toolbar or panel.

27. Select the area highlighted in Figure 2–237.

28. Set Extents to To Next and select the Join and OK buttons.

29. Select Sketch from the Command Bar toolbar and select the edge highlighted in Figure 2–238 to set up a new sketch plane.

30. Select Offset from the Sketch toolbar or panel and construct an offset circle.

31. Select General Dimension and add a dimension. (See Figure 2–239.)

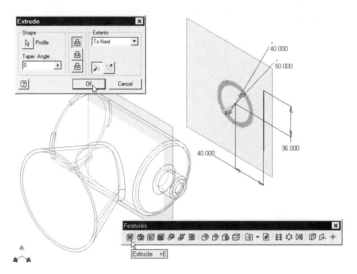

Figure 2–237   *Highlighted area being extruded*

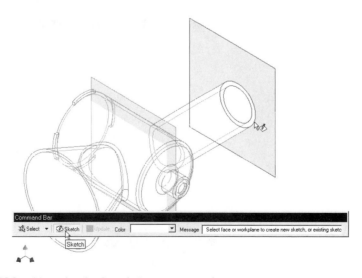

Figure 2–238   *New sketch plane being constructed*

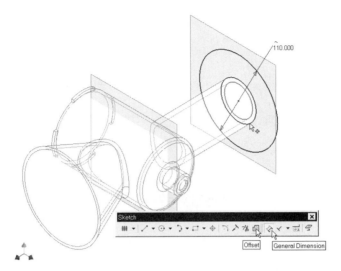

Figure 2–239 *Offset circle constructed and dimensioned*

32. Select Extrude from the Features toolbar or panel.

33. Select the area highlighted in Figure 2–240.

34. Set Extents to Distance of 4 units and select the Join and OK buttons.

35. Turn off the XY plane and the sketch plane by selecting them, right-clicking, and deselecting Visibility.

36. Select Sketch from the Command Bar toolbar and select the edge highlighted in Figure 2–241 to set up a new sketch plane.

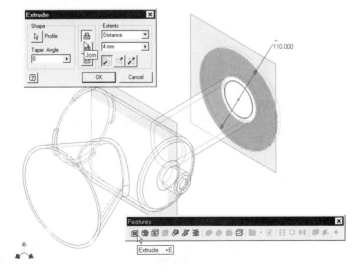

Figure 2–240 *Highlighted area being extruded*

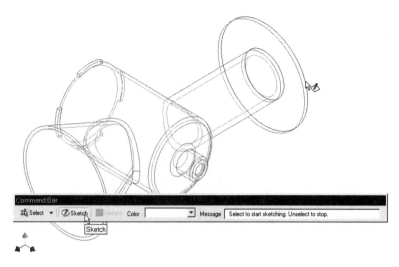

Figure 2–241   *Edge selected*

37. Select Offset from the Sketch toolbar or panel and construct an offset circle.

38. Select General Dimension to place a dimension. (See Figure 2–242.)

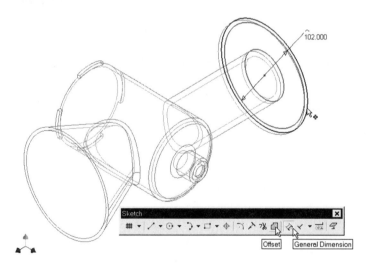

Figure 2–242   *Offset circle constructed and dimensions added*

39. Select Extrude from the Features toolbar or panel and select the area highlighted in Figure 2–243.

40. Set Extents to Distance of 4 units and select the Join and OK buttons.

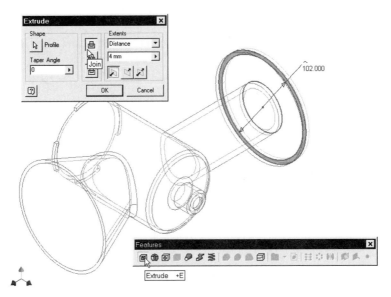

Figure 2–243 *Highlighted area being extruded*

Fillet

41. Select Fillet from the Features toolbar or panel.

42. Select the edges highlighted in Figure 2–244.

43. Set the fillet radius to 4 units.

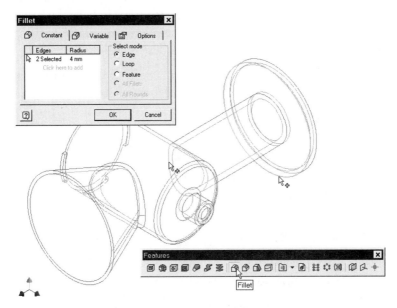

Figure 2–244 *Edges being filleted*

44. Select the OK button. (See Figure 2–245.)

45. The main body of the food grinder is complete. Save and close your file.

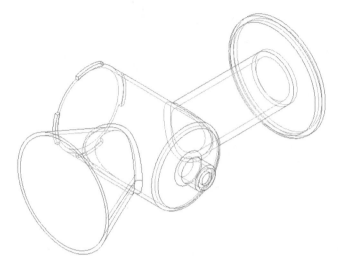

Figure 2–245   *Edges filleted*

DERIVED SOLID PARTS

Sometimes, an existing solid model can be a very good starting point for building a new solid model. Of course, you can save a new copy of a solid and from there continue to construct new features. However, if you want to have the new solid to be a mirror copy of the existing part and/or to maintain a reference with the existing solid from which it is made, you should make a derived solid.

A derived solid is referenced to the solid that it is derived from. If the original solid changes, the derived solid changes. You can add features to the derived part.

DERIVED FOOD GRINDER CUTTER

The food grinder cutter that you constructed has a right-hand helical cutting blade. Suppose you want to make a left-hand helix version of the food grinder cutter and you want to have the new solid part maintain a reference with the original cutter. This way, any modification you will make to the original cutter will be updated in the new cutter.

1. Start a new part file.

2. Then deselect Sketch on the Command Bar toolbar to exit sketch mode.

3. Select Derived Part from the Insert menu.

4. In the Open dialog box, select the file that you want to use as a source to make the derived solid. Here you select the file Cutter.ipt.

5. After you select a file as the referenced base solid part, the Derived Part dialog box displays. In the dialog box, set the scale to 1, select the Mirror part check box, and select XY Plane as the mirror plane. (See Figure 2–246.)

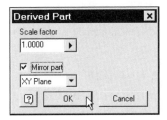

Figure 2–246   *Derived Part dialog box*

6. Select the OK button. A derived solid is constructed. (See Figure 2–247.)

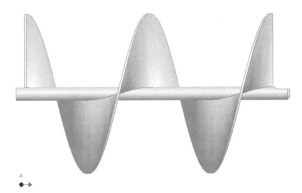

Figure 2–247   *Derived cutter constructed*

7. Set the display to an isometric view. (See Figure 2–248.)

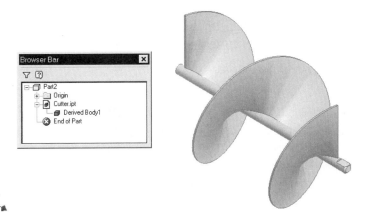

Figure 2–248   *Derived part in isometric view*

Compare Figure 2–248 with Figure 2–226. The original cutter has a right-hand helical cutting blade and the derived cutter has a left-hand helical cutting blade. In the expanded browser bar shown in Figure 2–248, you will find an item called Derived Body1. It denotes an external reference for the part with an external part.

8. The derived left-hand helical cutter is complete. Save and close the file (file name: CutterLH.ipt).

DERIVED FOOD GRINDER HANDLE

Suppose you want to design another version of the handle of the food grinder.

1. Start a new part file.

2. Then deselect Sketch on the Command Bar toolbar.

3. Select Derived Part from the Features toolbar or panel and select the part file Handle.ipt.

4. In the Derived Part dialog box, set the scale to 1 and do not select the Mirror button.

5. After constructing the derived part, set the display to an isometric view. (See Figure 2–249.)

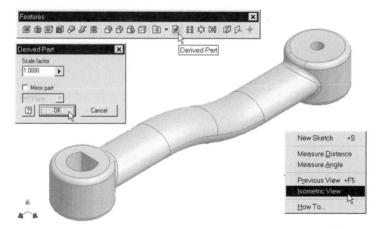

Figure 2–249 *Derived handle constructed*

Project Cut Edges

6. Now you will modify the derived handle. Select Sketch on the Command Bar toolbar and select the work plane highlighted in Figure 2–250 to construct a sketch plane.

7. Select Project Cut Edges from the Sketch toolbar or panel to project the cut edge at the intersection of the sketch plane and the solid part onto the current sketch plane. (See Figure 2–251.)

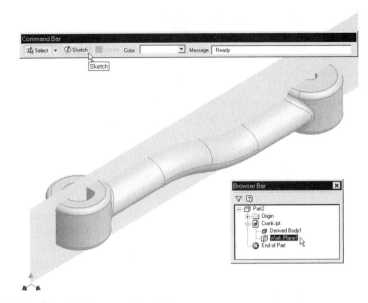

Figure 2–250 *Sketch plane constructed*

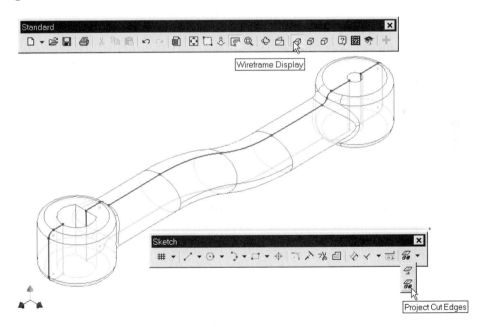

Figure 2–251 *Cut edge projected onto the current sketch plane*

8. Exit sketch mode.

9. Select Extrude from the Features toolbar or panel to extrude the area highlighted in Figure 2–252 a distance of 16 units from mid-plane.

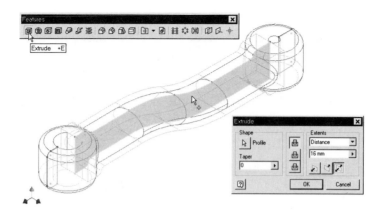

Figure 2–252 *Sketch being extruded*

10. Select the OK button. The sketch is extruded. Set the display to shaded mode. (See Figure 2–253.)

11. The derived handle is complete. Save and close your file (file name: Handle-A.ipt).

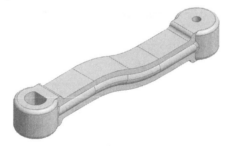

Figure 2–253 *Sketch extruded*

LIGHT, MATERIAL, COLOR, AND PART PROPERTIES

Now you will set the light style to manage the general lighting environment of the display, assign material and color to the solid part, and set properties to the solid part.

1. Open the file Cutter.ipt.

LIGHT

When you set the display to hidden edge display or shaded mode, the solid part is rendered and is illuminated by a default light.

2. To manage the color of the light, select Lighting from the Format menu to display the Lighting dialog box. (See Figure 2–254.)

3. In the Style Name box, add a new style.

4. Select the new style in the Active box.

5. Turn off the default light style and turn on the new style and select the new style in the Settings area.

Now you have a new light style.

6. For each light style, there are four lights (numbers 1 through 4). The On/ Off box enables you to turn on or off these lights. To modify the setting of a light, select the light in the Settings area to display the effect of the light.

7. Select and drag the vertical and horizontal scroll bar of view box to adjust the position of the light.

8. Select the Color swatch to display the Color dialog box (see Figure 2–255) to select a color.

9. Select the OK button after you complete the settings.

10. On returning to the Lighting dialog box, use the Brightness and Ambience scroll bars to adjust the brightness and the ambience of the light.

11. Then save and close the dialog box. A new lighting style is saved.

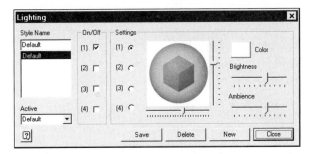

Figure 2–254   *Lighting dialog box*

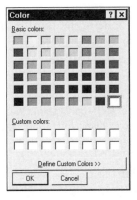

Figure 2–255   *Color dialog box*

MATERIAL

To incorporate mechanical and physical properties to the database of the 3D solid model, you assign material. Properties that you can include are Density, Young's

Modulus, Poisson's Ratio, Yield Strength, Ultimate Tensile Strength, Thermal Conductivity, Linear Expansion, Specify Heat, and Rendering Style.

12. To assign material to the solid part, select Materials from the Format menu to display the Materials dialog box. (See Figure 2–256.)

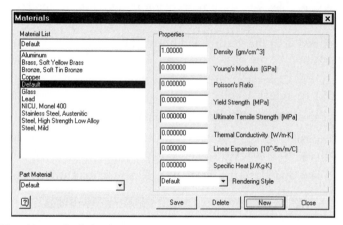

Figure 2–256 *Materials dialog box*

13. In the Materials list of the dialog box, select Stainless Steel, Austenitic.

14. Then save and close the dialog box.

COLOR

Color settings concern the appearance of the 3D solid part in your graphic screen. The faces of a 3D object have three color zones. The first zone is the portion of the 3D face under directed light. The second zone is the portion of the 3D face not under directed light. The third zone is the portion of the 3D face that reflects the light. In the shaded mode, they are represented by diffuse, ambient, and specular colors respectively.

To modify the color appearance, you set color.

15. To set the color of the solid part, select Colors from the Format menu. The Colors dialog box is displayed. (See Figure 2–257.)

16. In the dialog box, add a new style name.

There are two areas in the Color tab of the dialog box, Colors and Appearance. In the Colors area, you set the diffuse, emissive, specular, and ambient color of the 3D solid. Diffuse color sets the color of a face that is under directed light. Emissive color controls sets the color of an imaginary light that is put inside the 3D solid to make it illuminative. The default color is black, that is, no emissive color. Specular color sets the color of the portion of the face that is reflective to the light. Ambient color sets the color of the face that is not under directed light.

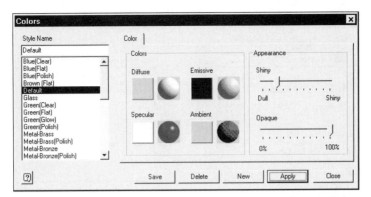

Figure 2–257   *Colors dialog box*

To further enhance the appearance of the 3D object, you modify shininess and opacity in the Appearance box. You change the reflective intensity by selecting and dragging the Shiny scroll bar. You change the opacity by selecting and dragging the Opaque scroll bar.

17. After setting the color, select the Save and Close buttons.

Note that the color display of the solid part is affected by the color of the light, the rendering style, and the color of the material. The resulting color can be very complex. Try various settings and discover the result.

18. Light, material, and color settings are complete. Save and close the file.

ORGANIZER

To copy the materials, color styles, and lighting styles from one solid part file to another solid part file, you use the organizer.

1. Open the file Blade.ipt.

2. Select Organizer from the Format menu to display the Organizer dialog box. (See Figure 2–258.)

In the dialog box, there are three tabs: Materials, Color Styles, and Lighting Styles. They enable you to copy the respective styles from a source file.

3. To open a source file for copying, select the Browse button and select a solid part file (Cutter.ipt) that has all the required materials, color, and light settings.

4. Then select the materials, color styles, or lighting styles.

5. Select the Copy>> button to copy to the current solid part file.

6. Now the materials, color, and light settings of the cutter are copied to the blade. Save and close your file.

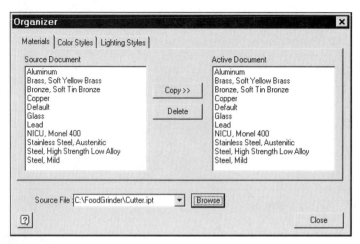

Figure 2–258 *Organizer dialog box*

PART PROPERTIES

In addition to the geometric data of the 3D solid model, you can incorporate various data in terms of properties in a 3D part file.

1. Open the file Cutter.ipt.

2. Select Properties from the File menu to see the file's Properties dialog box. (See Figure 2–259.)

In the Properties dialog box, there are seven tabs: Summary, Project, Status, Custom, Save, Physical, and Units.

- **Summary** Saves the general information about a 3D solid. It concerns title, subject, author, manager, company, category, keywords, and comments of the solid.

- **Project** Saves the information about the project. It concerns part number, description, revision number, project, cost center, estimated cost, creation date, vendor, and web link.

- **Status** Saves the status of the 3D solid. It concerns status, checked by, checked date, engineer approved by, engineer approved date, manufacture approved by, manufacture approved date, reserved by, reserved, last reserved by, and reserve removed.

- **Custom** Saves any customized data. You specify data name, type, and value.

- **Save** Saves save information. It concerns save preview picture, active window on save, active window, and import from file.

- **Physical** Saves physical properties assigned to the solid part. It concerns material, density, accuracy, mass, area, volume, center of gravity, and inertia properties.

- **Units**   Saves units information. It concerns length units, angle units, time units, mass units, linear dimension display precision, and angular dimension display precision.

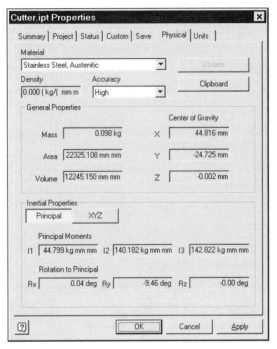

Figure 2–259   *Properties dialog box*

3. Select the Physical tab.

4. Change the material to Stainless Steel, Austenitic and select the OK button.

5. The properties are set. Save and close your file.

TEMPLATES, SAVE AS, OPEN, INSERT, AND COMPATIBILITY

TEMPLATES

Template files are computer files having all the necessary settings that may be required in repetitive work. When you need to construct a series of files that have a similar kind of settings, you can make use of a template file.

When you start a new file, you select a template from the New dialog box. The template file is saved in a template folder that was created when you installed Inventor in your computer. To save a file to become a template, select Save Copy As from the File menu. Then save the file in the Inventor Template directory. The saved file becomes a template. The next time you start a new file, you will find an icon representing the file in the New dialog box.

By default, there are three tabs in the New dialog box depicting Default, English, and Metric. To add a tab to the New dialog box, simply construct a folder in the Inventor template directory and put template files in that directory.

SAVE AS

It is very important that you can use an Inventor solid part file in other computerized activities without having to reconstruct the parts again. To output an Inventor file, you save it in various file formats: BMP, IGES, SAT, STEP, STL, XGL, and ZGL.

- **BMP** Windows Bitmap file translates the current screen display to a 2D raster image.

- **IGES** Initial Graphics Exchange Specification file translates lines and surfaces of the current 3D solid.

- **SAT** Save As Text file translates lines, surfaces, and solid information about the current 3D solid.

- **STEP** Standard for the Exchange of Product Model Data file translates wires, surfaces, and solids.

- **STL** Stereolithography file translates the current 3D solid model to a set of triangular flat surfaces.

- **XGL** X Windows Graphics Library file translates the 3D surface, texture mapping, and transparency of the current 3D model

- **ZGL** ZGL is a compressed XGL format.

These file formats are described more fully in Table 1–1 in Chapter 1.

1. Open the file Cutter.ipt.

2. Select Save Copy As from the File menu. (See Figure 2–260.)

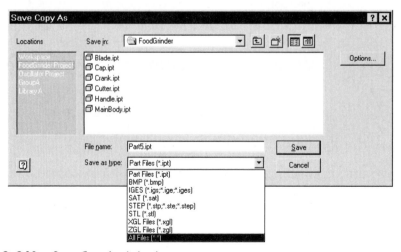

Figure 2–260 *Save Copy As dialog box*

3. In the Save Copy As dialog box, select a file format from the Save as type box.

4. Then type a file name and select the OK button. An Inventor solid part is exported to the selected file format.

OPEN

Reusing existing solid parts from other computer applications can speed up your design work tremendously. In Inventor, you can open SAT, STEP, and DWG files. To open such files, select Open from the File menu, select the file types (SAT, STEP, or DWG) in the Files of types box, and select a file.

When you open an SAT or STEP file, the solid part is read into Inventor and becomes a base solid part. You can construct additional sketched features, work features, and placed features as if you are working on an Inventor file. To edit the base solid feature, you use the tools available in the Solids panel or toolbar. (See Appendix A.)

If you have Mechanical Desktop R4 installed in your computer, you can open a Mechanical Desktop DWG file. After the file is read, all the parametric features of the original Mechanical Desktop file will be imported and displayed as objects in the browser bar.

COMPATIBILITY

Inventor is compatible with AutoCAD, Mechanical Desktop, and 3D Studio as shown in the following table.

Table 2–6 Inventor compatibility with other programs

Program	Compatibility
AutoCAD Files	You can open and read an AutoCAD file in Inventor.
Mechanical Desktop (MDT) Files	You can open and link Mechanical Desktop part files and retain an associative link to the relevant MDT files.
Mechanical Desktop R4/R5 (MDT4/5) Files	You can open and translate MDT4/5 part and assembly files and retain the part histories, assembly hierarchy, and constraint information.
3D Studio Max/Viz	By installing Inventor and 3D Studio Max/Viz in the same computer, you can use the 3D Studio rendering capabilities to output photo-realistic renderings of inventor designs.

Exercises

1. FOOD GRINDER COLORS

Set the color of the components of the food grinder as follows:

Main Body	Blue (clear)
Cap	Green (clear)
Crank	Red (clear)

2. MODEL CAR FOLDER

Construct a folder (C:\ModelCar) in your computer and use a text editor to construct a project path file (File name: ModelCar.ipj) as follows:

[Included Path File]
PathFile=C:\ModelCar\ModelCarProject.ipj
[Workspace]
Workspace=C:\Inventor New Projects
[Local Search Paths]
ModelCar=C:\ModelCar
[Workgroup Search Paths]
GroupA=C:\DesignGroupA
[Library Search Path]
Library A=C:\StandardParts

Select this project path file for Exercises 3 through 9 below.

3. MODEL CAR CIRCLIP

Figure 2–261 shows the circlip of a scale model car.

1. Start a new part file and construct a sketch in accordance with Figure 2–262.

2. Extrude the sketch a distance of 0.5 unit.

3. Save and close the solid part file (file name: carcir.ipt).

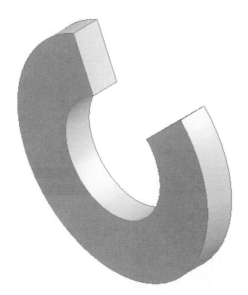

Figure 2–261 *Circlip*

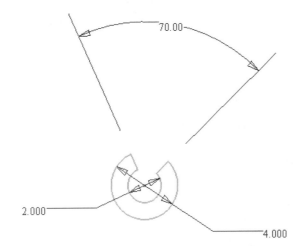

Figure 2–262 *Sketch*

4. MODEL CAR HUB

Figure 2–263 shows the hub of a scale model car.

1. Start a new part file and construct a rectangle (8 units by 9 units) on the XY plane as shown in Figure 2–264.

2. Extrude the sketch a distance of 7 units from mid-plane.

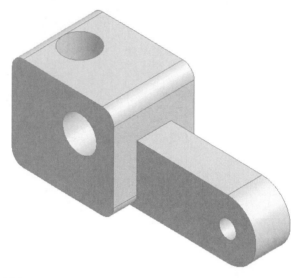

Figure 2–263   *Hub*

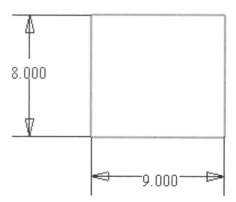

Figure 2–264   *First sketch*

3. Set the display to an isometric view and construct another sketch on the XY plane in accordance with Figure 2–265.

4. Extrude the sketch a distance of 3 units from mid-plane to join the solid part.

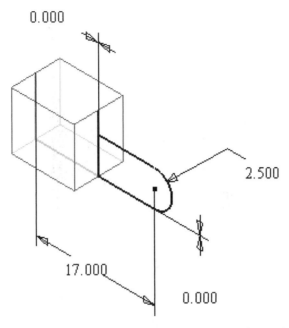

0.000

2.500

17.000

0.000

Figure 2–265 *First sketch extruded, display set to isometric view, and second sketch constructed*

5. Construct a third sketch as shown in Figure 2–266.
6. Extrude the sketch to cut through the solid.

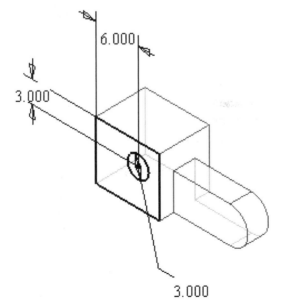

6.000

3.000

3.000

Figure 2–266 *Second sketch extruded to join the solid and third sketch constructed*

7. Construct a sketch as shown in Figure 2–267.

8. Extrude the sketch to cut through the solid.

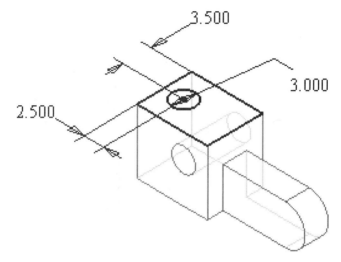

Figure 2–267   *Third sketch extruded to cut through, and fourth sketch constructed*

9. Fillet the edges in accordance with Figure 2–268. The fillet radius is 1 unit.

10. To complete the model, construct a hole with a diameter of 1.5 units in accordance with Figure 2–269.

11. The solid is complete. Save and close your file (file name: carhub.ipt).

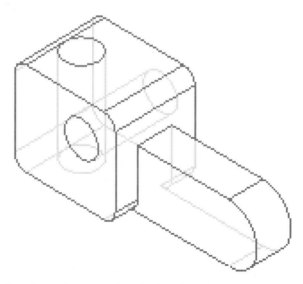

Figure 2–268   *Fourth sketch extruded and edges filleted*

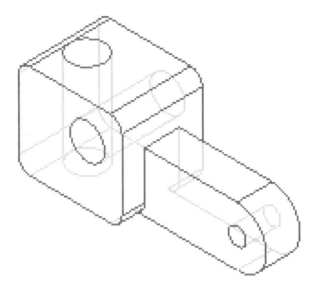

Figure 2–269 *Hole placed*

5. CAR MODEL FRONT AXLE

Figure 2–270 shows the front axle of a scale model car.

1. Start a new part file and construct a sketch in accordance with Figure 2–271.

2. Revolve the sketch to construct a revolved solid. Select the lower horizontal edge of the sketch as the axis for revolving.

3. Save and close the file (file name: caraxf.ipt).

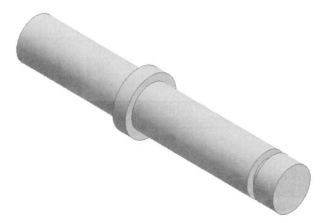

Figure 2–270 *Front axle*

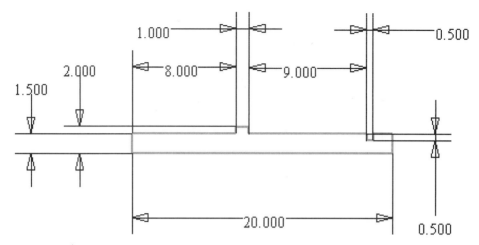

Figure 2–271 *Sketch*

6. CAR MODEL KING PIN

Figure 2–272 shows the king pin of a scale model car.

1. Start a new part file and construct a sketch in accordance with Figure 2–273.

2. Revolve the sketch about the lower edge.

3. Set the display to an isometric view and place a tapped hole in accordance with Figure 2–274. Thread size is M2. Depth of thread is 6 units.

4. The solid part is complete. Save and close your file (file name: carkpn.ipt).

Figure 2–272 *King pin*

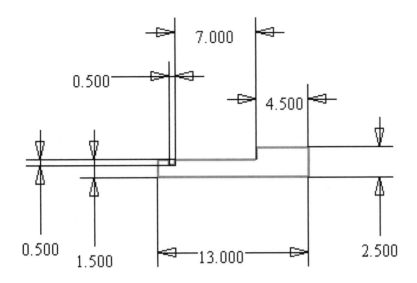

Figure 2–273   *Sketch*

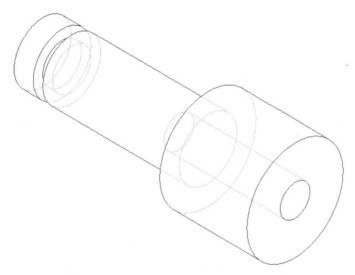

Figure 2–274   *Threaded hole placed*

7. CAR MODEL REAR AXLE

Figure 2–275 shows the rear axle of a scale model car.

Figure 2–275   *Rear axle*

1. Start a new part file and construct a sketch as shown in Figure 2–276.

2. Revolve the sketch about its lower edge.

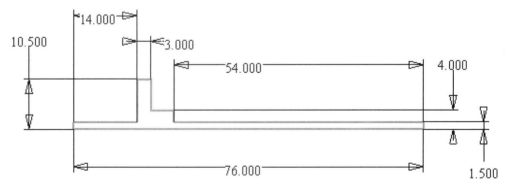

Figure 2–276   *Sketch*

3. Set the display to an isometric view and set up a sketch plane on the plane selected in Figure 2–277.

4. Select Look At from the Standard toolbar.

5. Construct a sketch in accordance with Figure 2–278. The sketch has two offset circles, two lines from the center to one of the offset circles, and two arcs. The radii of the arcs are the same and each of the arcs should be tangential to its adjacent line.

6. Trim the sketch and add dimensions in accordance with Figure 2–279.

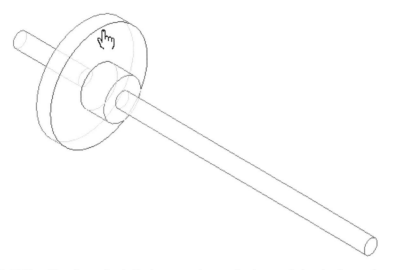

Figure 2–277 *Sketch revolved, display set to isometric view, and sketch plane selected*

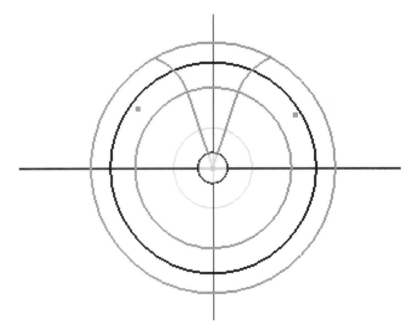

Figure 2–278 *Display set and sketch constructed*

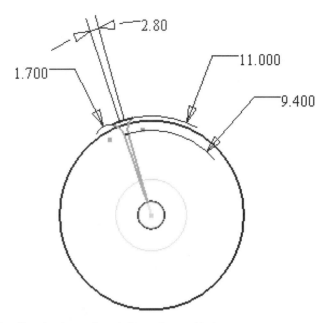

Figure 2–279   *Sketch trimmed, and dimensions added*

7. Set the display to an isometric view and extrude the sketch to cut through the solid. (See Figure 2–280.)

8. Select Circular Pattern from the Features toolbar or panel to construct a circular pattern of the extruded cut feature. The center of circular pattern is the centerline of the revolved solid. Placement count is **40** and the angle is **9** degrees. (See Figure 2–281.)

9. The solid part is complete. Save and close your file (file name: caraxr.ipt).

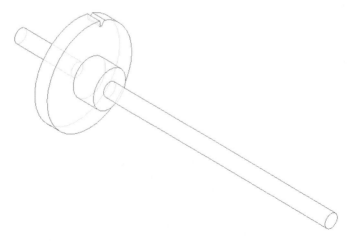

Figure 2–280   *Display set to an isometric view, and sketch extruded*

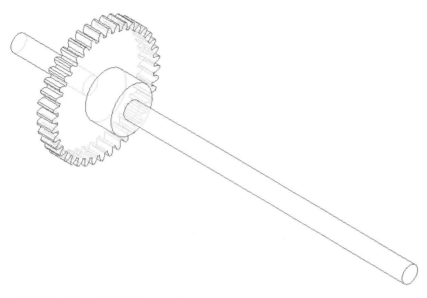

Figure 2–281 *Extruded solid arrayed*

8. CAR MODEL CHASSIS

Figure 2–282 shows the chassis of a scale model car.

1. Start a new part file and construct a sketch in accordance with Figure 2–283.

2. After constructing the sketch, add equal constraints to line segments not dimensioned in the figure to fully constrain the sketch.

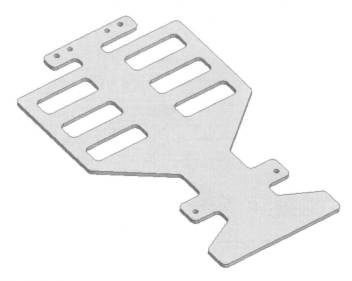

Figure 2–282 *Chassis*

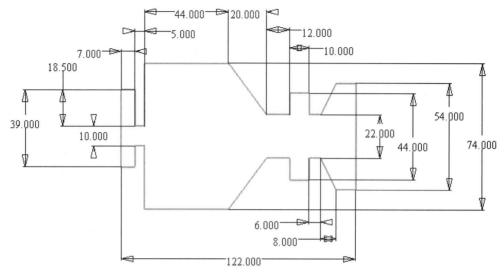

Figure 2–283   *Sketch*

3. Rotate the display in accordance with Figure 2–284.

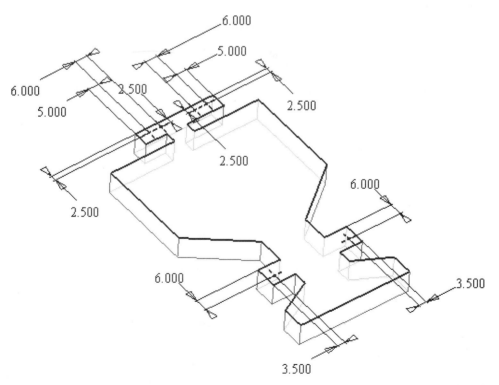

Figure 2–284   *Display rotated, sketch extruded, and hole center constructed and dimensioned*

4.Extrude the sketch a distance of 10 units.

5. Construct six hole centers and add dimensions. Construct six through holes of 2 units diameter on the hole centers. (See Figure 2–285.)

6. Place fillet features in accordance with Figure 2–286. The fillet radius is 2 units.

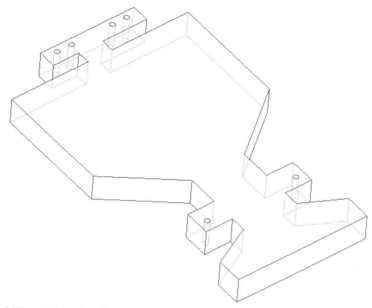

Figure 2–285   *Holes placed*

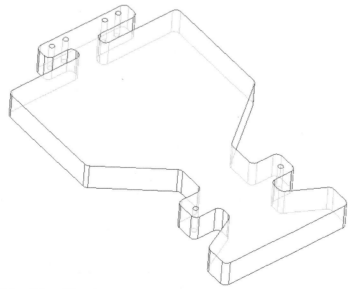

Figure 2–286   *Edges filleted*

7. Modify the extrusion thickness of the extruded solid to 1.5 units by selecting the extrusion in the browser panel, right-clicking, and selecting Edit Feature from the shortcut menu. (See Figure 2–287.)

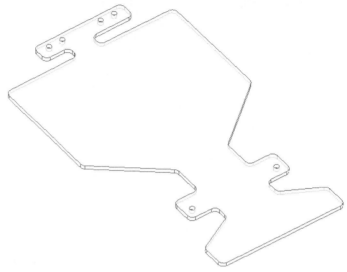

Figure 2–287   *Thickness changed*

8. Construct a rectangle and then fillet the corners of the rectangle. (See Figure 2–288.)

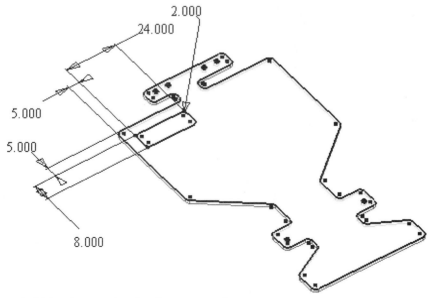

Figure 2–288   *Rectangular sketch constructed*

9. Extrude the sketch to cut through the solid. (See Figure 2–289.)

10. Construct a rectangular pattern of the extruded cut feature. The number of rows is 2 and the number of columns is 3. The distance between the rows is 40 units and the distance between the column is 14.5 units. (See Figure 2–290.)

11. The solid part is complete. Save and close your file (file name: carchs.ipt).

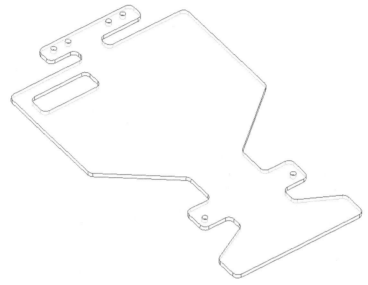

Figure 2–289   *Rectangular sketch extruded*

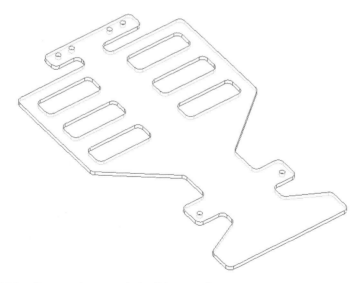

Figure 2–290   *Rectangular extruded solid arrayed*

9. CAR MODEL GEAR BOX

Figure 2–291 shows the gearbox casing of a scale model car.

1. Start a new part file and construct a sketch in accordance with Figure 2–292.

2. Set the display to an isometric view.

3. Extrude the sketch a distance of 39 units. (See Figure 2–293.)

4. Fillet four edges in accordance with Figure 2–294. The fillet radius is 2 units.

5. Make the solid hollow and remove some of the faces in accordance with Figure 2–295. The thickness of the shell is 1 unit.

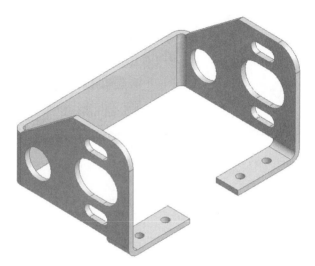

Figure 2–291   *Gear box*

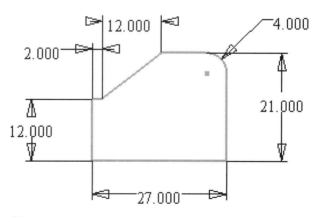

Figure 2–292   *Sketch*

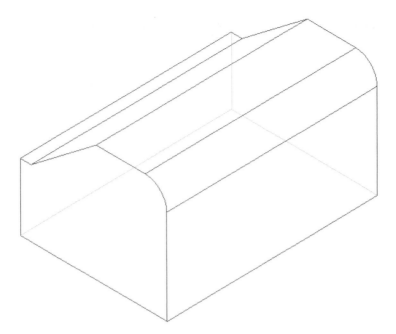

Figure 2–293 *Sketch extruded, and sketch constructed*

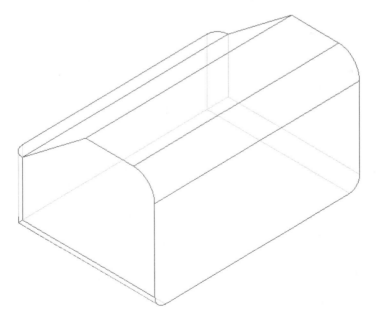

Figure 2–294 *Edges filleted*

6. Rotate the display and construct a sketch in accordance with Figure 2–296.

7. Set the display to an isometric view.

8. Extrude the sketch to cut through the solid. (See Figure 2–297.)

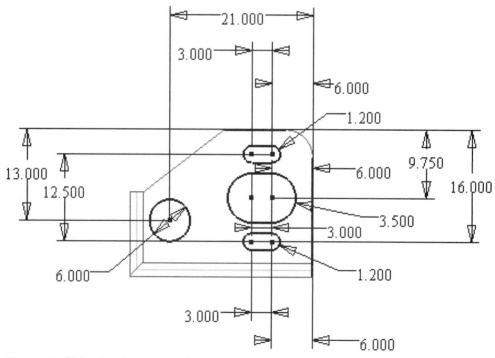

Figure 2–295 *Solid made hollow with faces removed*

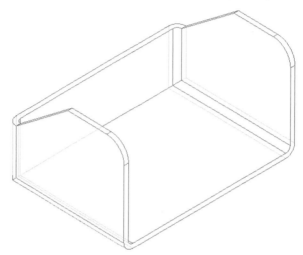

Figure 2–296 *Display set and sketch constructed*

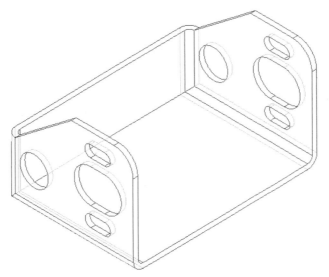

Figure 2–297   *Sketch extruded*

   9. Rotate the display and construct a sketch in accordance with Figure 2–298.

10. Add dimensions to the sketch in accordance with Figure 2–299.

11. Extrude the sketch a distance of 2 units to cut the solid. (See Figure 2–300.)

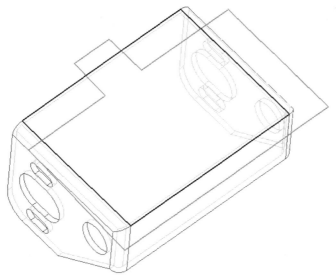

Figure 2–298   *Sketch constructed*

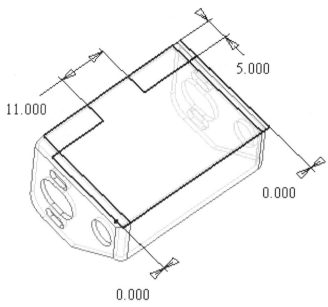

Figure 2–299 *Sketch dimensioned*

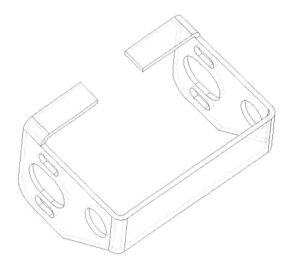

Figure 2–300 *Sketch extruded to cut*

12. Construct four hole centers in accordance with Figure 2–301.

13. Place four tapped through holes in accordance with Figure 2–302. Thread size is M2.

14. The solid part is complete. Save and close your file (file name: box.ipt).

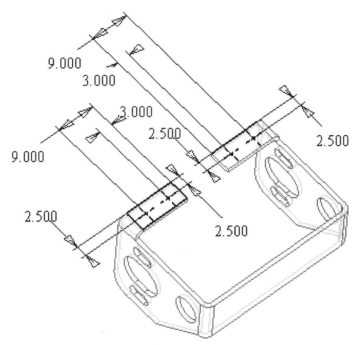

Figure 2–301 *Hole centers constructed*

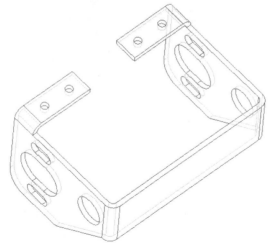

Figure 2–302 *Tapped holes placed*

10. MOBILE PHONE

1. Construct a folder (C:\MobilePhone) in your computer and use a text editor to construct a project path file (file name: MobilePhone.ipj).

[Included Path File]

PathFile=C:\MobilePhone\mblphoneProject.ipj

[Workspace]

Workspace=C:\Inventor New Projects

[Local Search Paths]

Mobile Phone Project=C:\MobilePhone

[Workgroup Search Paths]

GroupA=C:\DesignGroupA

[Library Search Path]

Library A=C:\StandardParts

Now design the upper and lower casing of a mobile phone. (See Figures 2–303 and 2–304.)

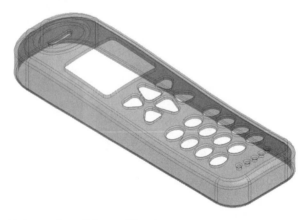

Figure 2–303   *Upper casing of the mobile phone*

Figure 2–304   *Lower casing of the mobile phone*

2. The base solid of the upper casing is a sweep solid. Figure 2–305 shows the dimensions of the sketches for your reference.

3. The second feature is an extruded solid. It intersects with the base sweep solid. Figure 2–306 shows the sketch.

4. Fillet the corners (radii of 16 units and 10 units) in accordance with Figure 2–307.

5. Place variable fillet features (radii of 3 units and 6 units) in accordance with Figure 2–308.)

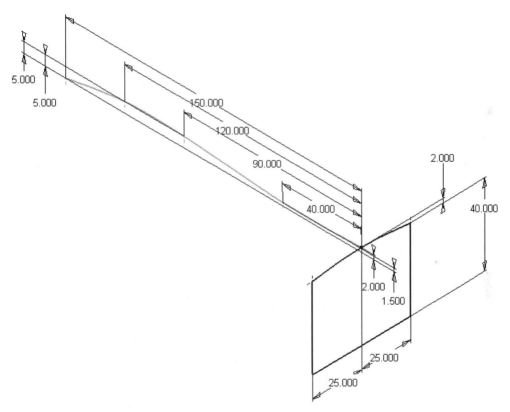

Figure 2–305 *Sketches for making the sweep solid*

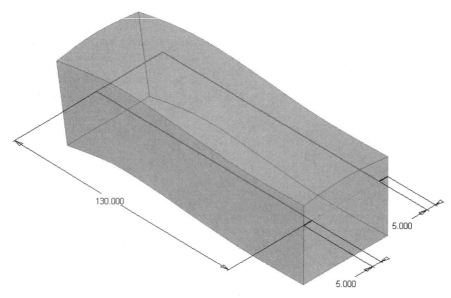

Figure 2–306 *Sketch for making the extruded solid*

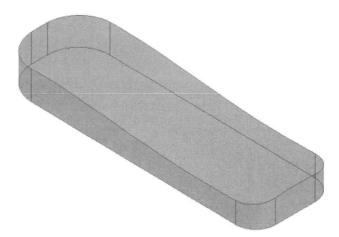

Figure 2–307 *Fillet features placed*

6. Construct a work plane that is 115 units offset from a vertical face.

7. Set up a sketch plane on the new work plane.

8. Construct a sketch. (See Figure 2–309.)

9. Revolve the sketch to cut the solid.

10. Fillet the edge (radius is 6 units).

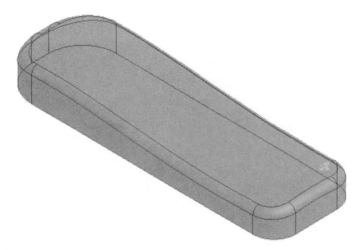

Figure 2–308   *Variable fillet feature placed*

11. Place a shell feature (thickness of the shell is 1.5 units). (See Figure 2–310.)

12. Construct a sketch for making a button opening. (See Figure 2–311.)

13. Extrude the sketch to make a button opening.

14. Fillet the edge (radius = 0.5 unit).

15. Place a rectangular pattern. (See Figure 2–312.)

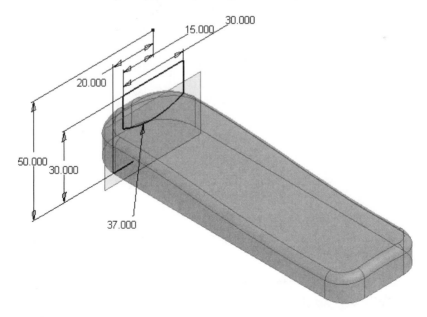

Figure 2–309   *Sketch on the new work plane*

202

Figure 2–310 *Fillet and shell features placed*

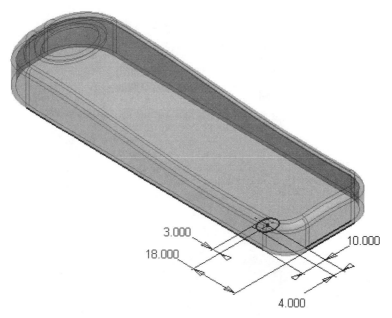

3.000

18.000

10.000

4.000

Figure 2–311 *Sketch for a button*

16. Design three openings by making a sketch and extruding the sketch to cut the solid. (See Figure 2–313.)

17. Mirror the three openings to make three more openings. (See Figure 2–314.)

18. Design and cut the openings. (See Figures 2–315 and 2–316.)

19. The upper casing is complete. Save and close your file (file name: MphoneUpper.ipt).

Figure 2–312   *Rectangular pattern placed*

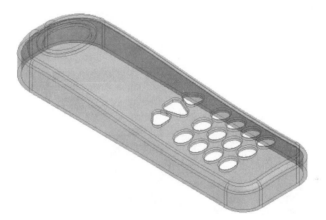

Figure 2–313   *Three openings constructed*

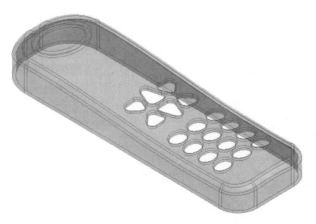

Figure 2–314   *Three openings mirrored*

204

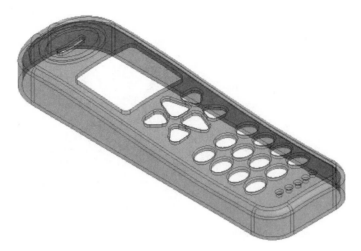

Figure 2–315 *Openings cut*

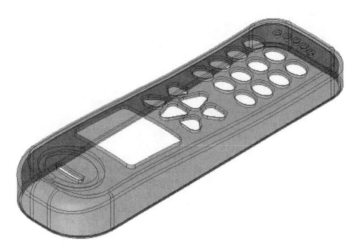

Figure 2–316 *Part completed*

20. Start a new part file and construct three work planes that are offset from the XY plane. (Offset distances are 40 units, 40 units, and 30 units.)

21. Construct four sketches on the XY plane and the work planes. (See Figure 2–317.)

22. Construct a loft solid from the sketches. (See Figure 2–318.)

23. Fillet the edges. (See Figure 2–319.)

24. Place a shell feature and cut an extruded feature. (See Figure 2–320.)

25. The solid part is complete. Save and close your file (file name: MphoneLower.ipt).

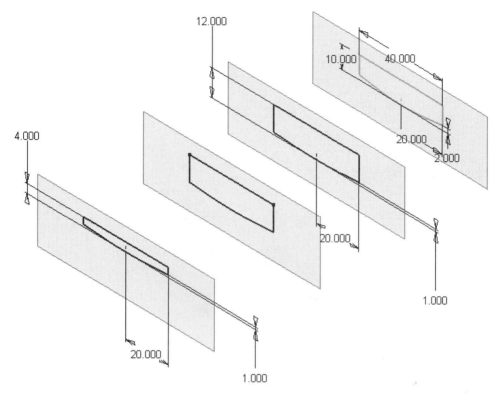

Figure 2–317 *Four sketches*

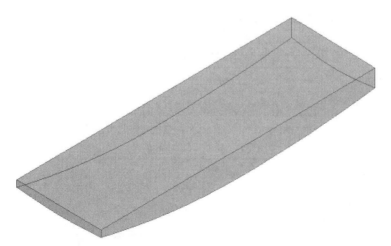

Figure 2–318 *Loft solid constructed*

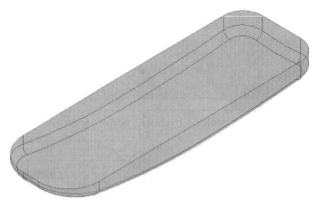

Figure 2–319   *Edges filleted*

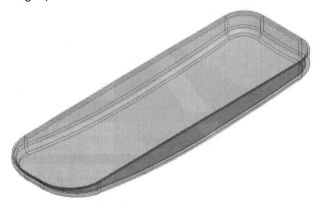

Figure 2–320   *Completed lower casing*

II. SKATE SCOOTER FOLDER

Construct a folder in your computer (C:\SkateBoard) and use a text editor to construct a project path file as follows (file name: SkateBoard.ipj).

[Included Path File]
PathFile=C:\SkateBoard\SkateBoardProject.ipj
[Workspace]
Workspace=C:\Inventor New Projects
[Local Search Paths]
Skate Board Project=C:\SkateBoard
[Workgroup Search Paths]
GroupA=C:\DesignGroupA
[Library Search Path]
Library A=C:\StandardParts

Use this project path file for Exercises 12 through 19 below.

12. SKATE SCOOTER STEERING ARM

Figures 2–321 and 2–322 show the component parts for the assembly of a steering arm of a skate scooter. Construct the solid parts as shown in the dimensions and data given.

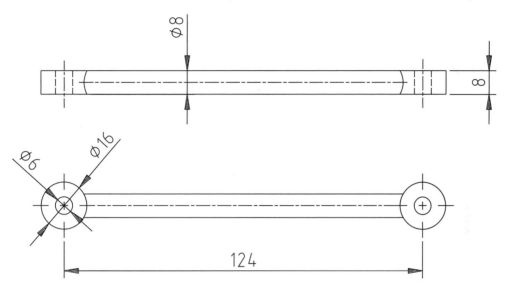

Figure 2–321   *Steering bar (File name: SteerBar.ipt)*

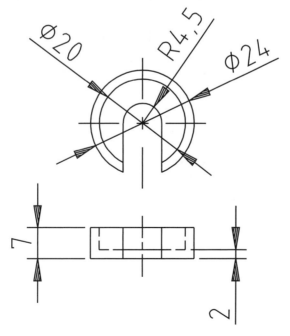

Figure 2–322   *Steering spring cap (file name: SteerSpringCap.ipt)*

13. SKATE SCOOTER STEERING SPRING

1. Construct a cylindrical solid part of 40 mm diameter and 47 mm height (file name: SteeringSpring.ipt).

2. Construct a coil to intersect. The coil has a mean diameter of 18 mm and a wire diameter of 2 mm.

Coil Size

Type:	Revolution and Height
Height:	47 mm
Revolution:	8

Coil Ends

Start:	Flat
Transition Angle:	90°
Flat Angle:	90°
End:	Flat
Transition Angle:	90°
Flat Angle:	90°

14. SKATE SCOOTER SHANK SPRING

1. Construct a spring by saving the part file that you constructed in Exercise 13 to a new file name (file name: ShankSpring.ipt).

2. Change the base cylindrical solid to 66 mm diameter and 120 mm height.

3. Modify the parameters of the coil as follows:

Mean Diameter:	44 mm
Wire Diameter:	4 mm

Coil Size

Type:	Revolution and Height
Height:	120 mm
Revolution:	12

Coil Ends

Start:	Flat
Transition Angle:	90°
Flat Angle:	90°
End:	Flat
Transition Angle:	90°
Flat Angle:	90°

15. SKATE SCOOTER SPRING SLEEVE

Figures 2–323 through 2–324 show the component parts of the spring sleeve assembly of a skate scooter. Construct the solid parts of the components in accordance with the dimensions shown.

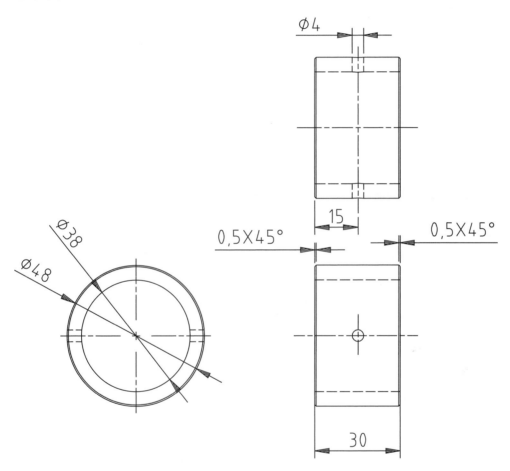

Figure 2–323 *Shank ring (file name: ShankRing.ipt)*

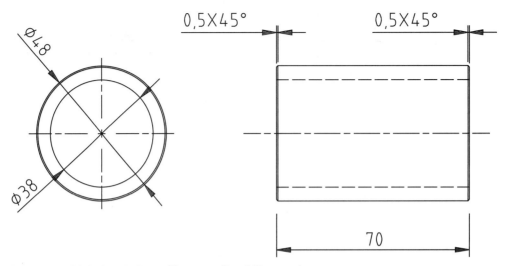

Figure 2–324 *Shank sleeve (file name: ShankSleeve.ipt)*

16. SKATE SCOOTER LOCKING MECHANISM

Figures 2–325 through 2–328 show the component parts for the assembly of a locking mechanism. Construct the solid parts in accordance with the dimensions given.

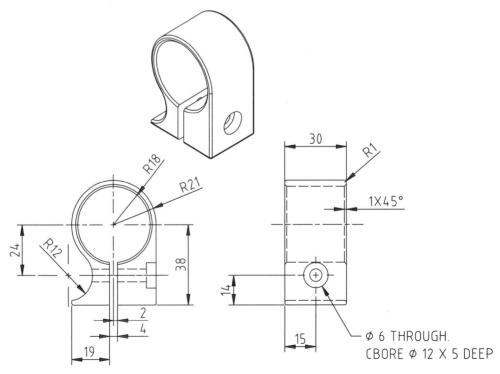

Figure 2–325 *Lock main body (file name: LockMain.ipt)*

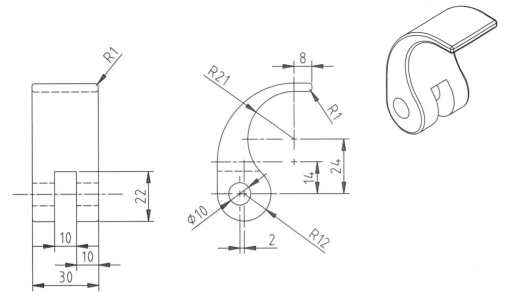

Figure 2–326 *Lock clamp (file name: LockClamp.ipt)*

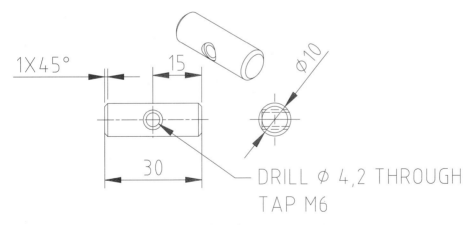

DRILL Ø 4,2 THROUGH
TAP M6

Figure 2–327 *Lock pin (file name: LockPin.ipt)*

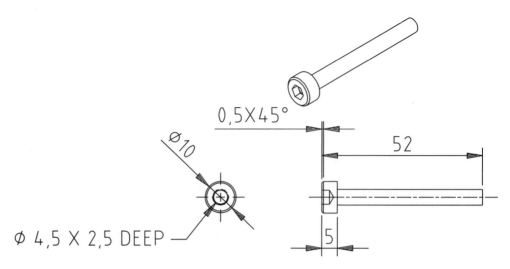

Figure 2–328 *Lock screw (file name: LockScrew.ipt)*

17. SKATE SCOOTER LEFT FRONT ARM

Figure 2–329 shows the left front arm of the skate scooter.

1. Construct a solid part of the component in accordance with the dimensions given.

2. Construct a mirror part of the component. (See Figure 2–330.)

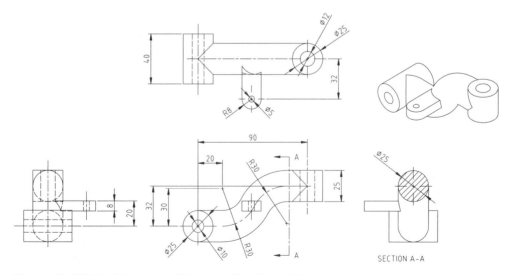

Figure 2–329 *Left front arm (file name: FrontArmLeft.ipt)*

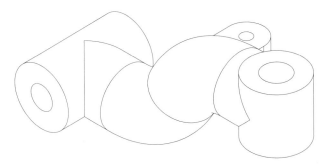

Figure 2–330 *Right front arm (file name: FrontArmRight.ipt)*

18. SKATE SCOOTER FRONT END

Figure 2–331 shows the front end of the skate scooter. Construct a solid part of the component.

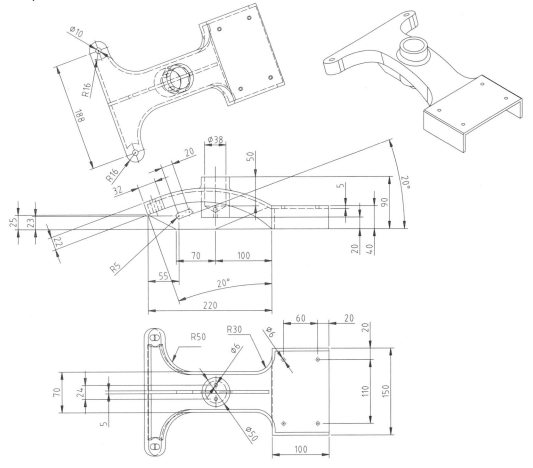

Figure 2–331 *Front end (file name: FrontEndBase.ipt)*

19. SKATE SCOOTER REAR END

Figure 2–332 shows the rear end of the skate scooter. Construct a solid part of the component.

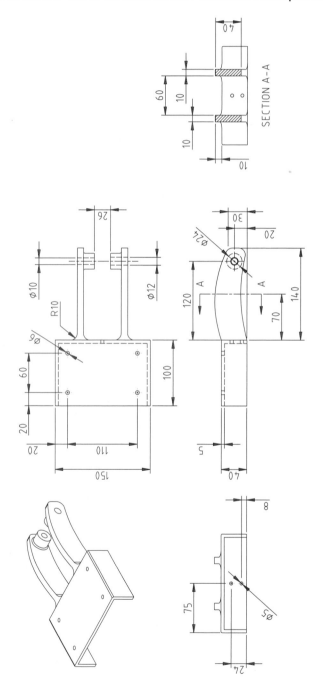

Figure 2–332 *Rear end (file name: RearEndBase.ipt)*

SUMMARY

You construct 3D solids through a feature-based approach: you deduce a complex 3D object into simple features and construct the features by making sketches. Sketches need not be precise; during the initial design stage, you should concentrate on the form and shape of the profiles. Then you apply geometric constraints to modify the geometric relationship among the objects in the sketch and set the size of the sketch by applying parametric dimensions.

After sketching, you construct solids from the sketches in several ways. You extrude a sketch to form an extruded solid. You revolve a sketch to form a revolved solid. You sweep a sketch along another sketch to form a sweep solid. You sweep a sketch along a helical sketch to form a coil solid. You loft along a number of sketches to form a loft solid. Furthermore, you use a sketch to split a face or a solid. Because these features derive from sketches, they are called sketched solid features.

When you construct more than one sketched solid feature in your solid, you can combine them in any of three ways: join, cut, and intersect.

To make a sketch, you need a sketch plane. You use the default planes (XY plane, XZ plane, and YZ plane) and the existing faces of a solid as sketch planes. In addition, you can construct work planes and use them as sketch planes. To help establish a proper relationship among the features, you use work axes and work points. Work planes, work axes, and work points are called work features.

Apart from making sketches to construct solid features, you can incorporate pre-constructed solid features in your solid model: hole, shell, fillet, chamfer, rectangular pattern, circular pattern, mirror, and face draft. Unlike working with sketched solid features, you do not need to construct any sketch—you simply place them on the solid part. These pre-constructed solid features are called placed solid features.

Altogether, there are three kinds of features in a solid: sketched solid features, placed solid features, and work features.

Because Autodesk Inventor is a parametric system, you can modify the parameters of the features (sketched solid features, placed solid features, and work features) any time during and after you construct the 3D solid and you can change the way sketched solid features are combined. For example, you can change a join operation to a cut or intersect operation.

In this chapter, you learned how to construct various kinds of features and constructed some of the solid parts of a food grinder. In Chapter 3, you will construct the remaining solid parts of the food grinder and put them together to form an assembly. In Chapter 4, you will learn advanced modeling techniques.

REVIEW QUESTIONS

1. What kinds of features can you construct by making sketches? With the aid of sketches, describe how the features are constructed.

2. Use simple examples to illustrate the three ways that you combine sketched solid features.

3. Use sketches to explain the ways you can construct work planes, work axes, and work points.

4. What kinds of features can you construct by specifying the locations and parameters? Use sketches to illustrate.

5. Differentiate between a face split feature and a face draft feature. Use examples to illustrate.

CHAPTER 3

Assembly Modeling

OBJECTIVES

The aims of this chapter are to explain the key concepts of assembly modeling and the three design approaches to constructing an assembly and to let you practice placing solid parts in an assembly drawing and constructing solid parts in an assembly. This chapter gives you an in-depth understanding about using assembly constraints, explains how to animate mechanical motion in an assembly and how to construct a bill of materials, and outlines the procedures for constructing presentation views of an assembly. After studying this chapter, you should be able to

- Describe the key concepts of assembly modeling
- State the three approaches in making an assembly
- Place existing components into an assembly
- Create new components in an assembly
- Apply assembly constraints to component parts of an assembly
- Animate mechanical motions in an assembly
- Construct a bill of materials in an assembly
- Produce a presentation view of assembly

OVERVIEW

An assembly is a collection of component parts, put together properly to form a device that serves a purpose. To construct an assembly in the computer, you use an assembly file. Because you have already used part files to construct the 3D solids and to store the information about the individual components, you will use the assembly file for two main purposes: linking the component parts together and keeping the information regarding how the component parts are put together. For complex devices that have many component parts, it is common practice to organize the component parts into a number of smaller sub-assemblies such that each sub-assembly has fewer parts. Therefore, an assembly set consists of an assembly file and a number of part files or an assembly file together with a number of sub-assembly files and part files.

To construct an assembly to link to a set of components, you use the bottom-up, top-down, or the hybrid approach. You construct an assembly file by placing components that you have already constructed in the assembly, by creating new components while you work on the assembly, or by placing existing components and creating new components simultaneously.

After you link the components to an assembly, they are free to translate. To establish a proper spatial relationship among them, you apply assembly constraints.

By varying the parameters of the assembly constraints that you set for the components of the assembly, you can animate mechanical motion. Using an assembly, you generate a bill of materials and exploded presentation views.

In the previous chapter, you constructed six components of a food grinder in part files. In this chapter, you will start an assembly file and place those component parts in an assembly. In addition, you will construct three more components in the assembly. In the assembly, you will move and rotate the components to their appropriate locations and you will apply assembly constraints to align the components properly. After making the assembly, you will construct a bill of materials and presentation views of the assembly.

ASSEMBLY MODELING CONCEPTS

With the exception of very simple objects, such as a ruler, most objects have more than one component part put together to form a useful assembly. When you design a set of components for an assembly, the relative dimensions and positions of parts, and how they fit together, are crucial. You need to know whether there is any interference among the mating parts. If there is interference, you need to find out where it occurs; then you can eliminate it. To shorten the design lead time, you construct virtual assemblies in the computer to validate the integrity of a set of component parts. To design a complex assembly that has a large number of parts, you organize the parts into smaller assemblies (sub-assemblies) and put the sub-assemblies into a larger assembly.

To construct an assembly in the computer, you start a new assembly file and connect a set of relevant part files and/or assembly files (sub-assemblies) to the assembly file. In the computer, the part files store the information about the 3D objects and the assembly files store the information about how the 3D objects are assembled together. Because the component part files hold the definitions of the solid parts and link to the assembly file, any change you make to the part files will be incorporated in the assembly automatically.

In an assembly file, the components are free to translate in three linear directions and three rotational directions. You move and rotate them as if you were manipulating a real object. To impose restriction to the movements and to align a component properly with another component in the assembly, you apply assembly constraints.

Construction of an assembly involves two major tasks: gathering a set of parts or sub-assemblies in an assembly file and applying appropriate assembly constraints on the components.

DESIGN APPROACHES

How you gather a set of components in an assembly depends on which design approach you take. As explained earlier, there are three approaches: bottom-up, top-down, (bottom being the parts, and top being the assembly) and hybrid. In a bottom-up approach, you construct all the component parts and then assemble them in an assembly file. In a top-down approach, you start an assembly file and construct the individual component parts while you are doing the assembly. The hybrid approach is a combination of the bottom-up and top-down approaches.

The Bottom-Up Approach

When you already have a good idea on the size and shape of the component parts of an assembly or you are working as a team on an assembly, you use the bottom-up approach. Through parametric solid modeling methods, you construct all the parts to appropriate sizes and shapes that best describe the component parts of the assembly. Then you start an assembly file and place the parts in the assembly. In the assembly, you align the component parts together by applying assembly constraints. After putting all the parts together, you analyze and make necessary changes to the parts.

The Top-Down Approach

Sometimes you have a concept in your mind, but you do not have any concrete ideas about the component parts. You use the top-down approach—you start an assembly file and design some component parts there. From the preliminary component parts, you improvise. The main advantages in using this approach are that you see other parts while working on an individual part and you can continuously switch from designing one part to another.

The Hybrid Approach

In reality, you seldom use one approach alone. You use the bottom-up approach for standard component parts and new parts that you already know about and you use the top-down approach to figure out new component parts with reference to the other component parts. This combined approach is the hybrid approach.

ASSEMBLY IN INVENTOR

To reiterate, Autodesk Inventor has four kinds of files: part, assembly, presentation, and drawing. You construct an assembly by using an assembly file.

Assembly File

To start an assembly file, select New from the File menu. In the New dialog box, select Standard.iam to start a new assembly file. Initially, the panel bar shows the Assembly panel. Place the cursor in the panel and right-click. You will find a shortcut menu showing five items: Sketch, Features, Solids, Sheet Metal, and Expert. (See Figure 3–1.)

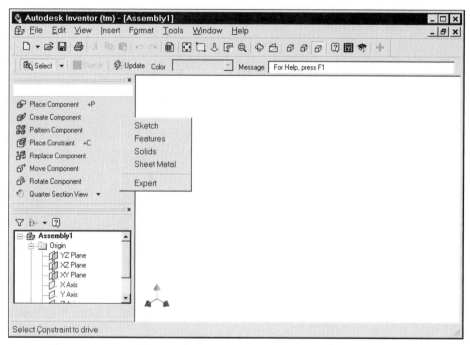

Figure 3–1   *New assembly file started*

To construct a solid part in the assembly, you use the Sketch and Features panels. To construct a sheet metal part in the assembly, you use the Sketch and Sheet Metal panels. To place a SAT or STEP format solid part in the assembly and edit the imported solid part, you use the Solids panel (see Appendix A). To hide the text describing the icons shown in the panel, select Expert.

Assembly Tools

Tools for constructing an assembly are available on the Assembly toolbar or panel. To display the Assembly toolbar, select View ▶Toolbar ▶Assembly. (See Figure 3–2.)

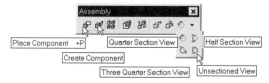

Figure 3–2   *Assembly toolbar*

The Assembly toolbar and panel has eight button areas. Table 3–1 describes the choices:

Table 3–1   Assembly toolbar and panel options

Button	Function
Place Component	Enables you to use the bottom-up approach to place existing parts or sub-assemblies to the assembly drawing.
Create Component	Enables you to use the top-down approach to start a new part or a new sub-assembly in the assembly drawing.
Pattern Component	Enables you to repeat a part or sub-assembly in a rectangular or circular pattern.
Place Constraint	Enables you to apply assembly constraints to selected pair of objects.
Replace Component	Enables you to replace a part or sub-assembly with another part or sub-assembly.
Move Component	Enables you to select an object and move it to a new location.
Rotate Component	Enables you to select an object and rotate it in 3D space.
Quarter Section View/ Half Section View/ Three Quarter Section View/ Unsectioned View	Enable you to set the display to a quarter section view, half section view, or three quarter section view through selected sketch planes, work planes, or faces and to reset the section view to a normal unsectioned view.

Placing a Component (Bottom-Up Approach)

In the bottom-up approach, you construct a number of 3D solid parts. Then you start an assembly file and place the components in the assembly file.

You select Place Component from the Assembly toolbar or panel and select a solid part or assembly from the Open dialog box. In the Open dialog box, you select a part file or assembly file and place the file in the assembly file.

Creating a Component (Top-Down Approach)

In the top-down approach, you start an assembly file and create a component in the assembly.

You select Create Component from the Assembly toolbar or panel and specify a new solid part or a new assembly. In the Create In-Place Component dialog box, you specify a part file or an assembly file that you will construct in the assembly. After starting a component part, you proceed to construct the component part in a way similar to what you learned in Chapter 2.

Degrees of Freedom

Initially, the component parts (except the first component) that you place or create in an assembly are free to translate in the 3D space, in three linear directions and three rotational directions. These free translations are called the six degrees of freedom (DOF). The DOF of a component is represented by a DOF symbol (see Figure 3–3).

You discover the DOF of the objects in an assembly by selecting Degrees of Freedom from the View menu.

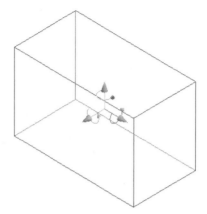

Figure 3–3 *Six degrees of freedom of a free object*

Grounding

By default, the first component part that you place or create in the assembly is fixed in the 3D space. A fixed object is a grounded object, with no degree of freedom—it cannot be moved. To free a grounded object, you select the object, right-click, and deselect Grounded. (See Figure 3–4.) On the other hand, if you want to fix an object in 3D space, you ground the object by right-clicking and selecting Grounded.

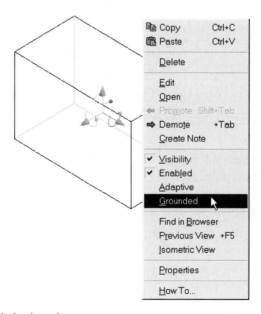

Figure 3–4 *Grounded selected*

Translation of Objects in 3D Space

Unless a component part is grounded, it is free to translate in the 3D space. To translate the component parts in an assembly to their appropriate locations, you move or rotate them.

To move an object, you select Move Component from the Assembly toolbar or panel, select the component part, and drag to move the component to a new position. To rotate an object, you select Rotate Component from the Assembly toolbar or panel, select the component part, and drag to rotate the component to a new orientation. (See Figure 3–5.)

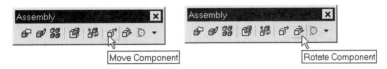

Figure 3–5 *Move and rotate*

Note that moving or rotating a component does not affect a component's DOF. You simply put it in a new position and new orientation.

Applying Assembly Constraints

To restrict the movement of a component part in 3D and to align it with another component in the assembly, you apply assembly constraints to selected faces, edges, and vertices of parts and the origin (with three work axes along the X, Y, and Z directions) of the part file or the sub-assembly file.

To apply assembly constraints, you select Place Constraint from the Assembly toolbar or panel. (See Figure 3–6.)

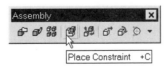

Figure 3–6 *Place assembly constraint*

There are four kinds of assembly constraints: mate, angle, tangent, and insert.

Mate causes two selected objects to mate or flush at a specified offset distance; you select points, edges, or faces. Figure 3–7 shows (from top to bottom) point B mated to point A, edge B mated to edge A, vertical face B mated to vertical face A, and vertical face B flush with vertical face A.

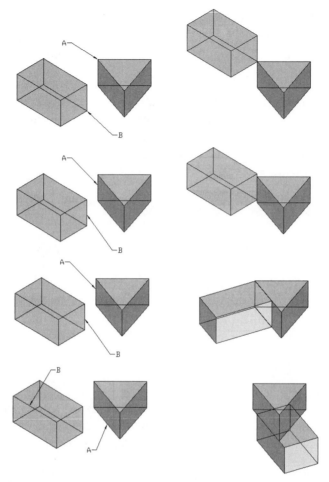

Figure 3–7 *Mate constraints*

Angle causes two selected objects to align at a specified angle; you select edges or faces. Figure 3–8 shows face B constrained at an angle with face A.

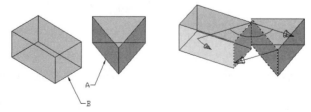

Figure 3–8 *Angle constraint*

Tangent causes selected faces, planes, cylinders, spheres, and cones to contact at their tangential point and at a specified offset distance; you select faces. (See Figure 3–9.)

Figure 3–9 *Tangent constraint*

Insert causes selected circular edges to align face to face and concentrically at a specified distance; you select circular edges. (See Figure 3–10.)

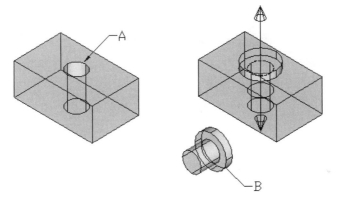

Figure 3–10 *Insert constraint*

Constraint Driving

You can animate mechanical motions in an assembly by varying the parameters of the assembly constraints that you set to the components of an assembly. This is called constraint driving.

Pattern of Components

When you need to set up an array of component parts in an assembly, it is very time consuming to place the components and apply assembly constraints to the component individually. To save time, you put a component part in the assembly, apply constraints to a component part to properly position it in the assembly, and construct a pattern. There are two kinds of patterns: rectangular and circular.

Figure 3–11 shows Pattern Component selected from the Assembly toolbar and a rectangular pattern of components placed in the assembly.

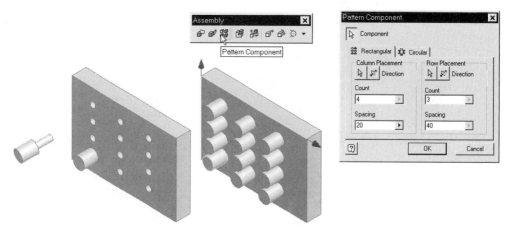

Figure 3–11 *Assembly pattern*

Toggling Assembly Mode and Part Modeling Mode

After you place components in an assembly or create components in an assembly, the component part files link to the assembly file. The component parts will display as objects in the assembly file browser bar. To edit a solid part, you double-click the component part in the browser bar to switch to part modeling mode. To return to assembly mode, you double-click the assembly in the browser bar.

Reordering Component Parts in the Assembly Hierarchy

Objects (components or sub-assemblies) placed or created in an assembly form a hierarchy in the browser. The hierarchy refers to the sequence of placement or creation. A component that is created or placed earlier in the assembly will be placed in a higher position in the hierarchy. When you apply assembly constraints to a pair of objects, an object located lower in the hierarchy will translate toward the object located higher in the hierarchy unless it is grounded. To reorder the hierarchy, you select the object in the browser bar and drag it to a new position. (See Figure 3–12.)

Restructuring the Assembly Hierarchy

In a large assembly that has many component parts, it is common practice to organize the components into sub-assemblies of smaller number of components. If you want to reorganize a component in an assembly into a sub-assembly, you select the part in the browser, right-click, and select Demote. If you want to move a component in a sub-assembly into the assembly, you select the part in the browser, right-click, and select Promote. (See Figure 3–13.)

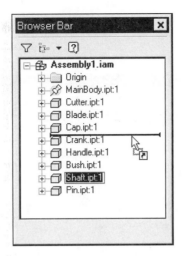

Figure 3–12 *Reordering the assembly hierarchy*

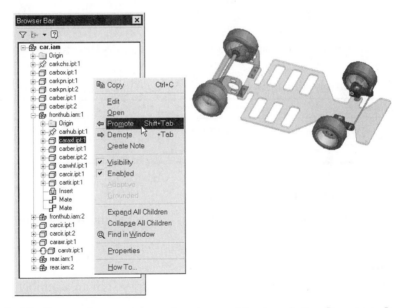

Figure 3–13 *Rearranging components in sub-assemblies by selecting Demote or Promote*

Replacing a Component in an Assembly

In the process of designing a new product, you may want to replace a component part in an assembly with another component. To save the effort needed to remove a component and re-link another component, you use a replacement. Select Replace Component from the Assembly toolbar or panel. (See Figure 3–14.)

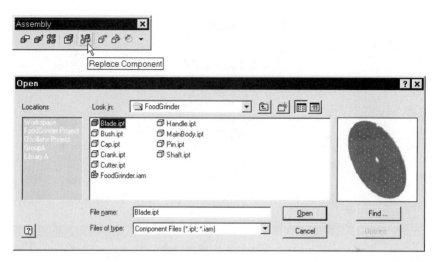

Figure 3–14 *Replacing a component*

Shortcut Keys

Apart from the appropriate toolbar and panel, Table 3–2 shows shortcut keys you can use:

Table 3–2 Assembly shortcut keys

Shortcut Key	Function
P	Place part
C	Assembly constraints

THE BOTTOM-UP APPROACH

Using the bottom-up approach, you start an assembly file and place existing parts or sub-assemblies in an assembly.

FOOD GRINDER ASSEMBLY

You will place the parts that you constructed in Chapter 2 in an assembly file.

1. Select New from the File menu.

2. In the New dialog box, select the Default tab.

3. Select Standard.iam and select the OK button.

Place Components

4. Select Place Component from the Assembly toolbar or panel.

5. Select Mainbody.ipt from the directory where you stored the part files of the food grinder.

6. Right-click and select Done. (See Figure 3–15.) A component is placed in the assembly.

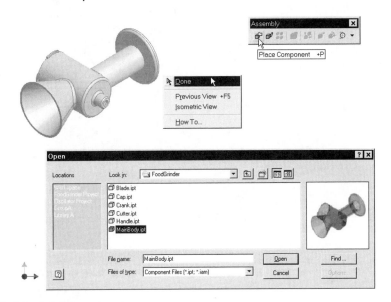

Figure 3–15 *Solid part placed*

By default, the first component placed or created in the assembly is grounded. Therefore, you find a pushpin symbol next to the Mainbody.ipt in the browser bar shown in Figure 3–16. It denotes that the object is grounded.

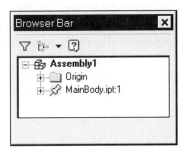

Figure 3–16 *Browser bar showing Mainbody.ipt pinned*

7. Now place three more components in the assembly. Select Place Component to place the solid parts (Cutter.ipt, Blade.ipt, and Cap.ipt) one by one. Remember to right-click and select Done each time after you place a component to the assembly.

8. To display the degrees of freedom symbol, select Degrees of Freedom from the View menu. (See Figure 3–17.)

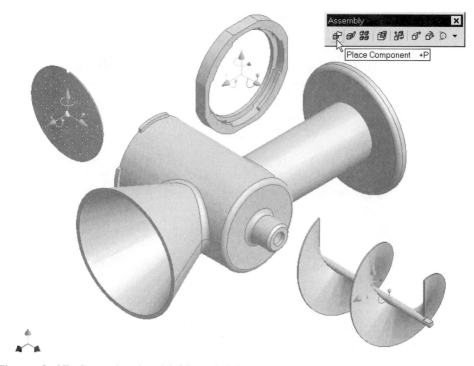

Figure 3–17 *Parts placed and DOF symbol displayed*

Apply Assembly Constraints

Now you have four components placed in the assembly. To restrict their relative movement and to align them to their proper position and orientation, you will apply assembly constraints to selected pairs of geometric references on the components.

9. You will align the axis of the blade to the axis of the cutter. Select Place Constraint from the Assembly toolbar or panel.

10. Select the Mate button in the Place Constraint dialog box.

11. Select the axis of the cutter indicated in Figure 3–18.

12. The axis of the cutter is selected. Now select Zoom Window from the Standard toolbar and zoom the display as shown in Figure 3–19.

13. Select the cylindrical face of the hole of the blade highlighted in Figure 3–19. The cutter translates toward the blade.

14. Right-click and select Previous View. (See Figure 3–19.)

15. Select the Apply button. The constraint is applied. (See Figure 3–20.)

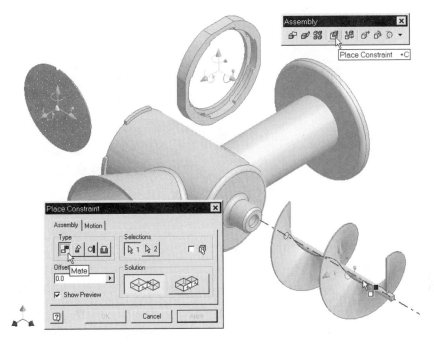

Figure 3–18 *Axis of the cutter selected*

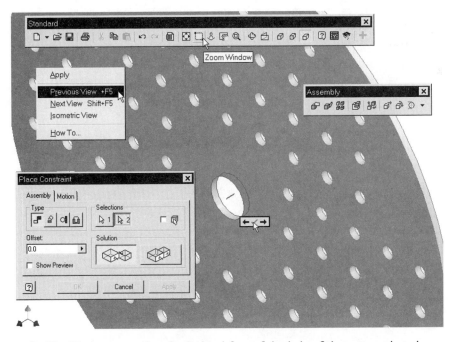

Figure 3–19 *Display zoomed and cylindrical face of the hole of the cutter selected*

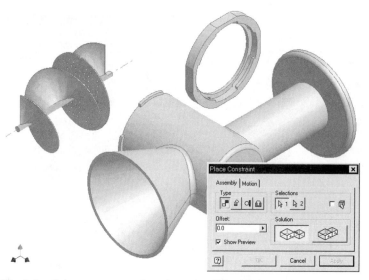

Figure 3–20 *Axis of the cutter mated to the axis of the blade*

Now you will align the edge of the cutter with a face of the blade.

16. Select Zoom Window from the Standard toolbar and zoom the display as shown in Figure 3–21.

17. Check to see that the Mate button of the Place Constraint dialog box is still selected.

18. Select the edge of the cutter indicated in Figure 3–21.

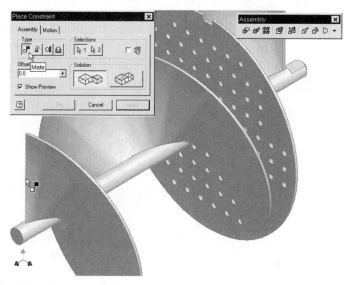

Figure 3–21 *Edge of the cutter selected*

19. Select the face of the blade indicated in Figure 3–22.

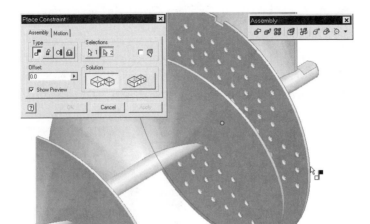

Figure 3–22 *Face of the blade selected*

20. Select the Apply button. The edge of the cutter is constrained to the face of the blade. (See Figure 3–23.)

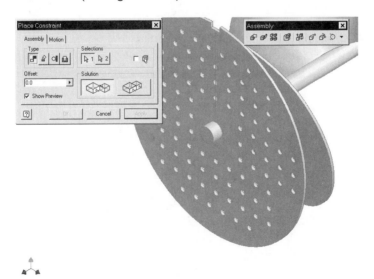

Figure 3–23 *Selected face of the blade mated with selected edge of the cutter*

Now you will orient the blade relative to the main body.

21. Zoom, pan, and rotate the display as shown in Figure 3–24.

22. Select the Angle button of the Place Constraint dialog box.

23. Select a face indicated in Figure 3–24.

24. Select Zoom Window from the Standard toolbar and zoom the display as shown in Figure 3–25.

25. Select an edge of the blade indicated in Figure 3–25.

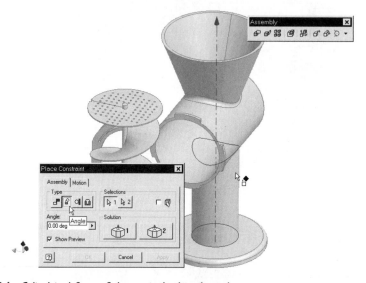

Figure 3–24 *Cylindrical face of the main body selected*

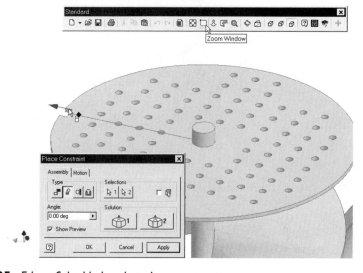

Figure 3–25 *Edge of the blade selected*

26. Right-click and select Previous View.

27. Select the Apply button. The direction of a face of the blade is aligned with the axis of the main body. (See Figure 3–26.)

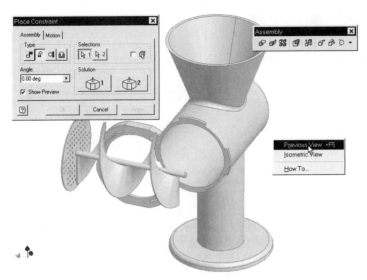

Figure 3–26 *Face of the blade aligned with the cylindrical face of the body*

28. Now you will align a face of the blade with a face of the main body. Select the Insert button in the Place Constraint dialog box.

29. Select a circular edge indicated in Figure 3–27.

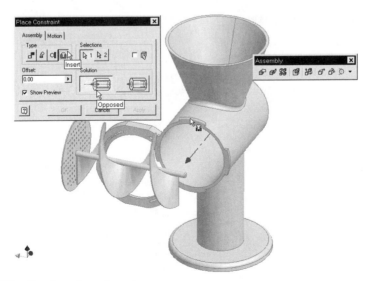

Figure 3–27 *Circular edge selected*

30. Select Zoom Window from the Standard toolbar to zoom the display as shown in Figure 3–28.

31. Select the circular edge of the blade indicated in Figure 3–28.

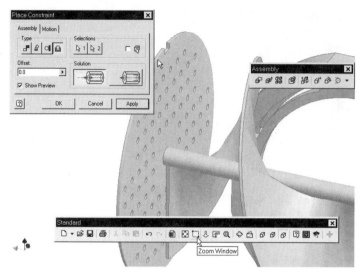

Figure 3–28 *Circular edge of the blade selected*

32. Right-click and select Previous View.

33. Select the Apply button to constrain the blade to the main body. (See Figure 3–29.)

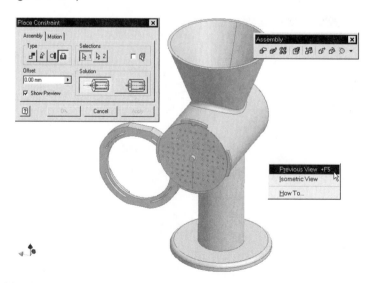

Figure 3–29 *Blade properly constrained*

34. Now you will constrain the cap. Select the Mate button of the Place Constraint dialog box.

35. Select the circular edges of the cap and the blade indicated in Figure 3–30.

Figure 3–30 *Circular edges of the cap and the blade selected*

36. Select the OK button. The cap is constrained to the main body. (See Figure 3–31.)

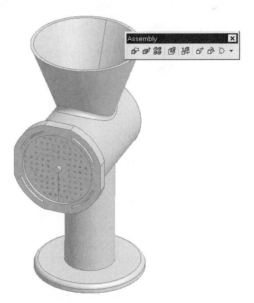

Figure 3–31 *Cap constrained*

Display Section View

While you construct an assembly, you have to select objects from individual components. To facilitate selection, you can set the display to either shaded mode or wireframe mode. In addition, you can set the display to three kinds of section view: quarter section, half section, and three quarter section.

To set the display to a half section view, you select Half Section View from the assembly toolbar or panel and select a plane, work plane, or face to define the section plane. To set the display to a quarter or three quarter section view, you select Quarter Section View or Three Quarter Section View from the assembly toolbar or panel. Then you select two mutually perpendicular planes, work planes, or faces to specify the section planes.

Now you will learn how to set the display to a half section view.

37. Select Half Section View from the Assembly toolbar or panel.

38. Select Workplane1 of the blade of the food grinder in the browser bar.

39. Right-click and select Done. (See Figure 3–32.)

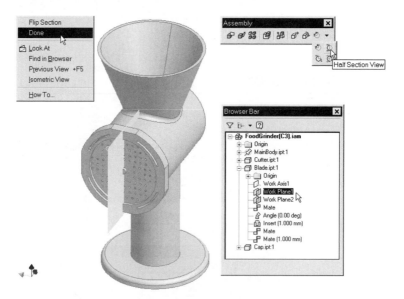

Figure 3–32 *Display being set to half section view*

40. The display is set. Now select Unsectioned View from the assembly toolbar or panel to set the display to a normal unsectioned view. (See Figure 3–33.)

41. Compare Figure 3–33 with Figure 3–31 and find out which kind of display you prefer to use.

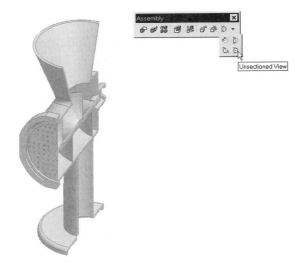

Figure 3–33 *Display being set to normal*

Place Two More Components

42. Now place two more solid parts, the crank and the handle, in the assembly.

43. Select Place Component from the Assembly toolbar or panel, select the files, and select the placement locations as shown in Figure 3–34.

Before you apply assembly constraints to them, you will construct three more parts.

44. Save your assembly (file name: Foodgrinder.iam).

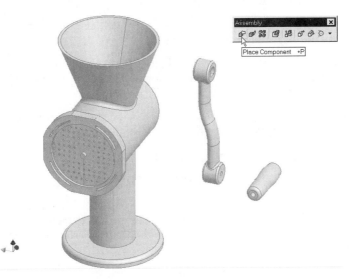

Figure 3–34 *Crank and handle placed*

THE TOP-DOWN APPROACH

In a top-down approach, you start an assembly and construct parts or sub-assemblies while working in the assembly environment. The advantages of constructing components in the assembly environment are that you see other parts and you make use of the features of other parts to construct the solid.

FOOD GRINDER ASSEMBLY

Now you will construct three more component parts in the assembly. (See Figure 3–35.)

Figure 3–35 *Component parts to be constructed in the assembly*

Because you used the bottom-up approach in working out six components and you use the top-down approach to work out the remaining components, you are, in fact, using the hybrid approach to construct the food grinder assembly.

Create Component (Bush of the Food Grinder)

Now you will construct the bush.

1. To create a new component in the assembly, select Create Component from the Assembly toolbar or panel.

2. Set the file name to Bush.ipt and file type to Part. The solid part that you are going to construct is saved in a part file and is referenced in the assembly file. (See Figure 3–36.)

3. Select the OK button and select a point on the screen to specify the location of the sketch plane of the new solid part.

In the browser bar and the graphics area, the assembly and other parts are grayed out. Now they are inactive, and you are working in part modeling mode. (See Figure 3–37.)

4. You will construct two circles in a sketch and extrude the sketch to form an extruded solid. Select Center Point Circle from the Sketch toolbar or panel.

5. Construct two concentric circles.

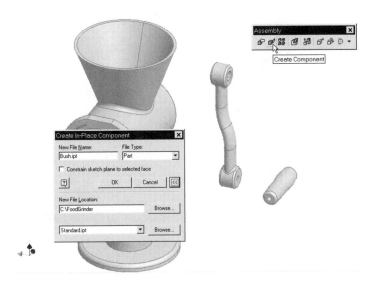

Figure 3–36   *New part initiated*

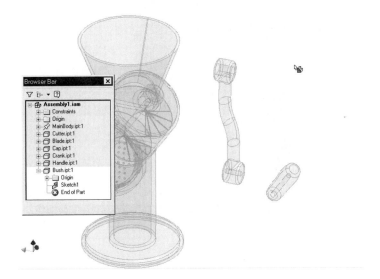

Figure 3–37   *Part modeling mode activated*

6. Select General Dimension from the Sketch toolbar or panel and add two dimensions (12 and 18 units) as shown in Figure 3–38.

7. Exit sketch mode and select Extrude from the Features toolbar or panel and select the area highlighted in Figure 3–39 to extrude it a height of 16 units.

8. Select the OK button.

242

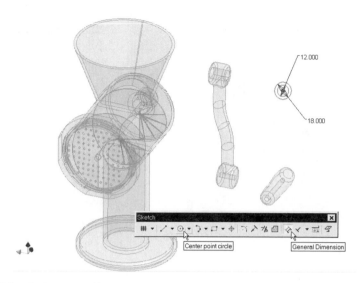

Figure 3–38 *Circles constructed and dimensioned*

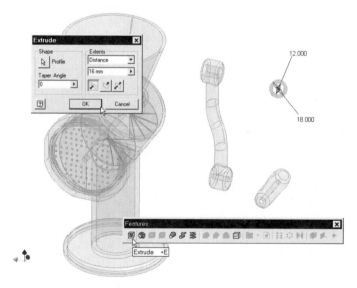

Figure 3–39 *Sketch being extruded*

An extruded solid is constructed. As you have seen, the way to construct a part in assembly mode is virtually the same as when you work in part modeling mode. The difference is that you see the other parts in the assembly.

9. Save the files.

Once a part is constructed, there is no difference in whether it is a placed part or a created part, because you can work in assembly mode or part modeling mode in an assembly. To translate parts in the 3D space or apply assembly constraints to parts, you work in assembly mode. To create or edit a part, you work in part modeling mode. To work in assembly mode, double-click the assembly in the browser bar. To switch to part modeling mode, double-click the part in the browser bar.

Translate Components

10. Because you will rotate, move, and constrain the newly constructed part, select the assembly in the browser to switch to assembly mode.

11. Now select Rotate from the Standard toolbar to set the display as shown in Figure 3–40.

12. Select Move or Rotate from the Assembly toolbar or panel to translate the bush to a new location.

13. Select Wireframe Display from the Standard toolbar to set the display to wireframe mode.

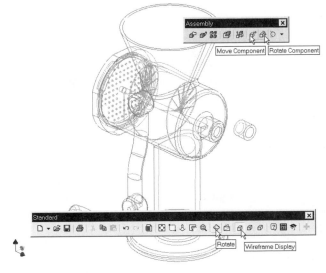

Figure 3–40 *Object translated, and display zoomed*

Apply Assembly Constraints

Now you will apply an insert constraint to align the bush to the main body.

14. Select Zoom Window from the Standard toolbar to set the display as shown in Figure 3–41.

15. Select Place Constraint from the Assembly toolbar or panel and select the Insert button.

16. Select the circular edge of the main body and the circular edge of the bush and set the constraint values as shown in Figure 3–41.

17. Select the OK button. The bush is constrained to the main body.

18. Right-click and select Previous View to set the display to the previous view.

19. Select the main body of the food grinder in the browser bar and right-click. (See Figure 3–42.)

20. Deselect Visibility from the shortcut menu. The main body is hidden. (See Figure 3–43.)

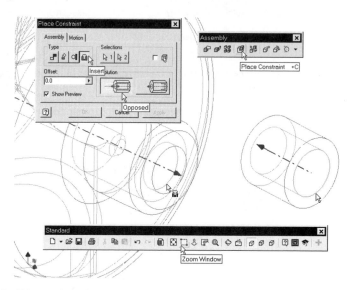

Figure 3–41 *Edges selected*

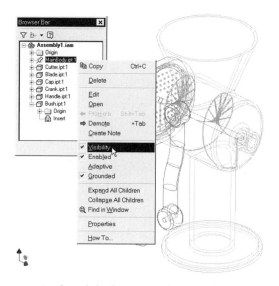

Figure 3–42 *Bush constrained, and display zoomed*

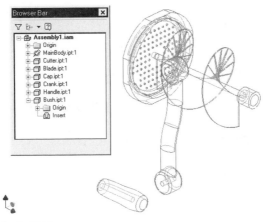

Figure 3–43 *Main body hidden*

Use Features of Existing Components (Shaft of the Food Grinder)

Now you will construct another part file in the assembly. While making the sketch for the new part, you will use the geometry of the another part in the assembly.

1. Select Create Component from the Assembly toolbar or panel to start a new solid part. The file name is Shaft.ipt.

On the default sketch plane, you will construct a sketch and revolve the sketch to make a revolved solid.

2. Select Line from the Sketch toolbar or panel to construct a series of horizontal and vertical lines and select General Dimension from the Sketch toolbar or panel to add dimensions as shown in Figure 3–44.

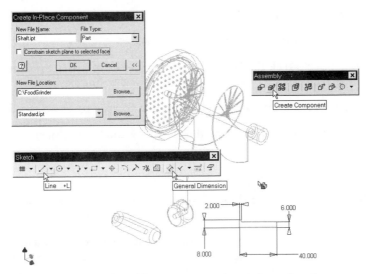

Figure 3–44 *New part started, and lines constructed and dimensioned*

3. The sketch is complete. Exit sketch mode and select Revolve from the Features toolbar or panel.

4. Select the area highlighted in Figure 3–45.

5. Select the lower horizontal line of the highlighted area as the centerline.

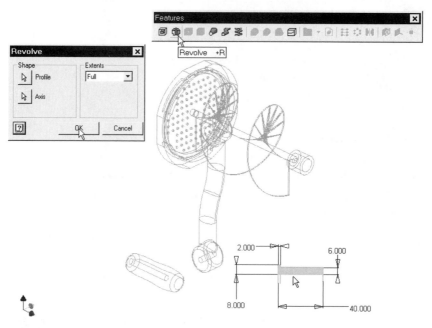

Figure 3–45 *Sketch being revolved*

6. Select the OK button. A revolved solid is constructed.

7. Save the files (shaft file name is Shaft.ipt).

8. Now double-click the assembly in the browser bar to activate assembly mode. You will apply an assembly constraint to the new part.

9. Select Zoom Window from the Standard toolbar to zoom the display.

10. Select Place Constraint from the Assembly toolbar or panel and select the Insert button.

11. Select the pair of circular edges highlighted in Figure 3–46.

12. Select the OK button. (See Figure 3–47.) An insert assembly constraint is applied.

13. Now you will make another sketch. Select Zoom Window and Rotate from the Standard toolbar to set the display as shown in Figure 3–48.

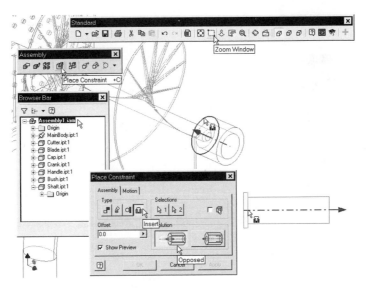

Figure 3–46 *Revolved solid constructed, assembly mode activated, and component being constrained*

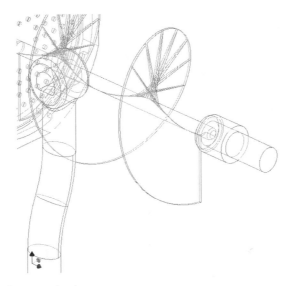

Figure 3–47 *Shaft constrained*

14. Select the Shaft in the browser bar by double-clicking to switch to part mode.

15. Select Sketch from the Command Bar toolbar and select the highlighted face to set up a new sketch plane.

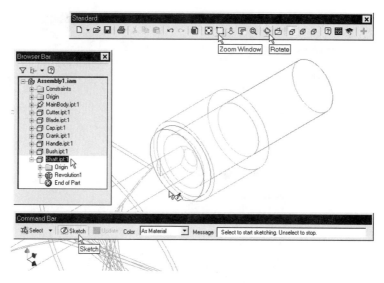

Figure 3–48 *Revolved solid constrained, display zoomed, and new sketch plane constructed*

16. To use the geometry of another part in the assembly, select Project Geometry from the Sketch toolbar or panel and select the edge highlighted in Figure 3–49 to project a line on the current sketch plane.

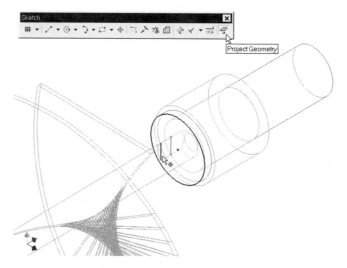

Figure 3–49 *Selected straight edge projected*

17. Now select the circular edge of the cutter that is highlighted in Figure 3–50.

18. Exit sketch mode and select Extrude from the Features toolbar or panel.

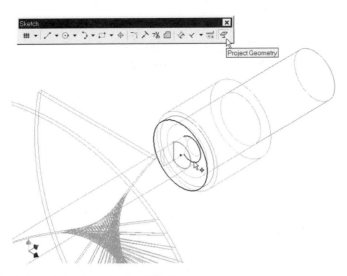

Figure 3–50 *Selected circular edge projected*

19. Select the area highlighted in Figure 3–51.

20. Extrude the area a distance of 10 units to cut the solid part and select the OK button.

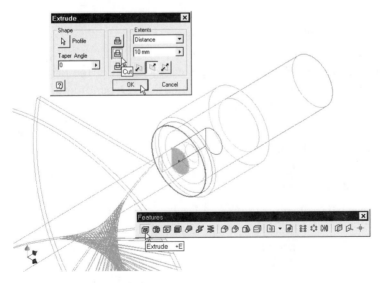

Figure 3–51 *Sketch being extruded*

In making this solid part, you constructed two sketched features. The sketch for making the second feature is projected from the geometry of the adjacent part.

Measure and Modify

While editing and constructing parts in assembly mode, you see all the parts put together and you can find out the distance between selected geometry of the parts selected.

21. Now select Measure Distance from the Tools menu.

22. Select the highlighted end points as shown in Figure 3–52.

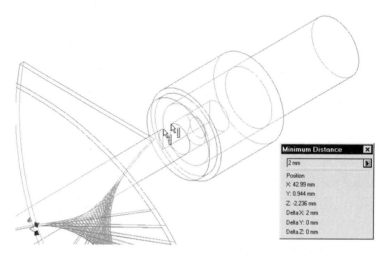

Figure 3–52 *End points selected*

The result of measurement indicates the exact locations of the first measuring point and the distance between the two selected points. (Depending on the exact size of the parts that you constructed, this measurement may not be the same as yours.) Here, the distance is 2 units. After you know the distance, modify the part.

23. Set the display as shown in Figure 3–53.

24. Select the shaft in the browser bar and right-click and select Edit Sketch to edit the sketch.

25. Select the dimension shown in Figure 3–53 and increase the dimension by 2 units.

26. Now update the change. Select Update from the Command Bar toolbar. (See Figure 3–54.)

27. Right-click and select Isometric View.

28. Select the assembly in the browser bar to update the assembly. (See Figure 3–55.)

In making this component in an assembly, you used the geometry of adjacent parts in sketching.

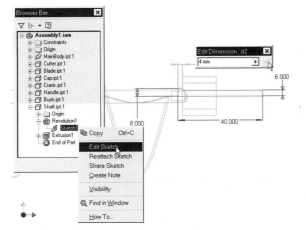

Figure 3–53 *Sketch being edited*

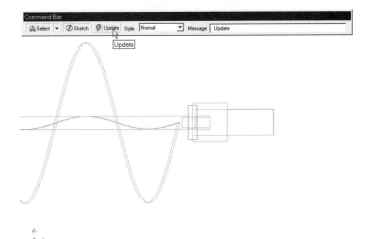

Figure 3–54 *Solid part updated*

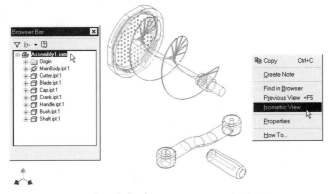

Figure 3–55 *Assembly updated and display set to isometric view*

Check Interference

Because it is difficult to tell from the display whether there is any interference among the parts in an assembly, you will let the computer find it out for you.

29. Select Analyze Interference from the Tools menu.

30. Select the shaft and the cutter to find out if there is any interference between the two selected parts. (See Figure 3–56.)

31. Select the OK button. After an interval, the result is displayed. (See Figure 3–57.)

The dialog box shown in Figure 3–57 indicates that there is no interference. If there is any interference, a message will tell you where the interference occurs.

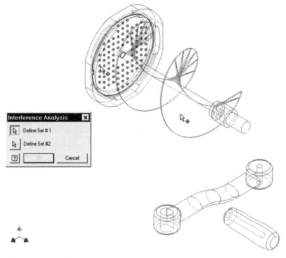

Figure 3–56 *Interference checking*

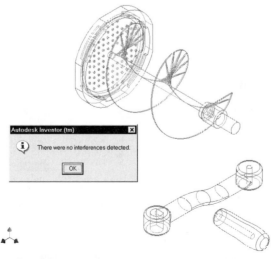

Figure 3–57 *Result of interference check*

Use Projected Geometry

Now you will use projected geometry to modify a part.

1. Select Mainbody in the browser.

2. Right-click and select Visibility to display it.

3. Select Rotate from the Standard toolbar and rotate the display as shown in Figure 3–58.

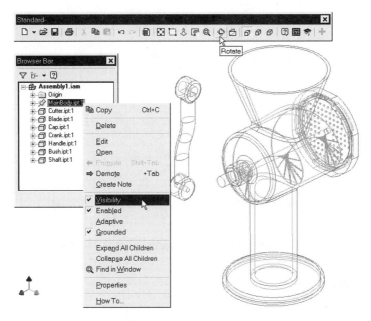

Figure 3–58 *Main body made visible and display rotated*

4. Select Zoom Window from the Standard toolbar to zoom the display as shown in Figure 3–59.

5. Select Place Constraint from the Assembly toolbar or panel and select the Insert button.

6. Select the pair of circular edges as highlighted in Figure 3–59 and select the OK button.

7. Now double-click the shaft in the browser bar to activate part modeling mode.

8. Select Zoom and Pan from the Standard toolbar to set the display as shown in Figure 3–60.

9. Select Sketch from the Command Bar toolbar and select the face high-lighted in the figure to set up a new sketch plane.

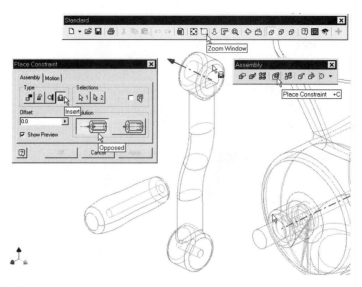

Figure 3–59 *Handle being constrained to the main body*

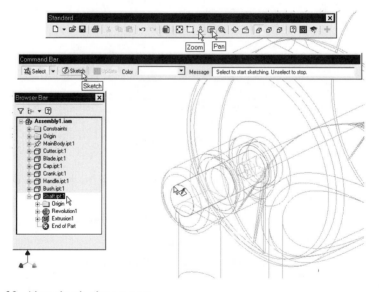

Figure 3–60 *New sketch plane set up*

10. Select Project Geometry from the Sketch toolbar or panel.

11. Select the straight edge as highlighted in Figure 3–61.

12. Exit from sketch mode and select Extrude from the Features toolbar or panel.

13. Select the area highlighted in Figure 3–62.

14. Extrude the area a distance of 18 units to cut the solid.

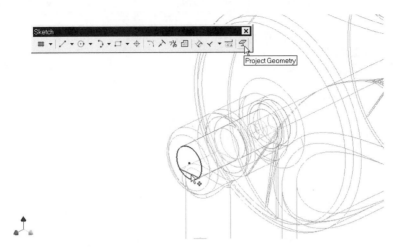

Figure 3–61 *Edges projected*

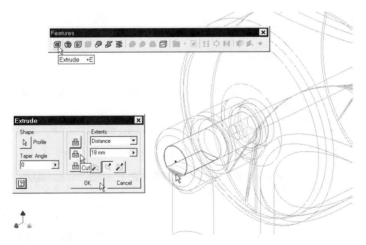

Figure 3–62 *Shaft being cut*

Geometry projected from the handle is used to cut the shaft.

Use Existing Geometry to Construct a New Solid Part

Now you will use existing geometry to construct a new solid part.

1. Select Zoom All from the Standard toolbar.

2. Then double-click the assembly in the browser bar to activate assembly mode.

3. Select Rotate from the Standard toolbar to rotate the display as shown in Figure 3–63.

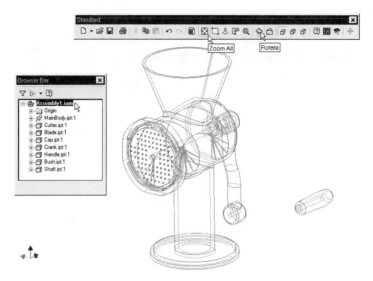

Figure 3–63 *Assembly mode set and display rotated*

4. Select Zoom Window from the Standard toolbar to set the display.

5. Select Create Component from the Assembly toolbar or panel to start a new solid part (file name is Pin.ipt).

6. In the Create In-Place Component dialog box, select the Constrain sketch plane to selected face button.

7. Select the face highlighted in Figure 3–64 to set up a new sketch plane for the new part.

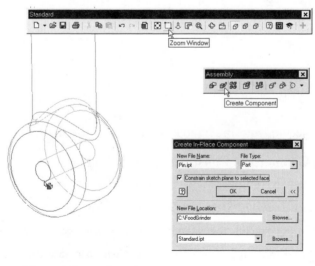

Figure 3–64 *New solid part being activated*

8. Select Project Geometry from the Sketch toolbar or panel.

9. Select the circular edge highlighted in Figure 3–65.

10. Exit sketch mode.

11. Select Extrude from the Features toolbar or panel and extrude the area a distance of 65 units to form an extruded solid. (See Figure 3–66.)

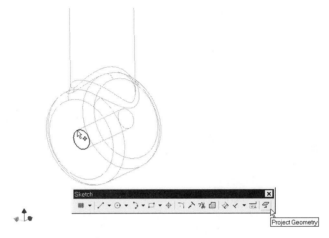

Figure 3–65 *Circular edge projected*

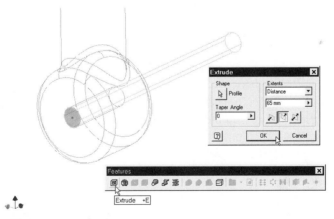

Figure 3–66 *Sketch being extruded*

12. Select Sketch from the Command Bar toolbar to set up a new sketch plane on the highlighted face shown in Figure 3–67.

13. Select Offset from the Sketch toolbar or panel and construct an offset circle.

14. Select General Dimension from the Sketch toolbar or panel to add a dimension (10 units). (See Figure 3–68.)

15. Select Extrude to extrude the area highlighted in Figure 3–69 to a distance of 5 units to join the solid part.

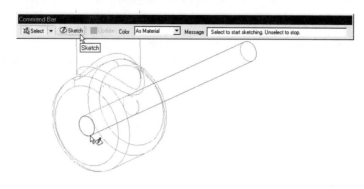

Figure 3–67 *Sketch plane constructed*

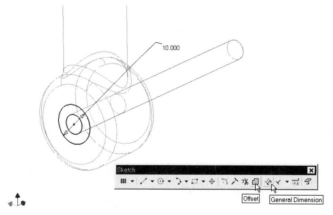

Figure 3–68 *Offset circle constructed, and dimension added*

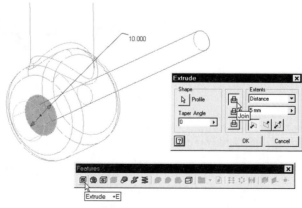

Figure 3–69 *Highlighted area being extruded*

The solid part is complete. Now you will apply assembly constraints.

16. Select Zoom All from the Standard toolbar to zoom the display.

17. Select the assembly in the browser bar to activate assembly mode. (See Figure 3–70.)

18. Select Zoom Window from the Standard toolbar to zoom the display.

19. Select Place Constraint from the Assembly toolbar or panel and select the Insert and the Opposed buttons.

20. Select the pair of the circular edges highlighted in Figure 3–71 and select the Apply button.

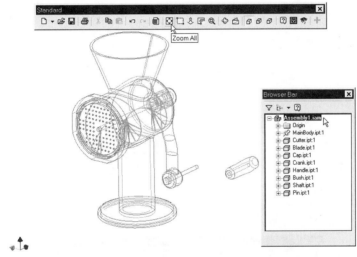

Figure 3–70 *Constraint being applied*

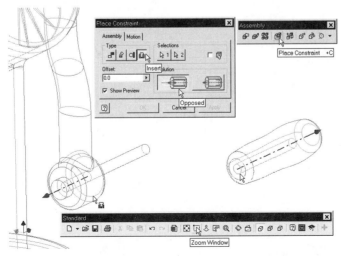

Figure 3–71 *Circular edges selected*

21. In the Place Constraint dialog box, select the Insert and the Opposed buttons.

22. Select the circular edges highlighted in Figure 3–72.

23. Select the OK button.

24. Select Zoom All from the Standard toolbar. (See Figure 3–73.)

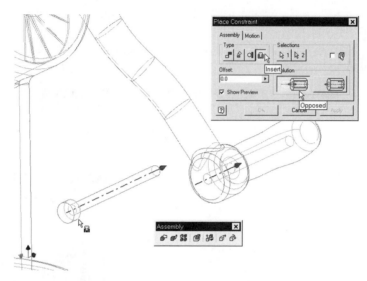

Figure 3–72 *Circular edges selected*

Figure 3–73 *Display rotated*

25. Select Rotate and Zoom Window from the Standard toolbar to set the display as shown in Figure 3–74.

26. Select Place Constraint from the Assembly toolbar or panel and select the Angle button.

27. Select the highlighted edges and select the Apply button.

28. Select Zoom from the Standard toolbar to set the display.

29. Select an edge of the crank and the axis of the main body indicated in 3–75.

30. Set the angle to 90 degrees and select the OK button.

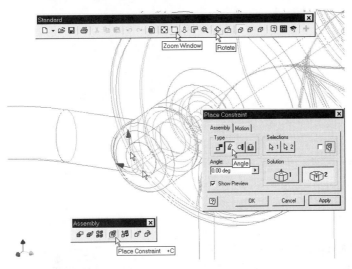

Figure 3–74 *Edges selected*

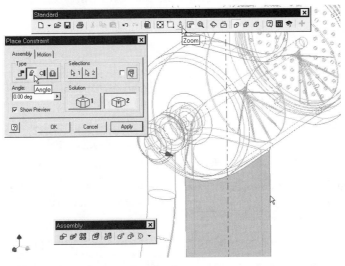

Figure 3–75 *Edge and axis being constrained*

31. Select Zoom All from the Standard toolbar to set the display. (See Figure 3–76.)

The parts of the food grinder are complete.

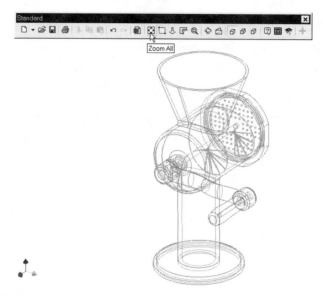

Figure 3–76 *Food grinder assembled*

CONSTRAINT DRIVING

You can animate mechanical motions in an assembly to help you evaluate your design. To animate the effect of mechanical motion, you drive an assembly constraint through a sequence of steps by selecting the constraint and setting the animation parameters.

Drive Angle Constraint

Now you will set the shaft of the food grinder in motion by driving the angle constraint that you applied to the shaft and the main body.

1. Select the Angle constraint (applied in the previous sequence of steps) in the browser bar.

2. Right-click and select Drive Constraint as shown in Figure 3–77.

3. Set Start to 0 degrees and End to 360 degrees to animate the angle between the shaft and the main body from an initial value of 0 degrees through 360 degrees.

(To save the animation to AVI file format, you can select the Record button.)

4. Now select the >> button.

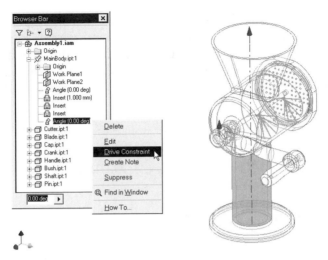

Figure 3–77 *Drive Constraint selected*

A Drive Constraint dialog box displays. (See Figure 3–78.)

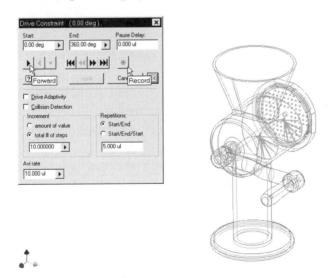

Figure 3–78 *Constraint being driven*

In the expanded dialog box, the Drive Adaptivity and Collision Detection check boxes concern adaptivity and collision detection, which you will learn about in Chapter 4, and the Avi rate box concerns the number of frames per second of an AVI file that you will save if you select the Record button.

    5. Select the Forward button.

You will see that the crank, the pin, and the handle are rotating together with the shaft because the crank is constrained to the shaft and the pin and the handle are constrained to the crank. However, if you examine the motion carefully, you will see that the cutter does not rotate, because the cutter it is not properly constrained to the shaft. To set the cutter in motion as well, you will apply a constraint between the cutter and the shaft.

6. Select the Step Forward button of the Drive Constraint dialog box to step the mechanism four steps forward.

7. Select the Apply button. (See Figure 3–79.) This will show where you need to apply constraints.

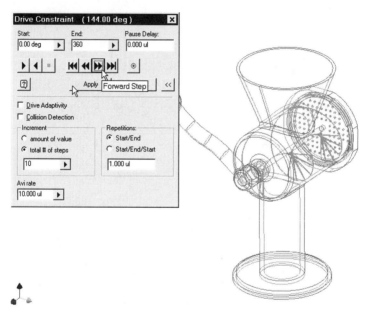

Figure 3–79 *Mechanism stepped forward*

8. Hide the main body, bush, and crank.

9. Zoom the display as shown in Figure 3–80.

10. Zoom the display as shown in Figure 3–81.

11. Select Place Constraint from the Assembly toolbar or panel.

12. Select the Angle button and select the edges that are highlighted with arrows in Figure 3–81.

13. Select the OK button.

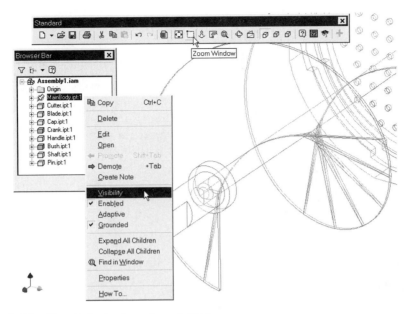

Figure 3–80 *Main body, bush, and crank hidden*

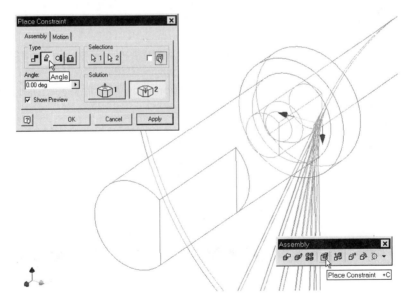

Figure 3–81 *Selected edges being constrained*

14. Now set the main body, bush, and crank visible again.

15. Select Zoom All from the Standard toolbar to set the display. (See Figure 3–82.)

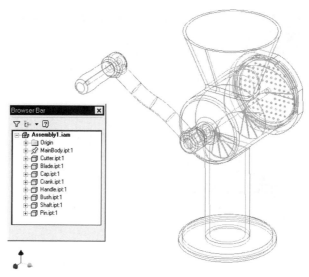

Figure 3–82 *Parts made visible*

Save Animation File

16. Select the angle constraint of the crank in the browser, right-click, and select Drive Constraint.

17. Select the Forward button to run the simulation. If the cutter is rotating together with the crank and handle, your assembly is complete.

18. You can save the animation of the motion in an AVI file. Select the Record button of the Drive Constraint dialog box.

19. Specify a file name and a compression ratio. (See Figure 3–83.)

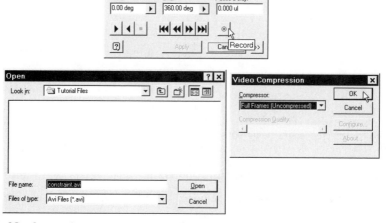

Figure 3–83 *Saving the animation in an AVI file*

Save Design View

Display views can be saved as design views.

20. Select Design Views from the View menu.

21. In the Design Views dialog box, type a design view name.

22. Select the Save button. The current display view is saved. (See Figure 3–84.)

23. Now save and close your file.

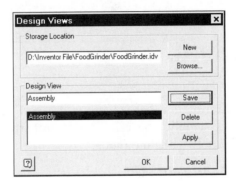

Figure 3–84 *Design Views dialog box*

BILL OF MATERIALS, ASSEMBLY PROPERTIES, AND DESIGN ASSISTANT

Now you will construct a bill of materials, set assembly properties, and use Design Assistant to manage design properties.

BILL OF MATERIALS

1. Open the food grinder assembly (file name: FoodGrinder.iam).

A bill of materials is a table detailing the properties and textual information regarding the component parts of an assembly.

2. To construct a bill of materials, select Bill of Materials from the Tools menu.

3. In the Bill of Materials dialog box, select the >> button to expand the dialog box and view the Column Properties area. (See Figure 3–85.)

4. Set the width of the columns and the alignment of the name and data. Select the OK button.

5. Now the assembly of the food grinder is complete. Select Save All from the File menu to save all your files.

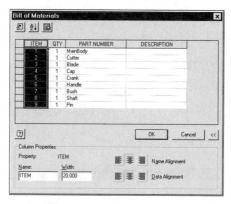

Figure 3–85 *Bill of Materials dialog box*

PROPERTIES

Properties are textual information that you store in a part file or an assembly file. When you work in an assembly, you can set the properties in two modes: part modeling mode and assembly modeling mode.

Set Properties in Part Modeling Mode

1. To set the properties of a solid part in part modeling mode, activate part modeling mode by double-clicking the part in the browser bar.

2. Select the part and right-click.

3. Select Properties. (See Figure 3–86.) The properties of the solid part in part modeling mode are displayed; there are six tabs.

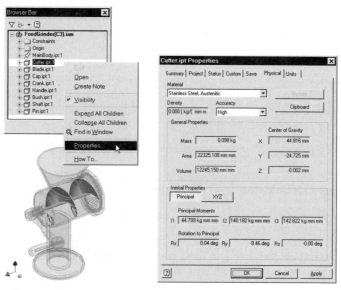

Figure 3–86 *Physical tab of the Properties dialog box in part modeling mode*

4. Select the OK button to close the Properties dialog box.

Set Properties in Assembly Modeling Mode

1. In assembly modeling mode, double-click the assembly in the browser bar to activate the assembly.

2. Select the solid part and right-click.

3. Select Properties. (See Figure 3–87.) As in the part modeling Properties dialog box, there are six tabs, but instead of the Units tab, there is an Occurrence tab. You set how the solid part occurs in the current assembly.

4. Select the OK button to close the Properties dialog box.

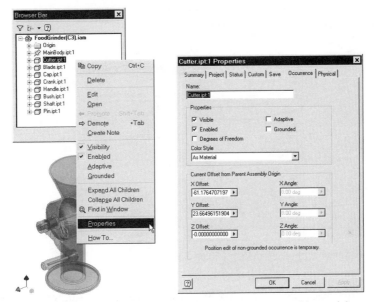

Figure 3–87 *Occurrence tab of the Properties dialog box in assembly modeling mode*

DESIGN ASSISTANT

To find, keep track of, maintain, and report on the solid part and assembly files, you use Design Assistant.

1. Open the file FoodGrinder.iam.

2. Select Design Assistant from the File menu.

In the Design Assistant dialog box, there are buttons enabling you to maintain the solid part and assembly files and a design properties tab enabling you to add, change, and report file properties.

3. Select the Select Properties To Display button; the Select Properties To View dialog box is displayed. (See Figures 3–88 and 3–89.)

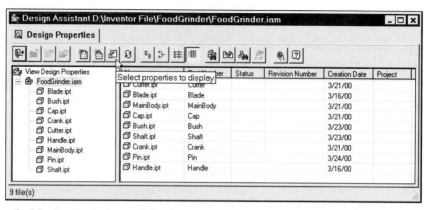

Figure 3–88 *Design Assistant dialog box*

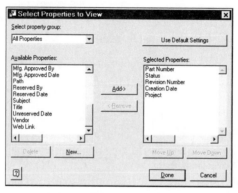

Figure 3–89 *Select Properties To View dialog box*

4. In the Select Properties To View dialog box, select Author from the Available Properties box, select the Add-> button, and select the Done button. The author of the part files is displayed.

5. Now close the Design Assistant dialog box.

PRESENTATION VIEWS

To illustrate how component parts of an assembly fit together, you use exploded views.

To construct presentation views, start a presentation file by selecting New from the File menu and selecting Standard.ipn. (See Figure 3–90.)

Tools for constructing presentation views of an assembly are available on the Presentation toolbar or panel. To display the Presentation View Management toolbar, select View ▶ Toolbar ▶ Presentation View Management. (See Figure 3–91.)

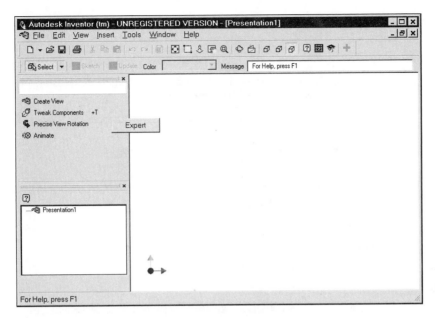

Figure 3–90 *New presentation file started*

Figure 3–91 *Presentation View Management toolbar*

The Presentation View Management toolbar has four button areas. Table 3–3 describes the choices:

Table 3–3   Presentation View Management toolbar and panel options

Button	Function
Create View	Enables you to construct presentation views.
Tweak Components	Enables you to tweak the components of an assembly apart.
Precise View Rotation	Enables you to precisely rotate the presentation view.
Animate	Enables you to animate the presentation.

There are two kinds of presentation views: automatic and manual. In an automatic presentation view, the component parts are exploded apart from the grounded component by a specified explosion amount. In a manual presentation view, you tweak the component parts apart.

CONSTRUCTING A PRESENTATION VIEW MANUALLY

1. Select New from the File menu and select Standard.ipn.

2. Select Create View from the Presentation View Management toolbar or panel.

3. In the Select Assembly dialog box, select Manual in the Explosion Method box and select the Explore Directories button.

4. Select the file Foodgrinder.iam and select the OK Buttons. (See Figure 3–92.)

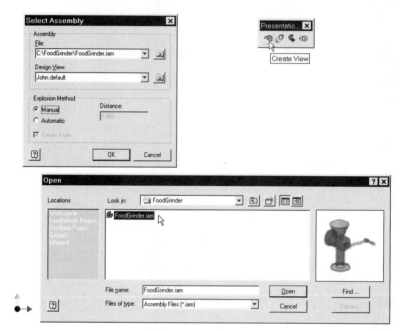

Figure 3–92 *Opening an assembly file in a presentation file*

Tweak the Components Apart

5. Select Tweak Components from the Presentation View Management toolbar or panel.

6. In the Tweak Component dialog box, select the Direction button and select the handle as shown in Figure 3–93 to use the coordinates of the handle for tweaking.

7. Select the Component button and select the handle to tweak the handle.

8. In the Transformations area, select the Translate and the X buttons and set the translation distance to 30 units. (See Figure 3–94.)

9. Select the Apply button. The handle is tweaked a distance of 30 units along the X axis of the selected coordinate system.

10. Select the Apply button again. The handle is tweaked a further distance of 30 units. (See Figure 3–95.)

In the browser bar, a Tweak item shows a distance of 60 units. In the presentation view, the handle is tweaked and a trail line is displayed showing the path of tweaking.

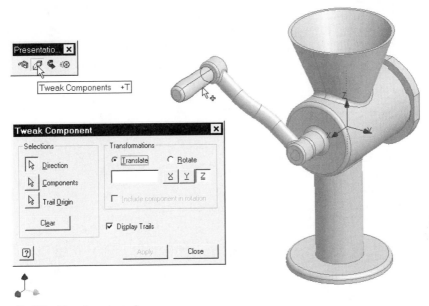

Figure 3–93 *Handle selected*

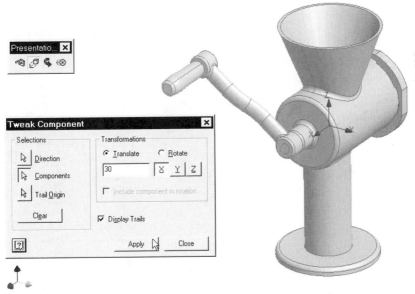

Figure 3–94 *Translation parameters specified*

11. Set the display to wireframe mode and then manually tweak the components apart as shown in Figure 3–96.

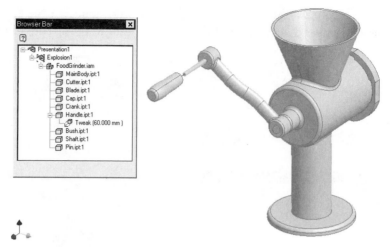

Figure 3–95 *Handle tweaked for a distance of 60 units*

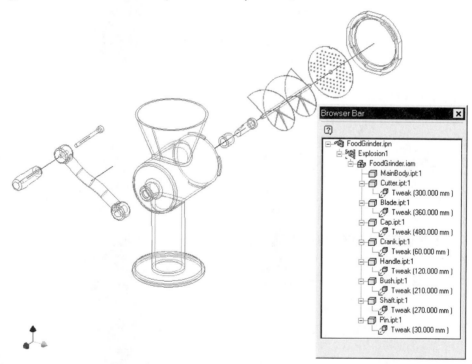

Figure 3–96 *Components tweaked apart*

12. The presentation view is complete. Save the file (file name: FoodGrinder.ipn).

CONSTRUCTING A PRESENTATION VIEW AUTOMATICALLY

In a presentation file, you can construct more than one presentation view.

1. Select Create View from the Presentation View Management toolbar or panel.

2. In the Explosion Method area, select the Automatic button and set the distance to 60 units. (See Figure 3–97.)

3. Select the OK button. A presentation view is constructed. (See Figure 3–98.) Note that there are two objects representing the presentation views in the browser bar.

4. Whether tweaking is manual or automatic, you can modify the amount of tweak. Select Tweak of Handle.ipt in the browser bar.

5. Then modify the tweak amount to 100 units. (See Figure 3–99.)

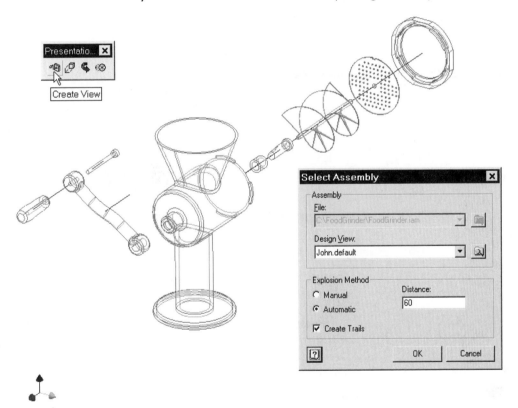

Figure 3–97 *Automatic explosion selected*

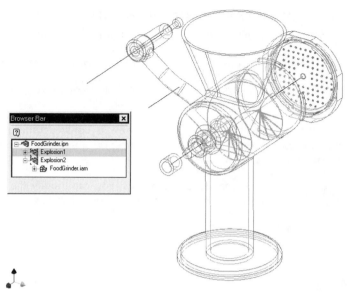

Figure 3–98 *Second presentation view*

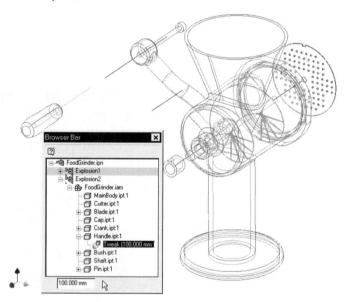

Figure 3–99 *Amount of tweak modified*

PRECISE VIEW ROTATION

1. Double-click Explosion1 in the browser bar to activate the first exploded view.

2. Select Precise View Rotation from the Presentation View Management toolbar or panel.

3. In the Incremental View Rotate dialog box, there are six buttons: Rotate Down, Rotate Up, Rotate Left, Rotate Right, Roll Counter Clockwise, and Roll Clockwise. Select these buttons to set the display as shown in Figure 3–100.

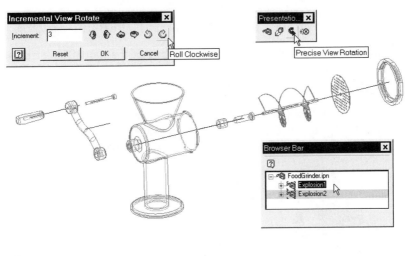

Figure 3–100 *Precise view setting*

ANIMATING A PRESENTATION

Tweaking can be animated.

1. Select Animate from the Presentation View Management toolbar or panel. (See Figure 3–101.)

2. Select the Play Forward button to play the animation.

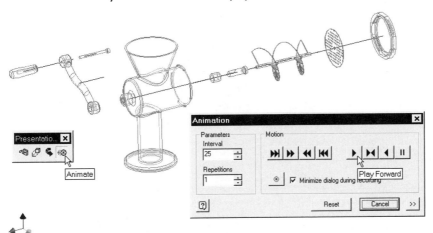

Figure 3–101 *Tweak animation*

1. MODEL CAR ASSEMBLY

In chapter 2, you constructed seven solid parts of a scale model car. Now you will place them in an assembly and create additional components. Figure 3–102 shows the assembly.

1. Start an assembly file and place the chassis of the model car (carchs.ipt) in the assembly.

2. Rotate the display in accordance with Figure 3–103.

3. Place a gear box (carbox.ipt) and two king pins (carkpn.ipt) in the assembly file in accordance with Figure 3–104, and turn on the degrees of freedom symbols. Note that the chassis is grounded and has no freedom at all.

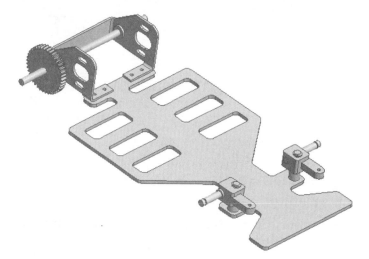

Figure 3–102   *Assembly of the model car*

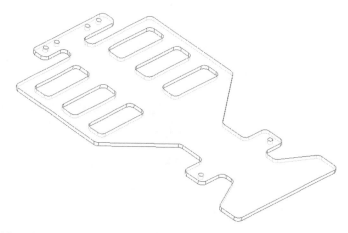

Figure 3–103   *Chassis placed in the assembly*

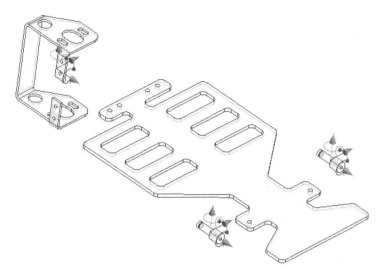

Figure 3–104 *Gear box and king pins placed in the assembly*

4. Apply an insert constraint on the pair of circular edges shown in Figure 3–105.

5. Repeat the insert constraint on a second pair of edges shown in Figure 3–106. The gear box is properly constrained.

6. Apply an insert constraint on a king pin and the chassis as shown in Figure 3–107.

7. Repeat the insert constraint on the second king pin. (See Figure 3–108.)

8. Save your file (file name: Car.iam).

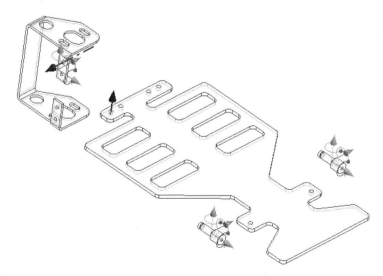

Figure 3–105 *Circular edges of the gear box and the chassis selected*

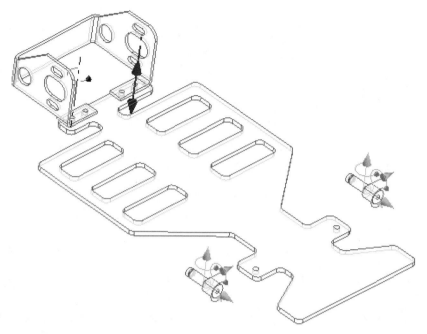

Figure 3–106   *Second pair of circular edges selected*

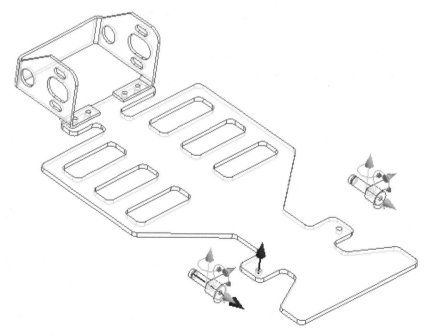

Figure 3–107   *King pin being constrained*

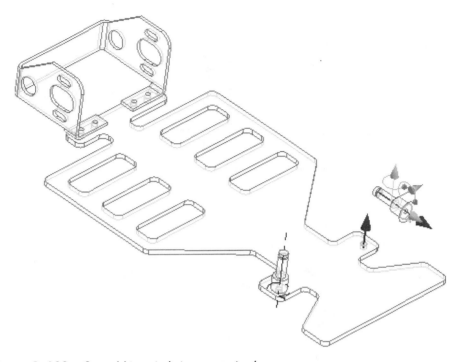

Figure 3–108   *Second king pin being constrained*

2. MODEL CAR BEARINGS

1. Use the assembly car.iam to continue from Exercise 1. Select Create Component to create a new component¾carber.ipt.

2. Construct a sketch as shown in Figure 3–109.

3. Revolve the sketch and use the lower edge of the sketch as the centerline. A revolved solid is constructed. This is the bearing of the scale model car. (See Figure 3–110.)

4. Select the assembly in the browser bar to change to assembly mode and save the assembly (file name: car.iam) and the new part file (file name: carber.ipt).

5. Select the bearing, right-click, and select Copy.

6. Right-click and select Paste. The bearing is copied. (See Figure 3–111.)

7. Apply the insert constraint on the bearings to assemble them to the gear box as shown in Figure 3–112.

8. Save your file.

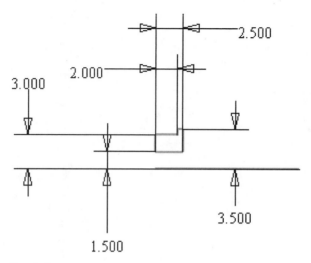

Figure 3–109 *Sketch for a new part*

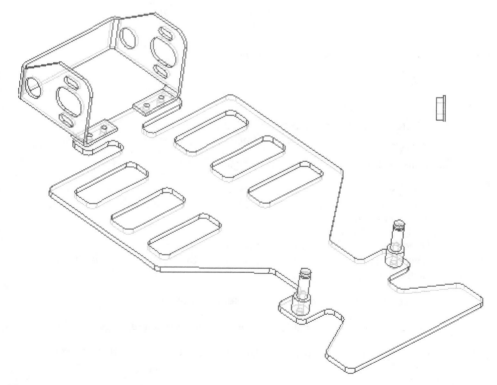

Figure 3–110 *Sketch revolved*

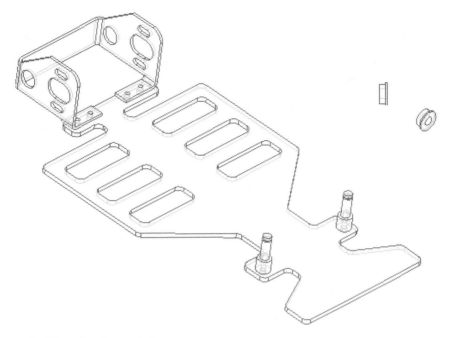

Figure 3–111   *Bearing copied*

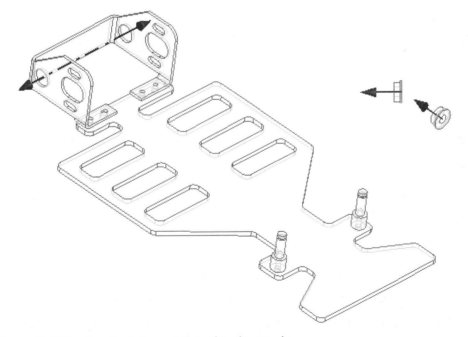

Figure 3–112   *Bearing being constrained to the gear box*

3. MODEL CAR SUB-ASSEMBLIES

1. Working with the Model Car assembly file (file name: car.iam) and continuing from Exercise 2, select Create Component from the Assembly toolbar or panel. The file type is assembly and the file name is fronthub.iam. Note that an assembly symbol is placed in the browser.

2. Now select Place Component from the Assembly toolbar or panel.

3. Then place two solid parts (carhub.ipt and caraxf.ipt) in the sub-assembly.

4. Rotate the parts as shown in Figure 3–113.

5. Save your file.

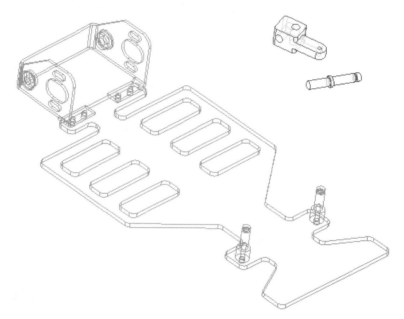

Figure 3–113   *New sub-assembly constructed and two solid parts placed in the sub-assembly*

6. Apply the insert constraint to assemble the front axle to the hub. (See Figure 3–114.)

7. Select the main assembly in the browser bar and select Save All in the File menu.

8. Select the sub-assembly, right-click, and select Copy.

9. Right-click and select Paste. The sub-assembly is copied. (See Figure 3–115.)

10. Rotate and move the sub-assemblies in accordance with Figure 3–116. Apply an insert constraint to assemble a sub-assembly to the main assembly.

11. Repeat the insert constraint on the second sub-assembly and the main assembly in accordance with Figure 3–117.

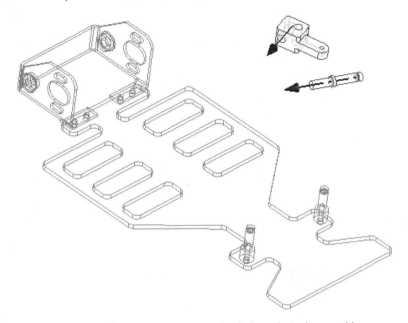

Figure 3–114   *Front axle being constrained to the hub in the sub-assembly*

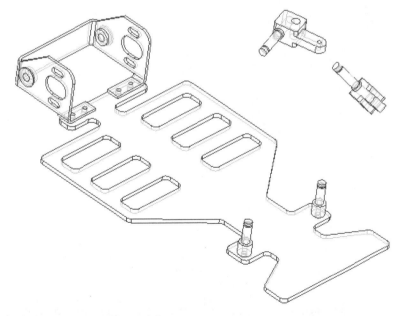

Figure 3–115   *Sub-assembly copied*

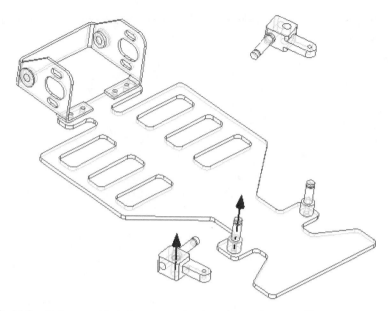

Figure 3–116   *Sub-assembly being constrained to the main assembly*

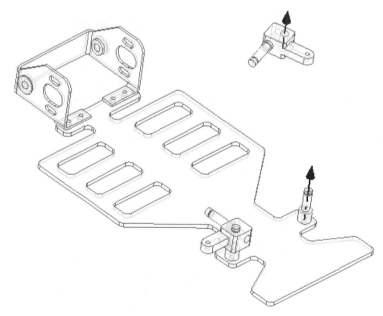

Figure 3–117   *Second sub-assembly being constrained*

12. Apply the angle constraint in accordance with Figure 3–118 to align the sub-assembly.

13. Repeat the angle constraint on the second sub-assembly in accordance with Figure 3–119.

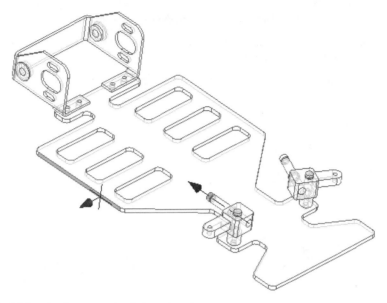

Figure 3–118   *Angle constraint being placed*

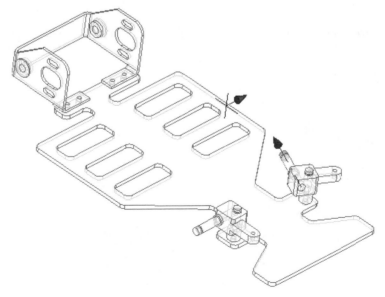

Figure 3–119   *Sub-assembly being constrained*

4. MODEL CAR ASSEMBLY COMPLETION

1. Working with the Model Car assembly file (file name: car.iam) and continuing from Exercise 3, place two copies of the circlip (carcir.ipt) and a copy of the rear axle (caraxr.ipt) in accordance with Figure 3–120.

2. Apply insert constraint in accordance with Figure 3–121. The solid parts are assembled. (See Figure 3–122.)

3. Save and close your files (assembly file name: Car.iam).

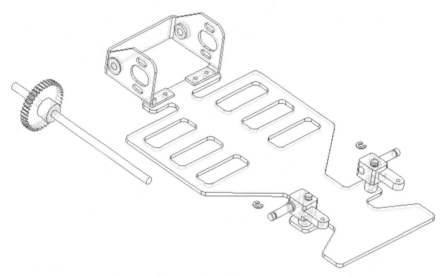

Figure 3–120   *Sub-assemblies constrained, and rear axle and circlips placed*

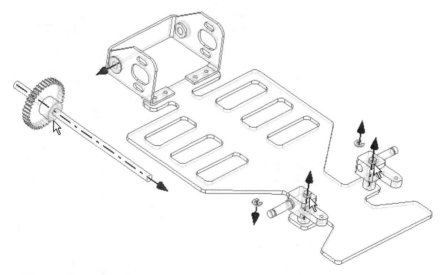

Figure 3–121   *Constraints being placed*

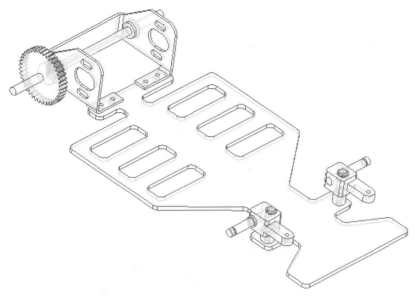

Figure 3–122   *Solid parts assembled*

5. MODEL CAR PRESENTATION

1. Start a presentation file and construct a presentation view of the model car.

2. Tweak the component parts apart manually in accordance with Figure 3–123.

3. Save and close your file (file name: Car.ipn).

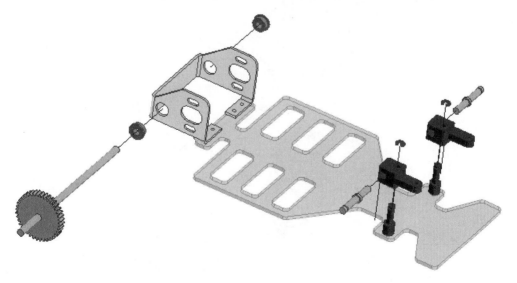

Figure 3–123   *Presentation view*

6. MOBILE PHONE ASSEMBLY AND PRESENTATION

1. Start an assembly and construct an assembly of the mobile phone upper and lower casings. (See Figure 3–124.)

2. Save and close your file (file name: Mphone.iam).

3. Start a new presentation file and tweak the parts in accordance with Figure 3–125.

4. Save and close the file (file name: Mphone.ipn).

Figure 3–124   *Assembly of mobile phone casing*

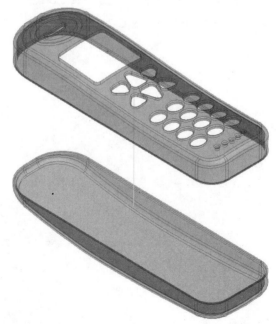

Figure 3–125   *Exploded view of the mobile phone*

7. SKATE SCOOTER STEERING SPRING ASSEMBLY AND PRESENTATION

1. Start a new assembly file and place a copy of the steering spring (file name: SteerSpring.ipt) and two copies of the steering spring cap (file name: SteerSpringCap.ipt).

2. Assemble the parts in accordance with Figure 3–126.

3. Construct a presentation file in accordance with Figure 3–127.

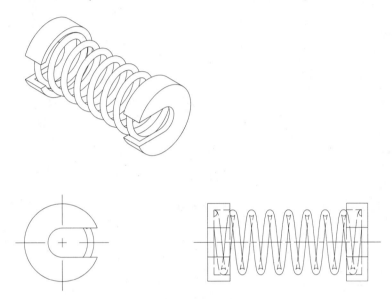

Figure 3–126   *Steer spring assembly (file name: SteerSpringSet.iam)*

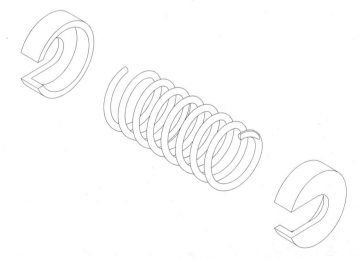

Figure 3–127   *Exploded view of the steering spring assembly (file name: SteerSpringSet.ipn)*

8. SKATE SCOOTER STEERING ARM ASSEMBLY AND PRESENTATION

1. Start a new assembly file and place a copy of the steering arm (file name: SteerArm.ipt) and two copies of the steer spring assembly (file name: SteerSpringSet.iam).

2. Then assemble the components in accordance with Figure 3–128.

3. Construct a presentation file in accordance with Figure 3–129.

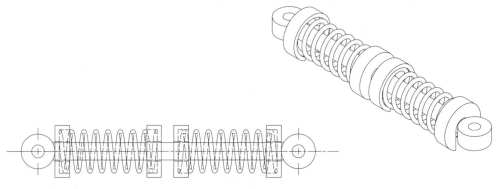

Figure 3–128   *Steering mechanism (file name: Steer.iam)*

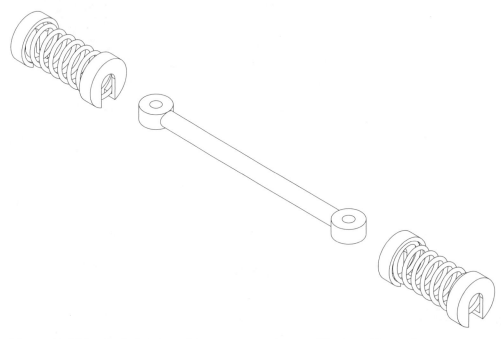

Figure 3–129   *Exploded view of the steering mechanism (file name: Steer.ipn)*

9. SKATE SCOOTER SHANK LOCK ASSEMBLY AND PRESENTATION

Construct an assembly (Figure 3–130) and an exploded view (Figure 3–131) of the shank lock assembly by placing a shank ring (file name: ShankRing.ipt), a shank sleeve (file name: ShankSleeve.ipt), and a shank spring (file name: ShankSpring.ipt) that you constructed in Chapter 2.

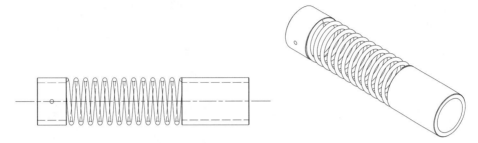

Figure 3–130 *Shank lock assembly (file name: ShankLockSet.iam)*

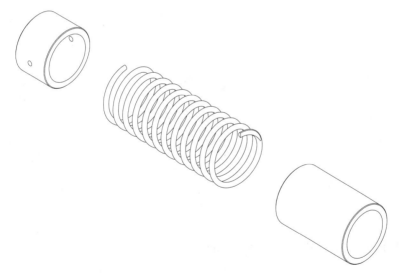

Figure 3–131 *Exploded view (file name: ShankLockSet.ipn)*

10. SKATE SCOOTER LOCKING MECHANISM ASSEMBLY AND PRESENTATION

Construct an assembly (Figure 3–132) and an exploded view (Figure 3–133) of the locking mechanism of the skate scooter by placing a lock main body (file name: LockMain.ipt), a lock clamp (file name: LockClamp.ipt), a lock pin (file name: LockPin.ipt), and a lock screw (file name: LockScrew.ipt) that you constructed in Chapter 2.

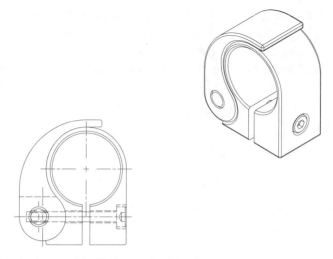

Figure 3–132   *Lock assembly (file name: Lock.iam)*

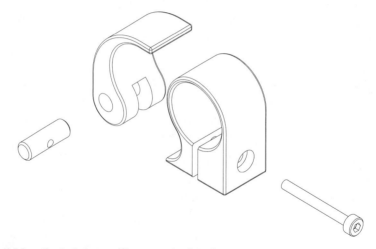

Figure 3–133   *Exploded view (file name: Lock.ipn)*

11. SKATE SCOOTER SHANK ASSEMBLY AND PRESENTATION

1. Now you will construct the shank assembly of the skate scooter. Start a new assembly file.

2. Place the lock assembly (file name: Lock.iam) and the shank lock assembly (file name: ShankLockSet.iam) in the assembly.

3. Construct a solid part (Figure 3–134) in the assembly and assemble the components together in accordance with Figure 3–135.

4. Construct an exploded presentation. (See Figure 3–136.)

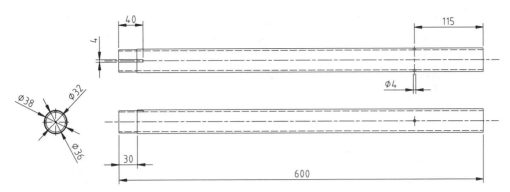

Figure 3–134   *Shank body (file name: ShankBody.iam)*

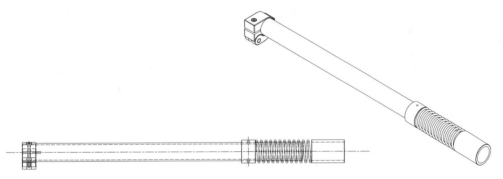

Figure 3–135   *Shank assembly (file name: Shank.iam)*

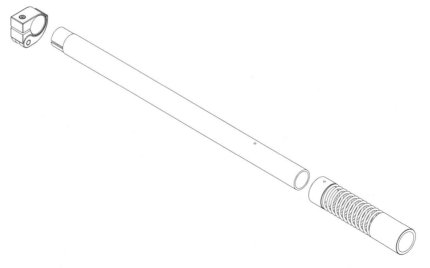

Figure 3–136   *Exploded view (file name: Shank.ipn)*

12. SKATE SCOOTER WHEEL ASSEMBLY AND PRESENTATION

1. Start a new assembly file and create two component parts in accordance with the dimensions shown in Figures 3–137 and 3–138.

2. Assemble the parts together and construct a presentation file. (See Figures 3–139 and 3–140.)

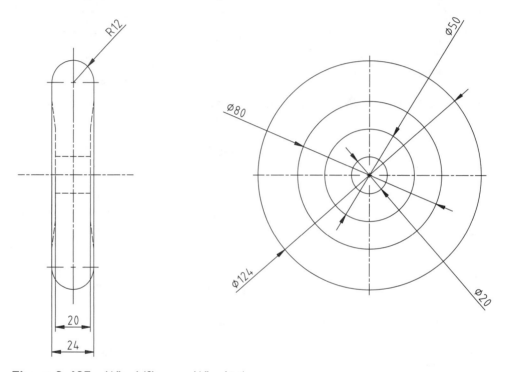

Figure 3–137   *Wheel (file name: Wheel.ipt)*

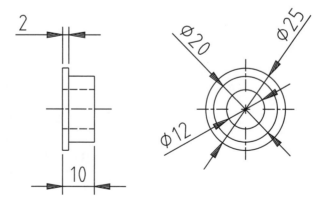

Figure 3–138   *Wheel bearing (file name: WheelBearing.ipt)*

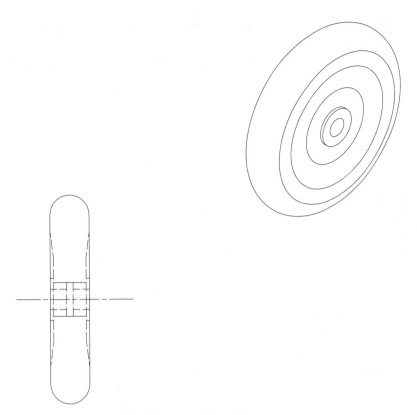

Figure 3–139   *Wheel assembly (file name: WheelSet.iam)*

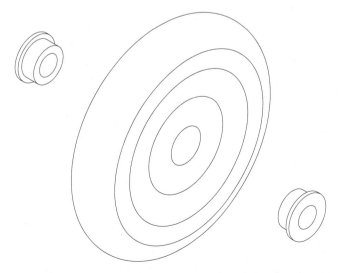

Figure 3–140   *Exploded view of the wheel assembly (file name: WheelSet.ipn)*

13. SKATE SCOOTER LOWER FRONT END ASSEMBLY AND PRESENTATION

Construct the lower front end assembly of the skate scooter as shown in Figures 3–141 and Figure 3–142.

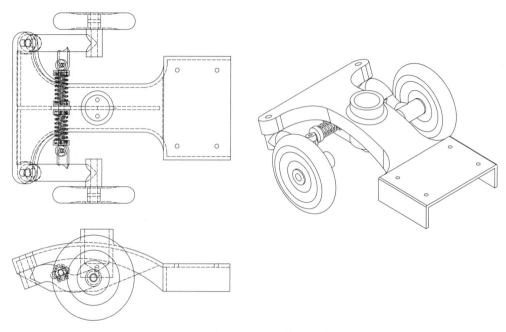

Figure 3–141   *Front end assembly (file name: FrontEndLower.iam)*

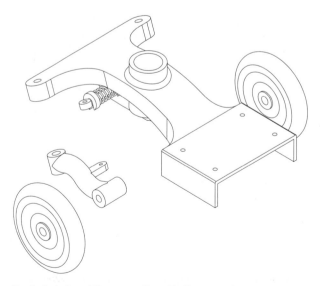

Figure 3–142   *Exploded view (file name: FrontEndLower.ipn)*

14. SKATE SCOOTER REAR END ASSEMBLY AND PRESENTATION

Construct the rear end assembly of the skate scooter. See Figures 3–143 and 3–144.

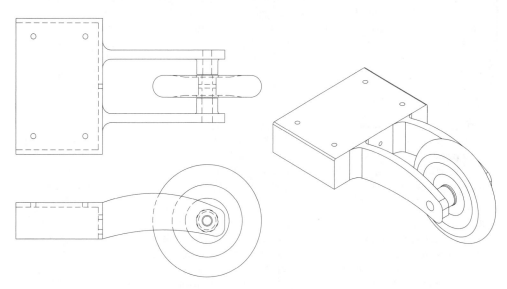

Figure 3–143   *Rear end assembly (file name: RearEnd.iam)*

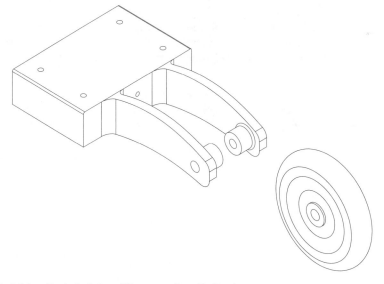

Figure 3–144   *Exploded view (file name: RearEnd.ipn)*

SUMMARY

An assembly is a collection of components put together properly to form a device. To construct an assembly, you use an assembly file, which links to a set of solid part files or sub-assembly files. In essence, an assembly file saves the assembly information, consisting of the location of the solid parts (or sub-assemblies) and the way the solid parts are assembled. Definitions of the individual solid parts are stored in the part files. Because the components of an assembly are not saved in the assembly file, every time you open an assembly drawing, the latest versions of the solid parts are loaded.

To design an assembly, you can use three approaches: bottom-up, top-down, and hybrid. In a bottom-up approach, you construct a set of solid parts and place the components in an assembly. In a top-down approach, you start an assembly file and create the components in the assembly. In a hybrid approach, you place some existing components in the assembly file and create some new components while working in the assembly environment

In 3D space, an object in the assembly can translate along three linear directions and rotate about three axes. The freedom of movement is called the six degrees of freedom (DOF). To restrict the movement of an object in 3D space, you can fix it by grounding. To maintain a proper positional relationship among a set of components in an assembly, you apply assembly constraints.

After you properly constrain the components in an assembly, you can animate the motion of individual or linked members of the assembly to illustrate mechanical motions.

For an assembly of large numbers of components, it is common practice to organize the components into sub-assemblies of smaller number of components. You can reorganize the components in an assembly into sub-assemblies and vice versa.

To display the components of an assembly in an exploded view, you construct a presentation file. In the presentation file, you tweak the components apart to illustrate how they are put together. Tweaking can be animated and saved as AVI files.

REVIEW QUESTIONS

1. Differentiate between a solid part file and an assembly file in terms of the data stored in the files.

2. Briefly describe the three approaches in constructing an assembly.

3. What are the six degrees of freedom of an object?

4. What kinds of assembly constraints can you apply to selected features of a pair of components?

CHAPTER 4

Advanced Modeling Techniques

OBJECTIVES

The aims of this chapter are to detail the use of 2D layout drawings in mechanism design, to let you gain an in-depth understanding of adaptive technology, to show you how to slice a shaded graphic display to show internal details of parts, and to let you practice using a design parameter spreadsheet. This chapter also outlines how to construct a design catalog and design elements, to use a design notebook, to set a relative motion among components in an assembly, and to design collaboratively. After studying this chapter, you should be able to

- Use 2D layout drawings to design mechanisms
- Apply adaptive technique in design
- Slice shaded graphics along the current sketch plane
- Construct and use a spreadsheet to control design parameters
- Make a design catalog and use design elements
- Use a design notebook
- Set relative motions in a mechanism
- Design collaboratively with other design team members

OVERVIEW

This chapter will address various advanced modeling techniques.

During the initial stage of designing a mechanism, you can construct a 2D layout drawing outlining the linkages in terms of line sketches. Using the sketches, you evaluate the validity of the mechanism and improve the design further. At a stage when you are satisfied with the mechanism, you build solid features on the 2D layout drawing.

In an assembly, the dimensions of the corresponding parts have to match in order to function properly. To ensure that the corresponding parts always match each other, you use adaptive technology to make a feature of one solid adapt to a feature of another solid part.

While designing in shaded mode, you can slice the graphics along the sketch plane to gain a better picture of the sketch.

To control the dimensions of a solid part or a set of solid parts globally, you can construct a spreadsheet and use the spreadsheet to guide the parameters of the solid part or the set of solid parts.

One of the ways to speed up the design process is to re-use solid features that are constructed in existing solid parts. You can export those reusable solid features as design elements to form a design catalog. Then you can insert the design elements into other solid parts.

You can use a design notebook to record design histories, intents, and information along with a part's graphical and geometric data.

To simulate mechanical motions in mechanical drive systems such as belt drive, friction drive, and gear drive, you can apply a relative motion among the members of the assembly.

You can work with a team of designers collaboratively on a design project by using various strategies. You can use NetMeeting to communicate and to work with others on the same file in real time, and you can use the Internet to transmit and share files.

2D DESIGN LAYOUT

When you start to design a mechanism, you have an idea of a set of linkages in your mind, but you do not know yet how the mechanism works. Before spending time designing the details of the component parts, you can quickly construct a set of 2D sketches to depict the linkages and assemble the 2D sketches together to form the mechanism. Because there are no solid features in the 2D sketches, you need to construct work features for the purpose of placing assembly constraints. You assign assembly constraints to selected work planes, work axes, and work points of the component parts to align the 2D component parts together. After validating and evaluating the simplified mechanism, you construct 3D solid models from the sketches.

FOUR-BAR MECHANISM

Now you will use a 2D layout assembly to design a four-bar mechanism: an oscillator. For the sake of simplicity in illustration, the standard engineering components such as bearings, seals, and fasteners are omitted. Figure 4–1 shows the completed oscillator. It has four members: a base, a crank, a lever, and a linkage. You will construct the assembly using the top-down approach.

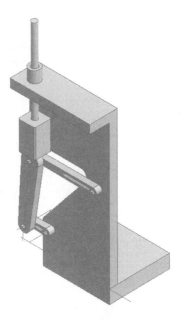

Figure 4–1 *Oscillator*

You will construct the individual components of this mechanism as simple 2D line sketches, construct work features, assemble the line sketches to form an assembly, validate the mechanism, and further refine the design. Figure 4–2 shows the layout of the mechanism.

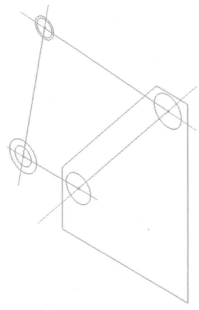

Figure 4–2 *2D layout of the oscillator*

Create the Project Path File

1. Now create a folder (C:\Oscillator) in your computer and use a text editor to construct a project path file as follows:

```
[Included Path File]
PathFile=C:\Oscillator\OscillatorProject.ipj
[Workspace]
Workspace=C:\Inventor New Projects
[Local Search Paths]
Oscillator Project=C:\Oscillator
[Workgroup Search Paths]
GroupA=C:\DesignGroupA
[Library Search Path]
Library A=C:\StandardParts
```

2. Save and close your file (file name: Oscillator.ipj).

Construct 2D Layout Parts

3. Select Projects from the File menu in Inventor and select the project file Oscillator.ipj to use the search paths depicted in the project path file.

4. Select New from the File menu, select Standard.iam, and select the OK button to start a new assembly drawing.

5. Select Properties from the File menu to display the Properties dialog box.

6. Set the Length Units to millimeter in the Units tab and select the OK button.

7. Select Create Component from the Assembly toolbar or panel to start a new part. (New file name is Base.ipt and file location is C:\Oscillator.)

8. Select Line from the Sketch toolbar or panel to construct a set of vertical, horizontal, and parallel lines as shown in Figure 4–3.

Although dimensions are unimportant at this preliminary design stage, putting major dimensions in the sketch enables you to gain a better control of the size and shape of the mechanism.

9. Select General Dimension from the Sketch toolbar or panel and add dimensions to the sketch as shown in Figure 4–4.

To assemble components together to form an assembly, you apply assembly constraints. Because there is no solid feature in the 2D layouts, you will use work features for placing assembly constraints.

10. Right-click and select Isometric View to set the display to an isometric view.

11. Deselect Sketch on the Command Bar toolbar to exit sketching mode.

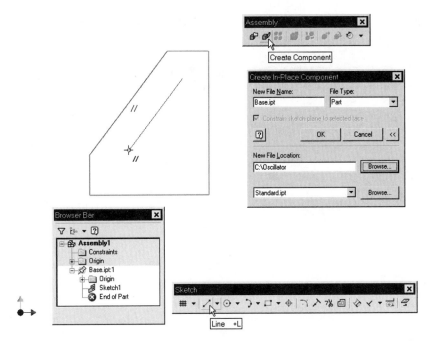

Figure 4–3   *Lines constructed on the sketch plane of the first part file of the assembly file*

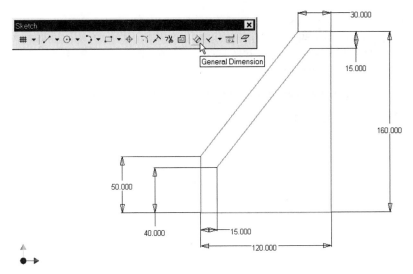

Figure 4–4   *Dimensions added to the sketch*

12. Select Work Axis from the Features toolbar or panel and select the lower edge of the sketch indicated in Figure 4–5 to construct a work axis.

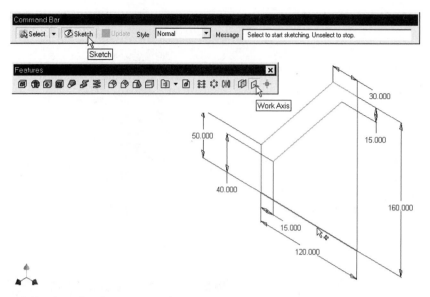

Figure 4–5 *A work axis constructed*

13. Select Work Points from the Features toolbar or panel and select the end points indicated in Figure 4–6 to construct two work points at the end points of the selected line.

14. Select Work Axis from the Features toolbar or panel, XY Plane in the browser bar, and the work points indicated in Figure 4–7 to construct two axes that are perpendicular to the XY plane and passing through the selected work points.

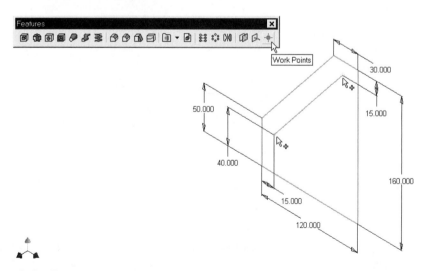

Figure 4–6 *Two work points constructed*

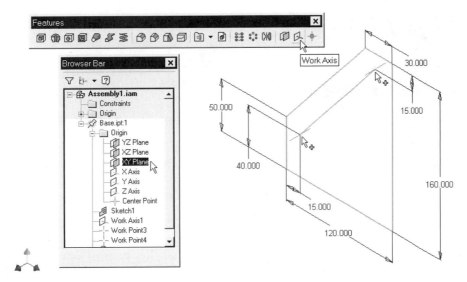

Figure 4–7 *Two work axes constructed*

The 2D layout drawing of the first component part of the oscillator, the base, is complete.

15. Double-click the assembly in the browser bar to toggle to assembly mode.

16. Select the XY Plane in the browser bar and select Look At from the Standard toolbar to set the display to the XY plane.

17. Now you will construct the crank of the oscillator. Select Create Component from the Assembly toolbar or panel to start another part (file name: Crank.ipt).

18. Select a point on the screen to establish the orientation of the sketch plane.

19. Use the Line and Circle tools from the Sketch toolbar or panel to construct a sketch as shown in Figure 4–8.

20. The sketch for the crank of the oscillator is complete. Deselect Sketch on the Command Bar toolbar to exit sketch mode.

21. Now you will construct the lever of the oscillator. Double-click the assembly in the browser bar to switch to assembly mode.

22. Select Create Component from the Assembly toolbar or panel to construct a new component (file name: Lever.ipt).

23. Select a point on the screen to locate a sketch plane and construct a line and two circles at the end points of the line as shown in Figure 4–9.

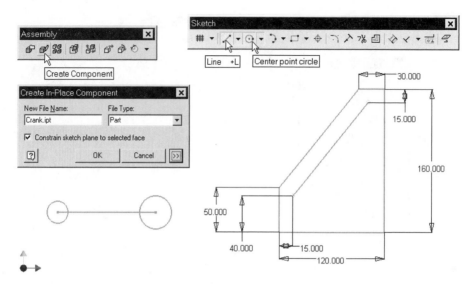

Figure 4–8 *Line and circle constructed on the sketch plane of the second part*

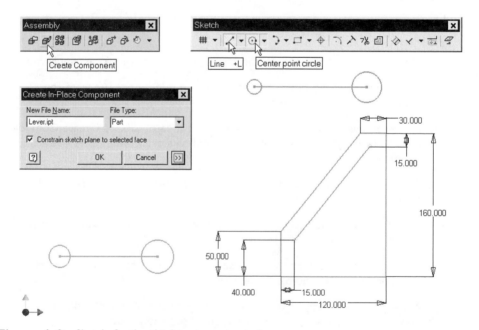

Figure 4–9 *Sketch for the third part constructed*

24. The sketch of lever is complete. Deselect Sketch on the Command Bar toolbar to exit sketch mode and double-click the assembly in the browser bar to switch to assembly mode.

25. Now construct the linkage of the oscillator. Select Create Component from the Assembly toolbar or panel to construct a new component (file name: Linkage.ipt).

26. Select a point on the screen to locate a sketch plane and construct a line and two circles at the end points of the line as shown in Figure 4–10.

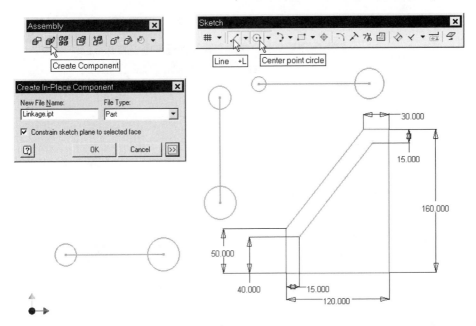

Figure 4–10 *Sketch for the fourth part constructed*

27. Deselect Sketch on the Command Bar toolbar to exit sketch mode. The sketches for the components of the oscillator are complete.

Now you will add work points and work axes to the crank, lever, and linkage in a way similar to what you did for the 2D layout drawing for the base of the oscillator.

28. Right-click and select Isometric View to set the display to isometric.

29. Double-click the crank in the browser bar to activate it.

30. Select Work Point from the Features toolbar or panel and select the end points of the line indicated in Figure 4–11 to construct two work points.

Now you will construct three work axes—two work axes to pass through the work points and perpendicular to the XY plane and a work axis to pass through two work points.

31. Select Work Axis from the Features toolbar or panel.

32. Select a work point and the XY plane of the crank in the browser bar to construct work axes that pass through the selected work points and are perpendicular to the selected XY plane. A work axis is constructed.

33. Repeat to construct another work axis passing through the other work point and perpendicular to the XY plane.

34. To construct the third work axis, select Work Axis from the Features toolbar or panel.

35. Select the two work points.

36. Select the XY plane of the crank in the browser bar. (See Figure 4–12.)

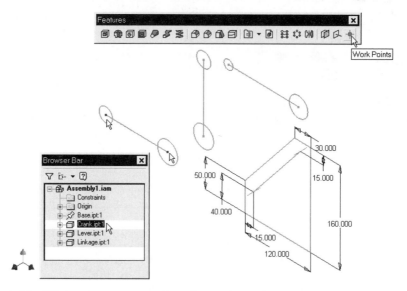

Figure 4–11 *Work points constructed on the crank*

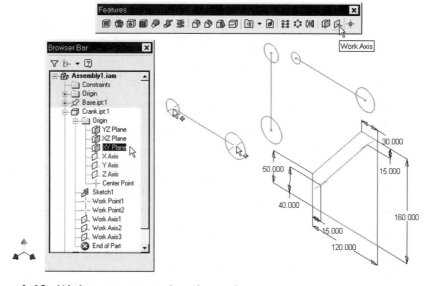

Figure 4–12 *Work axes constructed on the crank*

The work features for the crank are complete.

37. Now repeat the above steps, as outlined in Figures 4–11 and 4–12, to construct two work points and three work axes for the lever and the linkage. (See Figures 4–13 and 4–14.)

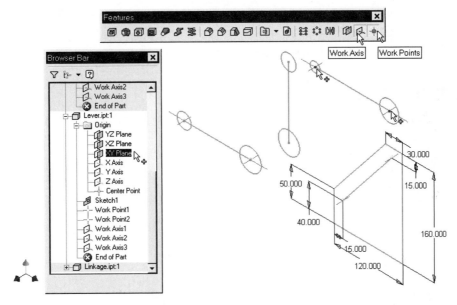

Figure 4–13 *Work points and work axes constructed on the lever*

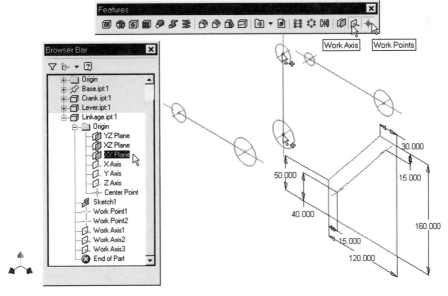

Figure 4–14 *Work points and work axes constructed on the linkage*

Set Adaptivity

You will learn about adaptivity in the next sequence of steps. Here you will set the sketch of the linkage to be adaptive and set the part to be adaptive so that the sketch can adapt to changes in its matching components.

38. Double-click the linkage in the browser bar to activate part editing.

39. Select the sketch of the linkage in the browser bar, right-click, and select Adaptive. (See Figure 4–15.) The sketch is now adaptive.

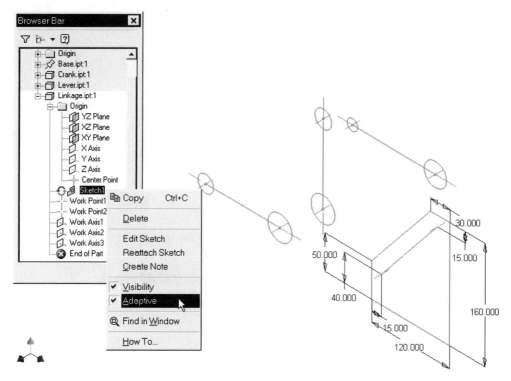

Figure 4–15 *Sketch of the linkage set to be adaptive*

40. Double-click the assembly in the browser bar to activate assembly mode.

41. Select the linkage in the browser bar, right-click, and select Adaptive. (See Figure 4–16.) The linkage is now adaptive.

42. Now the 2D layout drawings are complete. Select Save All from the File menu to save all the files (file name for the assembly is Oscillator.iam).

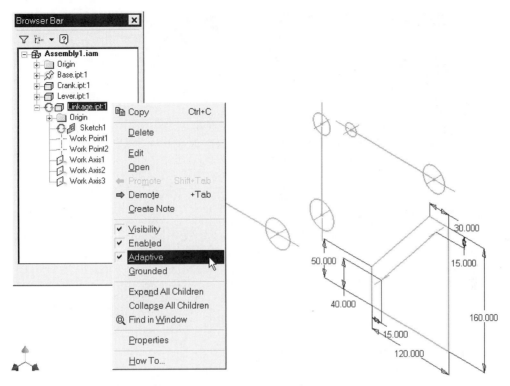

Figure 4–16 *Linkage set to be adaptive in the assembly*

ASSEMBLY OF 2D LAYOUTS

Now you will assemble the parts together to validate the design.

1. Select Place Constraint from the Assembly toolbar or panel.

2. Select the Mate button in the Type area and the Mate button in the Solution area.

3. Select the work axes indicated in Figure 4–17 and select the Apply button to mate the selected work axes.

4. Select the Angle button in the Type area of the Place Constraint dialog box.

5. Select the work axes indicated in Figure 4–18.

6. Select the Apply button to align the selected work axes.

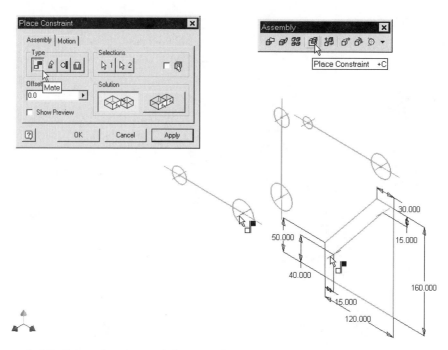

Figure 4–17 *Selected work axes being mated*

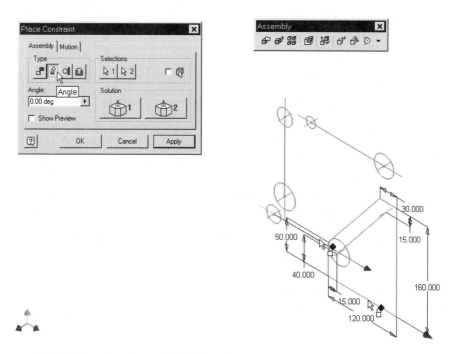

Figure 4–18 *Selected work axes being aligned*

7. Now you will assemble the lever and the linkage to the base. Select the Mate button in the Type area in the Place Constraint dialog box.

8. Select the work axes indicated in Figure 4–19.

9. Select the Apply button to mate the selected work axes.

10. Repeat the command to mate the work axes highlighted in Figures 4–20 and 4–21.

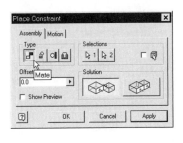

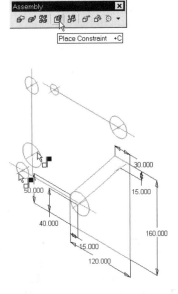

Figure 4–19 *Selected work axes being mated*

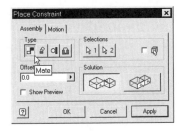

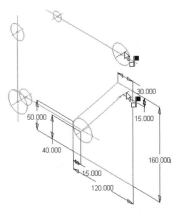

Figure 4–20 *Selected work axes being mated*

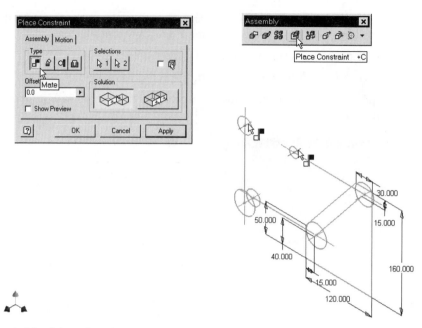

Figure 4–21 *Selected work axes being mated*

11. The constraint placement is complete. Select the XY plane in the browser bar.

12. Select Look At and Zoom All from the Standard toolbar to set the display as shown in Figure 4–22. Note that the figure will not look the same as yours because the size of the linkage, crank, and lever might not be exactly the same.

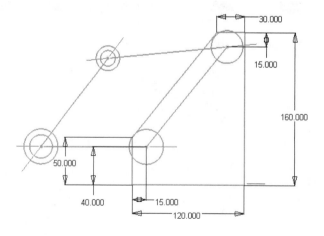

Figure 4–22 *Parts assembled*

13. Now you will refine your mechanism. Open the file Crank.ipt.

14. Add two lines that are tangential to the circles and three dimensions (10 units, 30 units, and 12 units) as shown in Figure 4–23.

15. Save and close the file.

16. Open the Lever.ipt.

17. Add two tangential lines and three dimensions (10 units, 105 units, and 12 units) as shown in Figure 4–24.

18. Save and close the file.

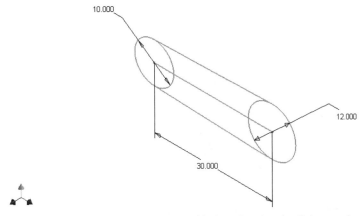

Figure 4–23 *Tangential lines and dimensions added to the sketch of the crank*

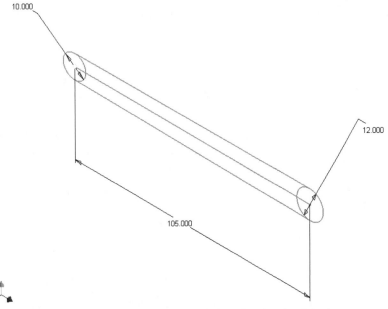

Figure 4–24 *Tangential lines and dimensions added to the sketch of the lever*

19. Select Update from the Command Bar toolbar. The assembly is updated. (See Figure 4–25.) Because the dimensions for the linkage have not been assigned yet, the figure might not look exactly the same as yours.

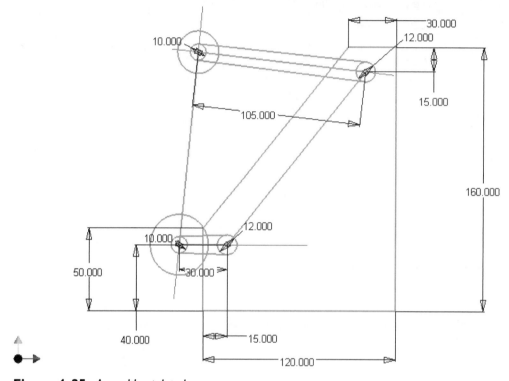

Figure 4–25 *Assembly updated*

20. Now validate your design by driving the crank. Select the angle constraint of the base in the browser bar, right-click, and select Drive Constraint. (See Figure 4–26.)

21. In the Drive Constraint dialog box, set the End value to 360 degrees, then select the >> button to expand the dialog box and set the total number of steps to 10 in the Increment area.

22. Select the Forward button to view the animation or select the Record button to save the animation to an AVI file.

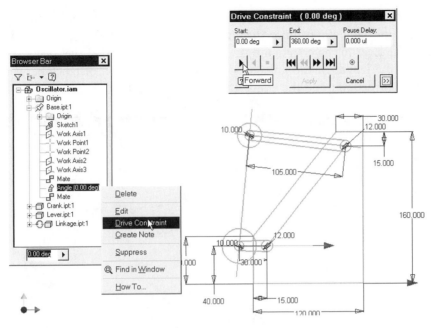

Figure 4–26 *Angle constraint being driven*

ADAPTIVE TECHNOLOGY

Adaptive technology is a solid modeling technique whereby a feature of a solid part changes in size to adapt to the feature of another solid part. For example, if the diameter of a shaft adapts to the diameter of a hole, changes in the hole diameter cause a corresponding change in the diameter of the shaft. For a solid part to be adaptive, both the solid part and the relevant feature of the solid part have to be adaptive. In addition, the adaptive feature must not be fully constrained (either geometrically or dimensionally).

To make a feature of a part adapt to the feature of another part in an assembly, you set the feature and the part to be adaptive. However, you can still change the adaptivity status while constructing the feature and the part.

SETTING ADAPTIVITY

In constructing the 2D layout of the oscillator, you set the linkage and its sketch to be adaptive. Now suppose we want to modify the length of the linkage such that the lever and the crank are horizontal in their initial position.

1. Select Place Constraint from the Assembly toolbar or panel.

2. Select the Angle button.

3. Select the edges highlighted in Figure 4–27. Select OK.

322

You have constrained the lever to be parallel to the horizontal base. Because there is already another angle constraint governing the position of the crank, the linkage connects the lever to the crank, and the linkage is adaptive, the lever becomes horizontal and the length of the linkage adapts to the change in distance between the end points of the lever and the crank. (See Figure 4–28.)

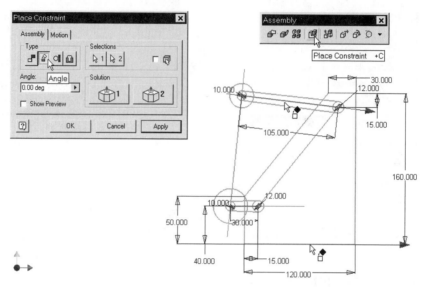

Figure 4–27 *Constraint being applied*

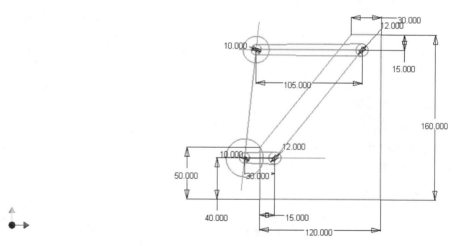

Figure 4–28 *Length of the linkage changed to adapt to change in its mating parts*

The purpose of applying the angle constraint to the lever is to set its position to cause the adaptive linkage to adapt to the change. However, this constraint will cause the mechanism to be inoperative.

4. Select the angle constraint applied to the lever in the browser bar, right-click, and select Delete to remove it. (See Figure 4–29.)

5. To validate the mechanism, select the angle constraint of the base in the browser bar, right-click, and select Drive Constraint to drive the animated constraint. (See Figure 4–30.)

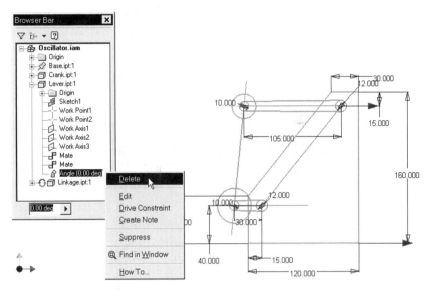

Figure 4–29 *Angle constraint of the lever being deleted*

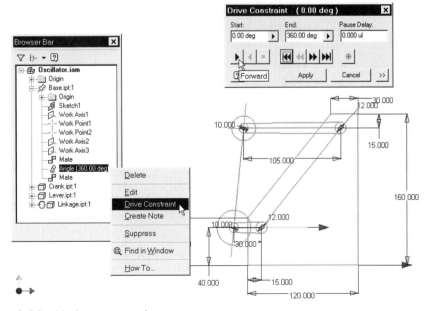

Figure 4–30 *Mechanism tested*

6. Select the Forward button of the Drive Constraint dialog box to validate the movement of the mechanism.

7. Close the Drive Constraint dialog box.

8. Save all files.

Construct the Parts

Now you will construct the parts from the 2D layout sketch.

1. Open the file Base.ipt.

2. Select Extrude from the Features toolbar or panel to extrude the sketch a distance of 15 units in a direction indicated in Figure 4–31.

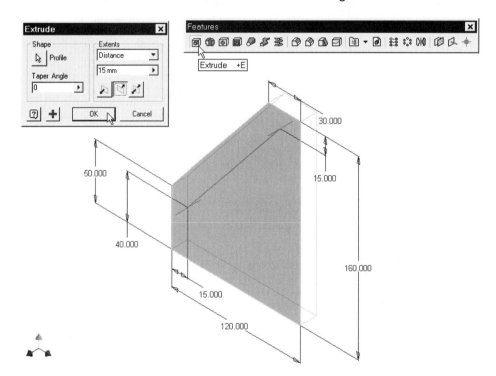

Figure 4–31 *Sketch being extruded*

3. Select Hole from the Features toolbar or panel and select the work point indicated in Figure 4–32 to construct a through hole with a diameter of 8 units.

4. Construct another through hole of 8 units diameter as shown in Figure 4–33.

5. Now the sketch is not required. Hide it by selecting the sketch in the browser bar, right-clicking, and deselecting Visibility. (See Figure 4–34.)

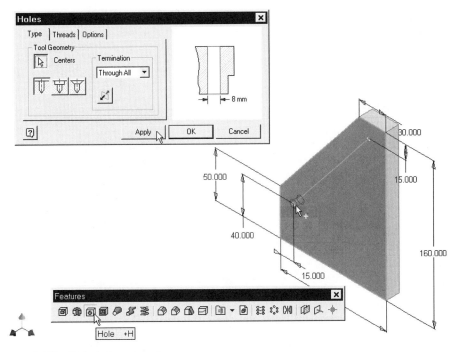

Figure 4–32 *Hole being placed*

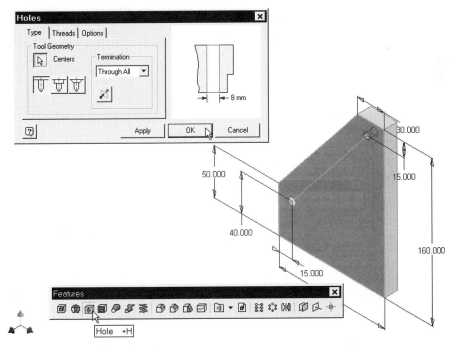

Figure 4–33 *Second hole being placed*

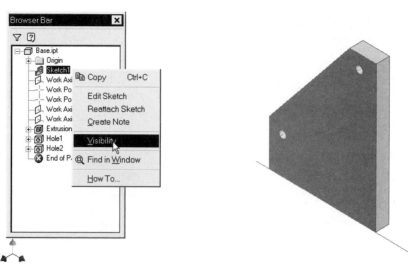

Figure 4–34 *Sketch hidden*

6. The base is complete. Save and close the file.

7. Now open the file Linkage.ipt and select the sketch to edit it.

8. Add two tangential lines and add two dimensions to set the size of the circles as shown in Figure 4–35. Remember that the sketch is adaptive and the part is adaptive in the oscillator assembly. Therefore, do not dimension the distance between the two circles.

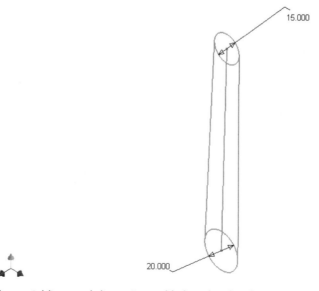

Figure 4–35 *Tangential lines and dimensions added to the sketch*

9. Exit sketch mode.

10. Select Extrude from the Features toolbar or panel to extrude the sketch a distance of 10 units. (See Figure 4–36.)

11. Select Hole from the Features toolbar or panel and select the work point indicated in Figure 4–37 to place a through hole with a diameter of 8 units.

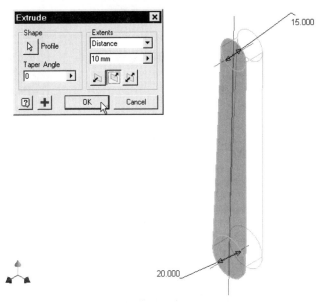

Figure 4–36 *Sketch being extruded*

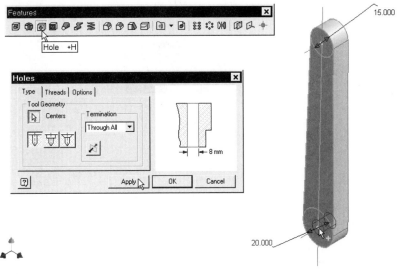

Figure 4–37 *Through hole being placed*

12. Now place another through hole of 8 units diameter on the other work point. (See Figure 4–38.)

13. The sketch is not required. Select it in the browser bar, right-click, and deselect Visibility. (See Figure 4–39)

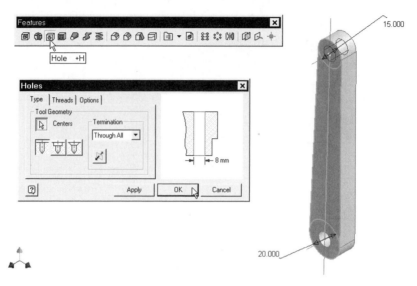

Figure 4–38 *Second through hole being placed*

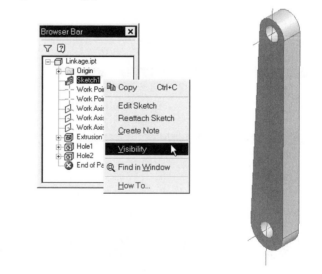

Figure 4–39 *Sketch hidden*

14. The linkage is complete. Save and close your file.

15. Now open the file Lever.ipt.

16. Extrude the sketch a distance of 10 units. (See Figure 4–40.)

17. Hide the sketch and save and close the file.

18. Open the file Crank.ipt.

19. Extrude it a distance of 10 units. (See Figure 4–41.)

20. Hide the sketch and save and close the file.

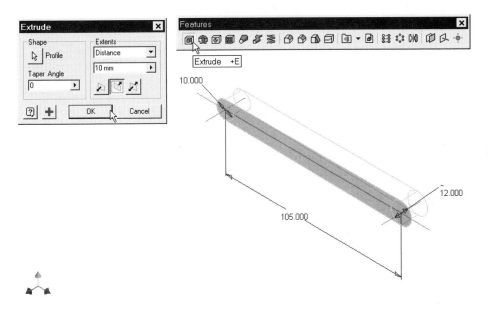

Figure 4–40 *Lever being extruded*

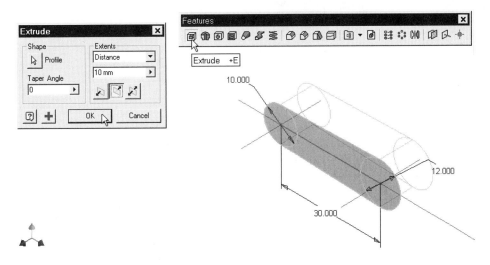

Figure 4–41 *Crank being extruded*

Place Constraints

1. Now you will work on the assembly Oscillator.iam. Select Update from the Command Bar toolbar to update the changes, right-click, and select Isometric View to set the display.

2. Select Wireframe Display from the Standard toolbar to set the display to wireframe mode. (See Figure 4–42.)

Note that the figure might not be the same as yours because the parts are not fully constrained yet; they can move along the work axes.

3. Before you proceed to add details to the oscillator, select the angle constraint in the browser bar, right-click, and select Drive Constraint to test the mechanism.

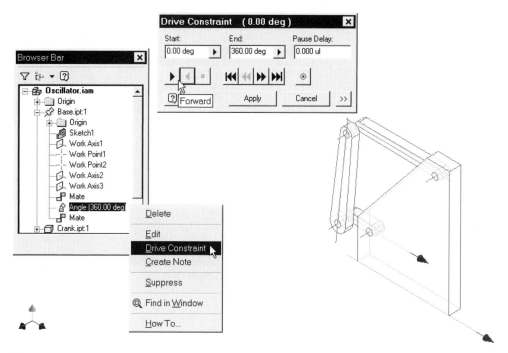

Figure 4–42 *Driving constraints*

4. Now select Place Constraint from the Assembly toolbar or panel.

5. Select the Mate button and select the faces highlighted in Figure 4–43.

6. Select the Apply button. The selected faces are mated.

7. Select the faces highlighted in Figure 4–44 to mate them.

8. Select the Apply button.

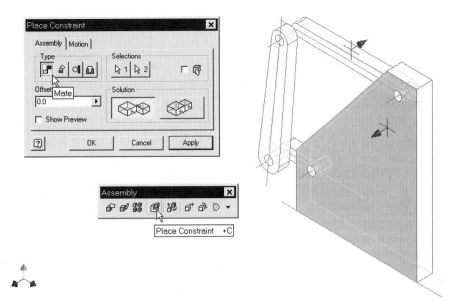

Figure 4–43 *Faces being mated*

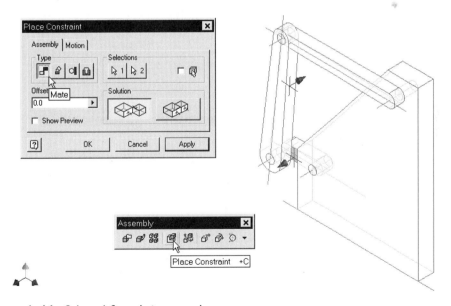

Figure 4–44 *Selected faces being mated*

9. Select the faces highlighted in Figure 4–45 to mate them.

10. Select the OK button. The faces are mated. (See Figure 4–46.)

332

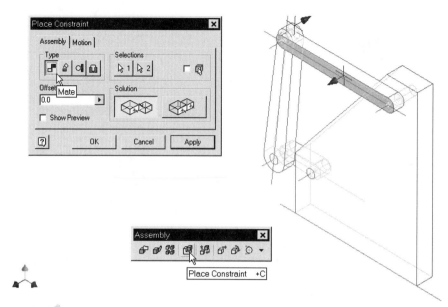

Figure 4–45 *Selected faces being mated*

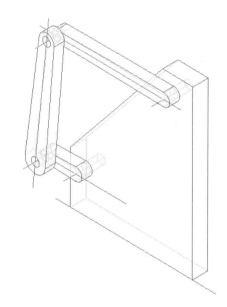

Figure 4–46 *Faces mated*

Set Adaptivity

Now you will work on the lever and the crank. On each of these parts, you will add two solid features extruded from circular sketches. You will not add any dimensions to these sketches and you will set them as adaptive. Being adaptive, they will adapt to the diameters of the holes placed in the linkage and the base.

1. Press and hold down CTRL and then select the crank, base, and linkage in the browser bar in turn, right-click, and deselect Visibility to hide them. (See Figure 4–47.)

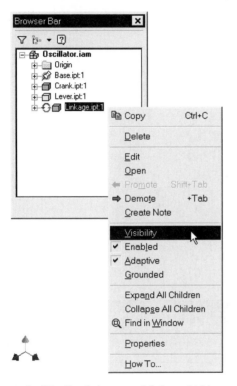

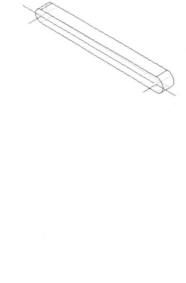

Figure 4–47 *Crank, base, and linkage hidden*

2. Double-click the lever in the browser bar to activate part modeling mode.

3. Select Sketch from the Command Bar toolbar and the face highlighted in Figure 4–48 to set up a new sketch plane.

4. Construct a circle at the work point. Do not add dimension to the circle.

5. Exit sketch mode and extrude the circle a distance of 10 units in the direction shown in Figure 4–49 to join the part.

6. Rotate the display as shown in Figure 4–50.

7. Select Sketch from the Command Bar toolbar and the face indicated to set up a sketch plane.

8. Construct a circle. Do not add dimension to the circle because you will make this feature adaptive.

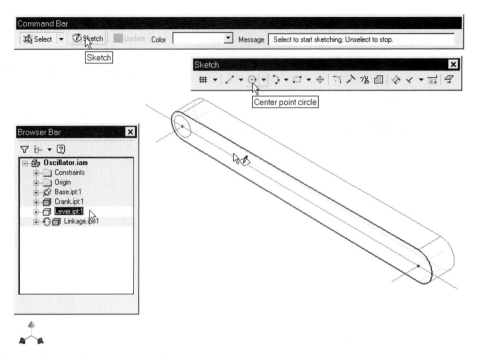

Figure 4–48 *Circle constructed*

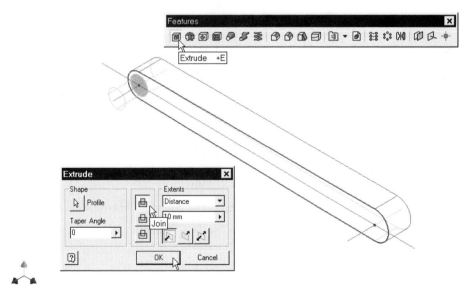

Figure 4–49 *Circle extruded*

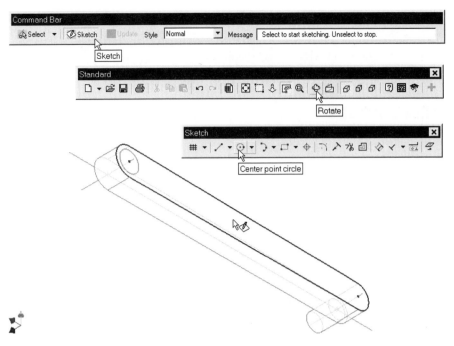

Figure 4–50 *Display rotated and circle constructed*

9. Exit sketch mode and extrude the circle a distance of 15 units as shown in Figure 4–51.

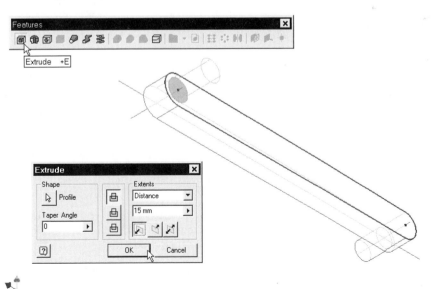

Figure 4–51 *Circle being extruded*

10. Double-click the assembly in the browser bar to switch to assembly mode.

11. Select the crank in the browser bar, right-click, and select Visibility to make the crank visible.

12. Set the display to an isometric view. (See Figure 4–52.)

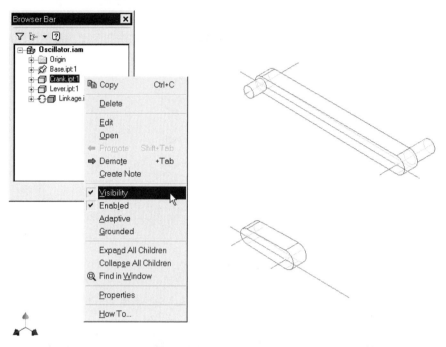

Figure 4–52 *Crank visible and display set*

13. Double-click the crank in the browser bar to switch to part modeling mode.

14. Follow the steps as shown in Figures 4–48 through 4–51 to construct two sketches and extrude them a distance of 10 units and 25 units, respectively. (See Figure 4–53.)

15. Double-click the assembly in the browser bar to switch to assembly mode.

16. Select the linkage and the base in turn, right-click, and select Visibility to make them visible. (See Figure 4–54.)

17. The crank, the lever, and the extruded features that you constructed on the crank and lever need to be set as adaptive. Double-click the crank in the browser bar to activate part modeling mode.

18. Select the two extruded features in turn, right-click, and select Adaptive to make them adaptive. (See Figure 4–55.)

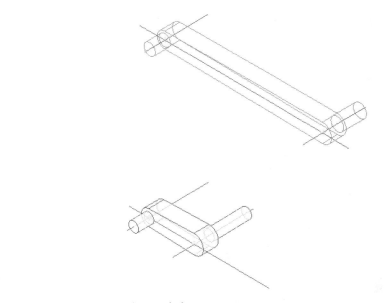

Figure 4–53 *Sketches constructed and extruded*

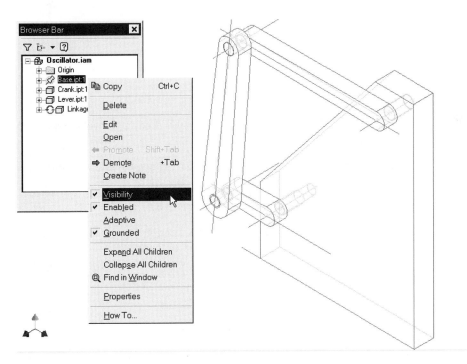

Figure 4–54 *Linkage and base made visible*

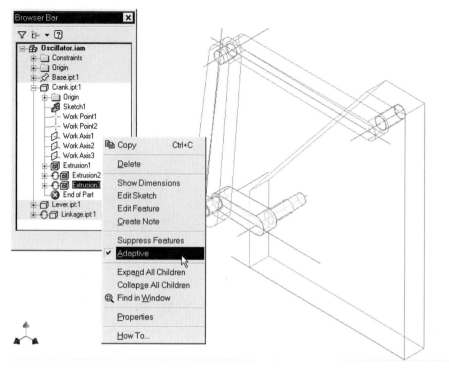

Figure 4–55 *Extruded solid features of the crank set to be adaptive*

19. Double-click the lever in the browser bar to activate it.

20. Select two extruded features in turn and make them adaptive as shown in Figure 4–56.

21. Double-click the assembly in the browser bar to set to assembly mode.

22. Select the lever and the crank in turn, and for each, right-click and select Adaptive to make them adaptive. (See Figure 4–57.)

23. Now you will mate the cylindrical faces indicated in Figure 4–58. Select Place Constraint from the Assembly toolbar or panel.

24. Select the cylindrical faces (not the circular edges).

25. Select the Apply button.

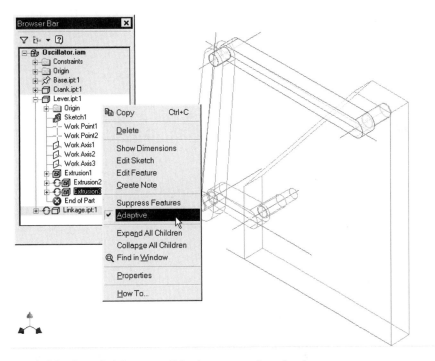

Figure 4–56 *Extruded features of the lever set to be adaptive*

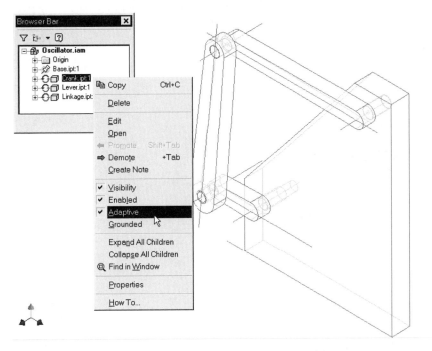

Figure 4–57 *Lever and crank set to be adaptive in the assembly*

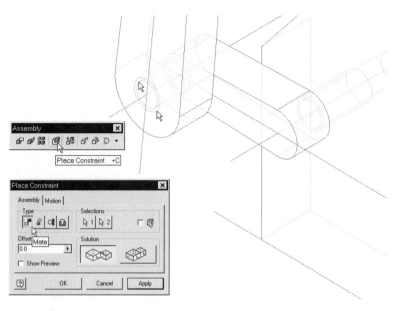

Figure 4–58 *Cylindrical faces selected*

26. The diameter of the extruded feature of the crank now changes and adapts to the diameter of the hole of the linkage. (If not, you probably did not select the cylindrical faces.) Now select the cylindrical faces indicated in Figure 4–59.

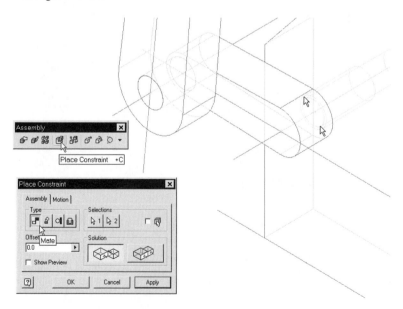

Figure 4–59 *Cylindrical faces selected*

27. Select the Apply button. The diameter of the selected extruded feature of the crank now adapts to the diameter of the selected hole of the base. (See Figure 4–60.)

28. Now zoom the display as shown in Figure 4–61 and repeat the steps delineated from Figure 4–58 to Figure 4–59 to apply mate constraints to the cylindrical faces of the extruded features of the lever and the holes of the linkage and the base.

The diameters of the extruded features adapt to the diameters of the holes of the linkage and the base.

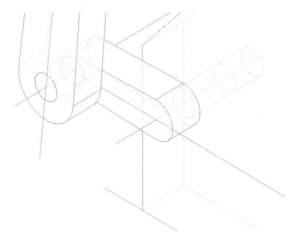

Figure 4–60 *Extruded features adapted and display changed*

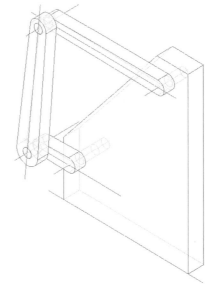

Figure 4–61 *Extruded features of the lever adapted to the holes of the linkage and the base*

After you apply the mate constraints to the four pairs of cylindrical faces, the mate constraint that you applied to the work axes becomes redundant.

29. Select the browser bar, right-click, and select Expand All. You will find constraints prefixed by a symbol "i" in the browser bar. (See Figure 4–62.)

30. Select these redundant constraints in turn, right-click, and select Delete to remove them.

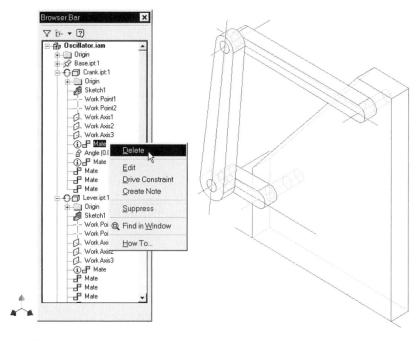

Figure 4–62 *Redundant constraint being deleted*

31. Now the mechanism is complete. Test the validity of your mechanism again by driving the angle constraint applied to the crank and the base.

32. Save and close all your files.

To reiterate, you must not fully constrain a sketch for it to be adaptive. Adaptivity should be applied to the relevant sketch of the part and to the part in the assembly.

GRAPHICS SLICING

In shaded mode or hidden edge display mode, the solid part is shaded. If you have a sketch plane that cuts through the solid, you can clip away a portion of the object from your display by means of graphics slicing. Graphics slicing is a technique in which the shaded solid part is sliced along the current sketch plane and the Z direction of the sketch plane is clipped.

BASE PLATE OF THE OSCILLATOR

1. Start a new solid part drawing. Construct a rectangle (80 units by 120 units).

2. Set the display to an isometric view and extrude the sketch a distance of 20 units. (See Figure 4–63.)

3. Select Shell from the Features toolbar or panel.

4. Set the shell thickness to 4 units and remove the vertical face highlighted in Figure 4–64.

5. Select the OK button. The part is made hollow.

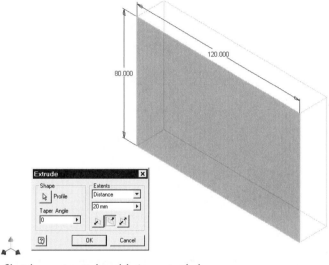

Figure 4–63 *Sketch constructed and being extruded*

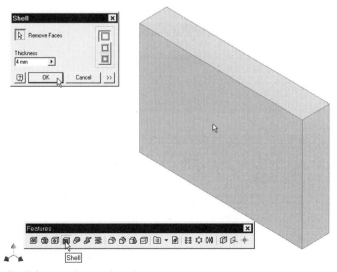

Figure 4–64 *Shell feature being placed*

6. Select Work Plane from the Features toolbar or panel.

7. Select the edges indicated in Figure 4–65 to construct a work plane.

8. Select Sketch from the Command Bar toolbar and select the new work plane from the browser bar to set up a new sketch plane. (See Figure 4–66.)

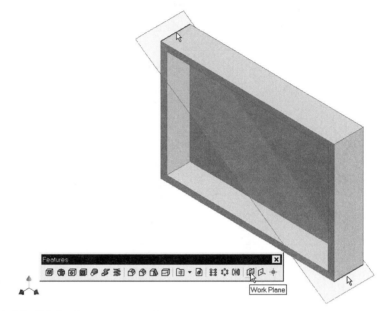

Figure 4–65 *Work plane constructed*

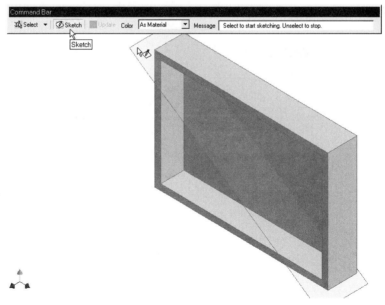

Figure 4–66 *Sketch plane constructed on the new work plane*

9. Select the new sketch plane from the browser bar, right-click, and select Slice Graphics. (See Figure 4–67.) The graphics display is sliced and you see the sketch plane.

10. Select Two point rectangle from the Sketch toolbar or panel.

11. Construct a rectangle as shown in Figure 4–68.

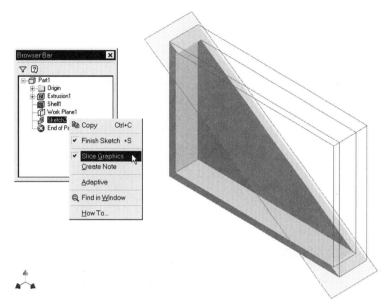

Figure 4–67 *Graphics sliced along the sketch plane*

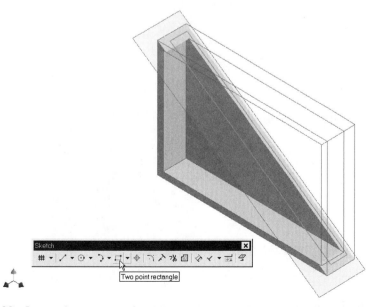

Figure 4–68 *Rectangle constructed*

12. Select General Dimension from the Sketch toolbar or panel and add dimensions as shown in Figure 4–69. The sketch is complete.

13. Deselect Sketch on the Command Bar toolbar.

14. Select Extrude to extrude the area a distance of 4 units from mid-plane. (See Figure 4–70.)

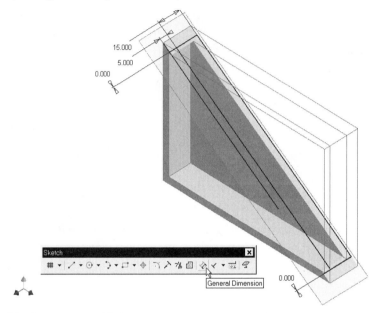

Figure 4–69 *Dimensions added*

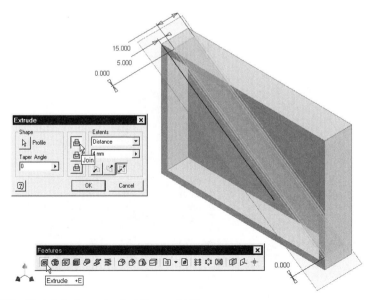

Figure 4–70 *Sketch being extruded from mid-plane*

15. Set the work plane to be invisible. (See Figure 4–71.)

16. Follow the same steps as shown in Figures 4–65 through 4–71 to construct another extruded solid. (See Figure 4–72.)

17. Save the file (file name: Plate.ipt).

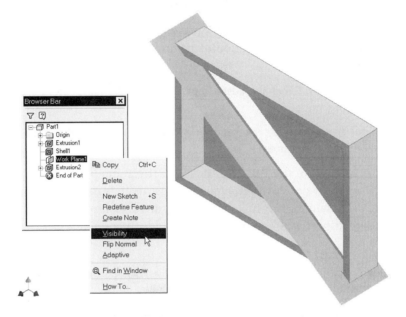

Figure 4–71 *Work plane made being invisible*

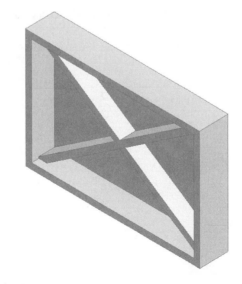

Figure 4–72 *Completed solid part*

DESIGN PARAMETERS

Each dimension that you placed in a solid part has a parameter name. By default, the name begins with a letter "d" followed by a number starting from zero. These are called model parameters. In addition, you can add user parameters, linked parameters, and embedded parameters and store them in a parameter table.

User parameters are parameters that you add to the table; you specify a parameter name and a value. To add a collection of parameters to the table, you can use an Excel spreadsheet.

- If you link the Excel spreadsheet to the table, the parameters are called linked parameters.

- If you embed the Excel spreadsheet in the table, the parameters are called embedded parameters.

Linked parameters link to the Excel spreadsheet. If you modify the spreadsheet, the parameters change as well. However, embedded parameters imported from the Excel spreadsheet do not maintain any link with the original Excel spreadsheet, and changes in the Excel spreadsheet will not affect the embedded parameters.

PARAMETERS TABLE

1. Now you will continue to work on the file Plate.ipt. Select Parameters from the Standard toolbar to display the parameter table. (See Figure 4–73.)

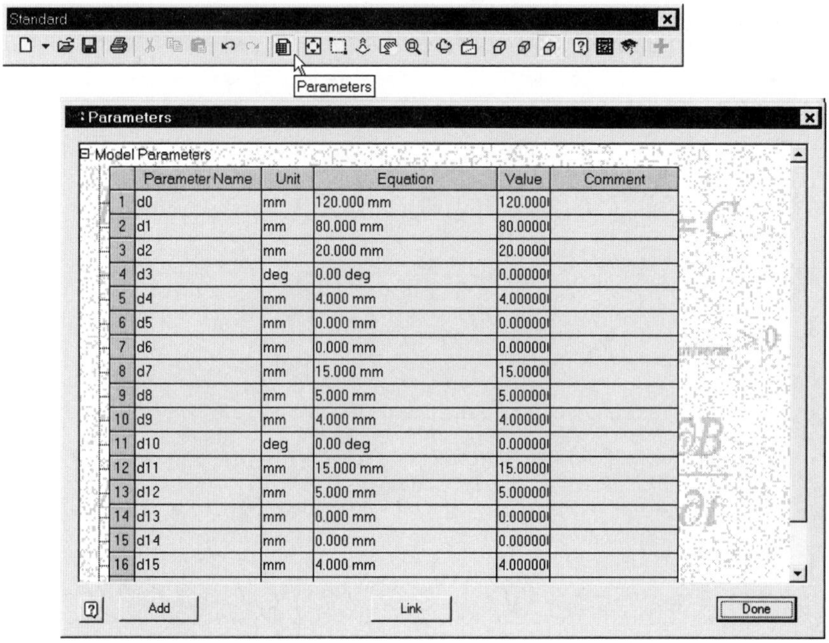

Figure 4–73 *Parameters dialog box*

Because your sequence of dimensioning the solid parts in your file may be different, the equation and value shown in Figure 4–73 might differ from yours.

Initially, you will find two tables: Model Parameters and User Parameters. In each table, there are six columns:

- Item number (unimportant)

- Parameter Name

- Units of measurement (you can use both Imperial size and metric size in a single table)

- Equation (you use a simple numeric value or an equation to express the dimension of the parameter)

- Value (this is the numeric value of the equation)

- Comment

The first table is a list of parameters that are assigned by Inventor to the solid part when you add dimensions to your solid part.

Add User Parameters

2. Now you will add a parameter in the User Parameters table. Select the Add button.

3. Type thickness in the Parameter Name column and 15 units in the Equation column. (See Figure 4–74.)

4. Select the Done button. A parameter is added to the User Parameters table.

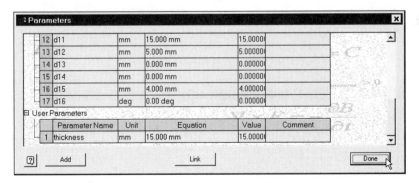

Figure 4–74 *Entry added to User Parameters table*

Edit the Part

5. Now you will use the user parameter to edit the solid part. Select the extruded solid feature indicated in Figure 4–75 from the browser bar, right-click, and select Edit Feature.

6. Change the distance of extrusion to "thickness." When you start typing the word thickness, the letters are initially red in color. They change to black after you type a name that is already in the user parameter table. (See Figure 4–75.)

7. Select the OK button. Now the extruded distance changes to the value specified by the parameter.

8. Select Parameters from the Standard toolbar to open the parameter table. (See Figure 4–76.)

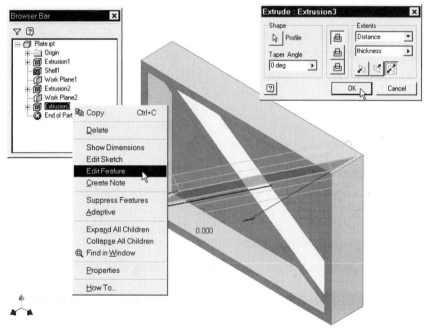

Figure 4–75 *Parameter name used to define extrusion distance*

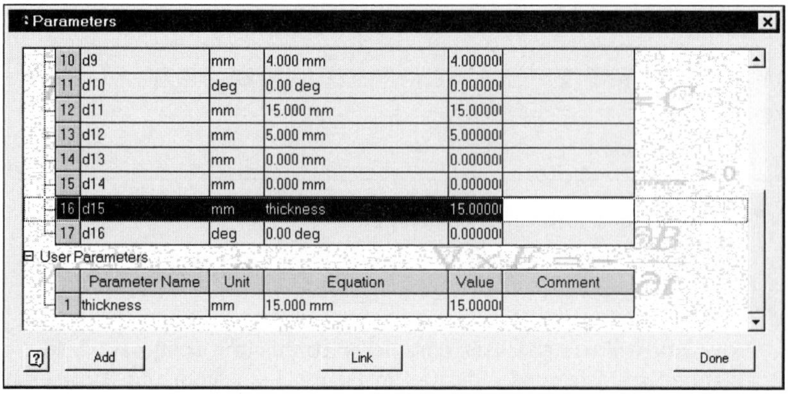

Figure 4–76 *Parameter table*

In Figure 4–76, the equation of the model parameter, d15, becomes "thickness." (In your table, the model parameter name may be different.)

Change the Parameter

9. Now you will modify the user parameter and see how the model changes. Select the Equation column of the user parameter table.

10. Change the equation to 4 mm and select the Done button.

11. Select Update from the Command Bar toolbar.

Because the user parameter is used to define the extrusion thickness, changing the user parameter causes the extrusion thickness to change as well.

SPREADSHEETS

To enter a set of parameters collectively, you can use an Excel spreadsheet. In the spreadsheet, construct two, three, or four columns. The first column is the parameter name, the second the value, the third the units of size, and the fourth the comment. If you do not specify the third column, the default units of size will be used. You should have Excel properly installed in your computer.

1. Start a new spreadsheet and fill in the sheet as shown in Figure 4–77.

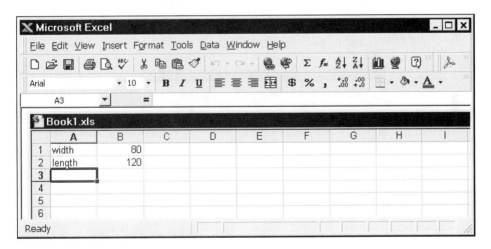

Figure 4–77 *Book1 spreadsheet*

2. Save your file (file name: Book1.xls).

3. Start another new spreadsheet and fill in the sheet as shown in Figure 4–78.

4. Save and close your file (file name: Book2.xls).

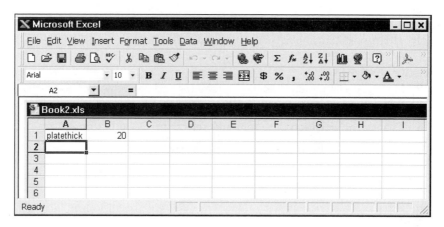

Figure 4–78 *Book2 spreadsheet*

Now you have two Excel spreadsheets. You will link the first and embed the second.

Link a Spreadsheet

5. In Inventor, select Parameters from the Standard toolbar.

6. Select the Link button.

7. In the Open dialog box, select Link, select the file Book1.xls, and select the Open button. (See Figure 4–79.)

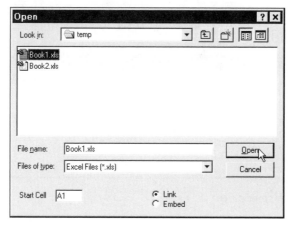

Figure 4–79 *Book1.xls linked to the solid part*

When you link a spreadsheet, the data in the spreadsheet is not saved in the solid part file. Every time you open a solid part file, the data in the linked spreadsheet is imported. As a result, you always get the latest version of the spreadsheet in your solid part.

Embed a Spreadsheet

Another way of using a spreadsheet is to embed the data from the spreadsheet in the solid part file. After embedding, there will be no more connection between your solid part file and the spreadsheet.

8. Select the Embed button, select Book2.xls, and select the Open button. (See Figure 4–80.)

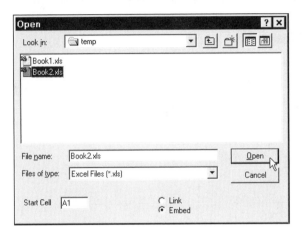

Figure 4–80 *Book2.xls embedded*

9. Now the parameter box has two more parameter tables. (See Figure 4–81.) Select the Done button.

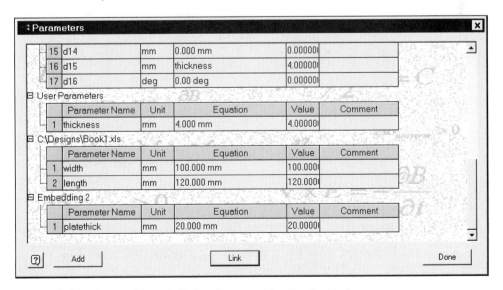

Figure 4–81 *A spreadsheet linked and a spreadsheet embedded*

Use Spreadsheet Parameters

Now you will use the parameters in both the linked table and the embedded table to dimension your solid part.

10. Select the first extruded feature from the browser bar, right-click, and select Edit Sketch.

11. Change the dimensions of the sketch as shown in Figure 4–82.

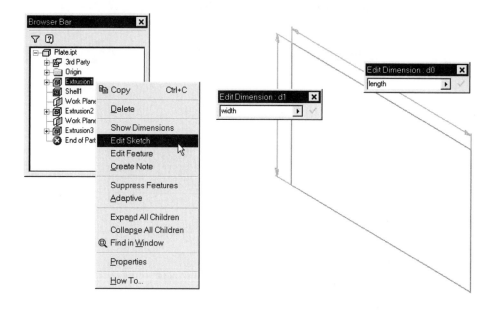

Figure 4–82 *Linked parameters used*

12. Select Update from the Command Bar toolbar.

13. Select the first extruded feature in the browser bar, right-click, and select Edit Feature.

14. Change the extruded distance to "platethick." (See Figure 4–83.)

15. Select Update from the Command Bar toolbar to update the change.

16. Select Embedding from the browser bar, right-click, and select Edit.

17. Change the value of B2 cell of the spreadsheet to 24. (See Figure 4–84.)

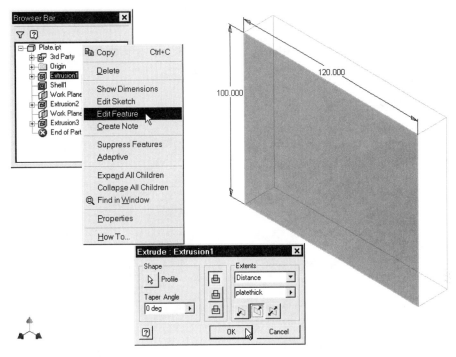

Figure 4–83   *Embedded parameter used*

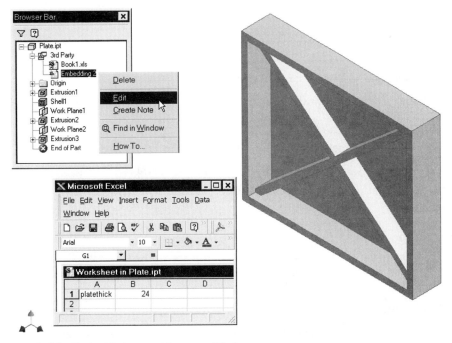

Figure 4–84   *Embedded spreadsheet modified*

18. Close the spreadsheet and select Update from the Command Bar toolbar. The solid is updated.

19. Now select Book1.xls (the linked spreadsheet), right-click, and select Edit.

20. Change cell B1 to 110. (See Figure 4–85.)

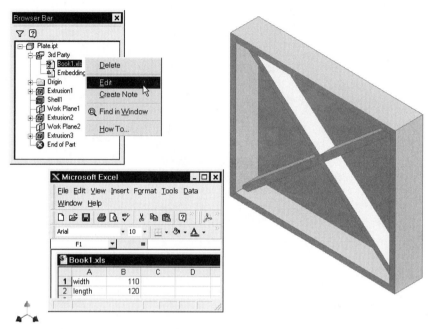

Figure 4–85 *Linked spreadsheet modified*

21. Save and close the spreadsheet.

22. Select Update from the Command Bar toolbar in Inventor. (See Figure 4–86.)

23. Save and close your file.

24. Now use Excel to open the spreadsheets Book1.xls and Book2.xls.

You will find that Book1.xls is modified (because it is linked) and Book2.xls is not changed (because it is not linked to the solid part and you changed only the embedded spreadsheet).

Link a Second Part

25. Open the file Base.ipt. You will link the spreadsheet (book1.xls) to this solid part. Now you use an Excel spreadsheet to control the dimensions of two solid parts.

26. Select Parameter from the Standard toolbar.

27. Select the Link button to link to the spreadsheet Book1.xls. (See Figure 4–87.)

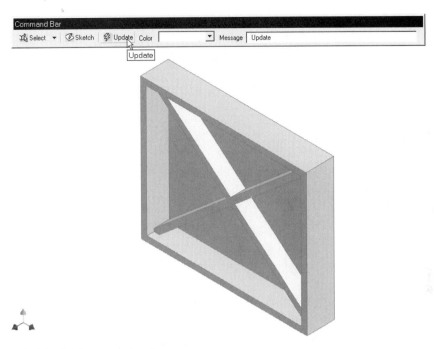

Figure 4–86   *Solid part updated*

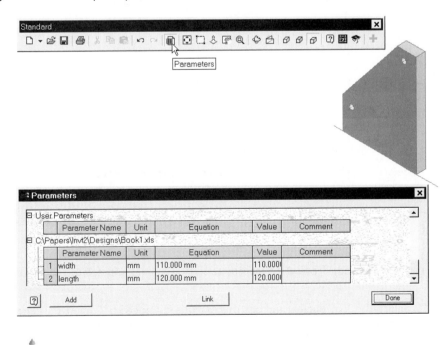

Figure 4–87   *Base linked to the spreadsheet Book1.xls*

28. Select the sketch of the first extruded feature from the browser bar, right-click, and select Edit Sketch.

29. Change the selected dimension value to the parameter "length." (See Figure 4–88.)

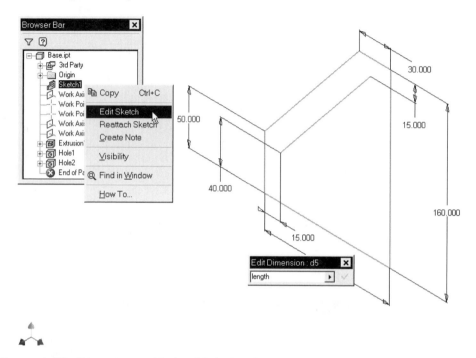

Figure 4–88 *Dimension modified and linked to the spreadsheet*

30. To update the change, select Update from the Command Bar toolbar.

Now link a dimension of the solid part to an external spreadsheet.

31. Select Sketch from the Command Bar toolbar and select the face indicated in Figure 4–89 to set up a new sketch plane.

32. Construct a sketch as shown in the figure.

33. Apply vertical and horizontal constraints to the lines and add three dimensions (15, 50, and 110 units).

34. Select Extrude from the Features toolbar or panel and select the sketch as the profile.

35. Select the Join button and set the extrude distance to 1.2*length. (The * symbol stands for multiply. Thus, 1.2*length means 1.2 multiples of the parameter name called length.)

36. Select the OK button. (See Figure 4–90.)

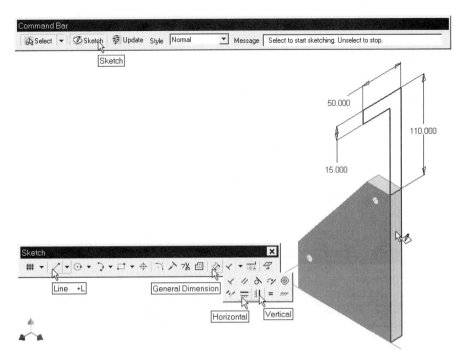

Figure 4–89   *Sketch constructed on new sketch plane*

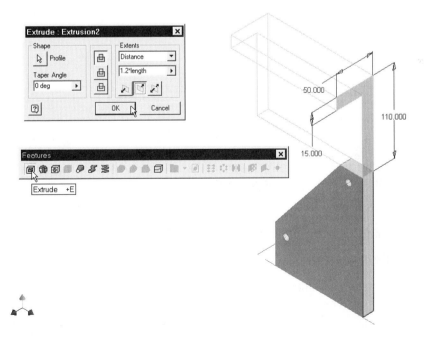

Figure 4–90   *Sketch being extruded*

37. Set the display to wireframe mode.

38. Select Fillet button from the Features toolbar or panel to place two fillet features.

39. Set the fillet radius to 10 units, select the edges highlighted in Figure 4–91, and select the OK button.

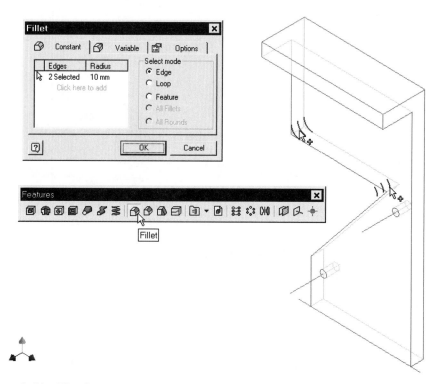

Figure 4–91 *Fillet feature being placed*

40. Select Sketch from the Command Bar toolbar and select the top face of the solid part to set up a sketch plane.

41. Construct a circle.

42. Add three dimensions (length, 10, and 20 units) as shown in Figure 4–92.

43. Extrude the sketch to cut through the solid. (See Figure 4–93.)

44. The base is complete. Save and close your file.

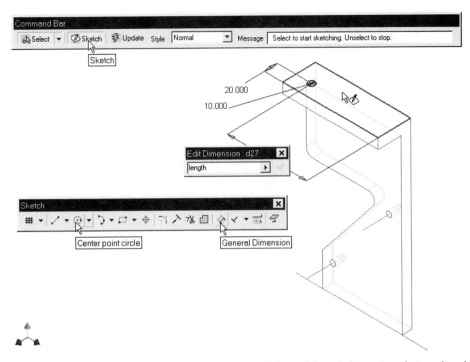

Figure 4–92 *Sketch constructed on the top face of the solid, and dimensions being placed*

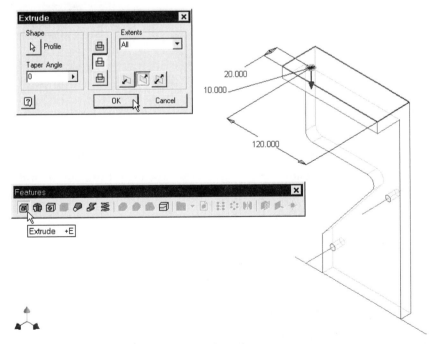

Figure 4–93 *Sketch being extruded to cut through*

Oscillator Assembly

1. Now open the assembly file Oscillator. You will place a component and add assembly constraints to complete the project.

2. Set the display to wireframe mode.

3. Select Place Component from the Assembly toolbar or panel and place the solid part file Plate.ipt in the assembly. (See Figure 4–94.)

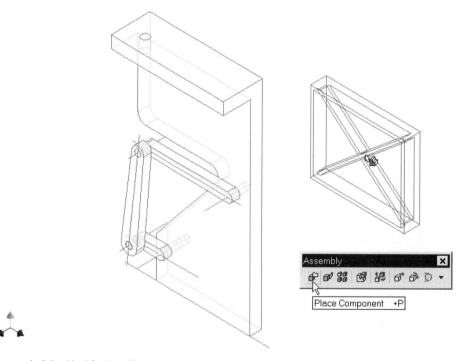

Figure 4–94 *Modified oscillator and the plate*

4. Select Place Constraint from the Assembly toolbar or panel.

5. Select the Mate button in the Type area, select the Mate button in the Solution area, and select the faces indicated in Figure 4–95.

6. Select the Apply button.

7. Select the Flush button in the Solution area.

8. Select the faces highlighted in Figure 4–96.

9. Select the Apply button.

10. Select the faces highlighted in Figure 4–97.

11. Select the OK button.

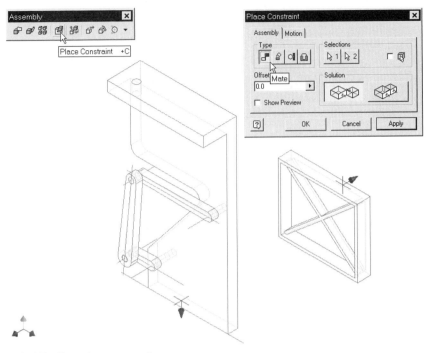

Figure 4–95 *Faces being mated*

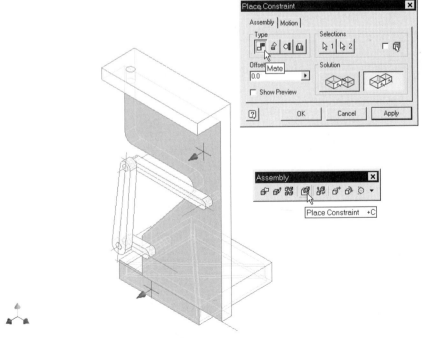

Figure 4–96 *Faces being constrainted flush*

The assembly is complete. (See Figure 4–98.) Note that two solid parts of this assembly are linked to a spreadsheet.

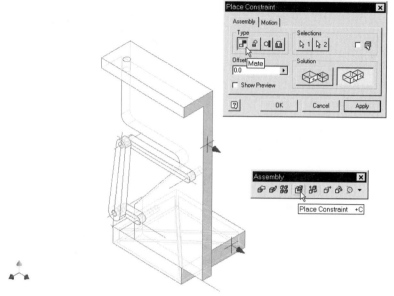

Figure 4–97 *Faces being flushed*

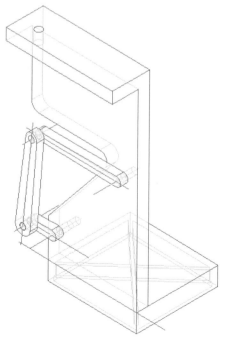

Figure 4–98 *Parts properly assembled*

DESIGN ELEMENTS

One of the major advantages of using the computer to design is that you can re-use existing data. For example, you can use standard fasteners, bearings, and seals in a design. You can construct these parts and put them in a library and retrieve them to put them in new assemblies.

Apart from using existing solid parts in new assemblies, you can make use of features from existing solid parts and incorporate them in new parts. You can select features to be design elements, put them in a design catalog, and insert them in new parts.

There are three kinds of design elements: join, cut, and intersect. When design elements are imported to a solid part, a join element joins to the part, a cut element cuts the part; and an intersect element intersects with the part. Now you will learn how to construct these three kinds of design elements.

JOIN DESIGN ELEMENT

Now you will construct a sketched solid feature on the part Plate.ipt and export it as a design element.

1. Select all the parts except Plate.ipt in turn, right-clicking, and deselecting Visibility to hide them.

2. Double-click Plate.ipt in the browser bar to activate part modeling mode.

3. Select Sketch from the Command Bar toolbar and select the top face of the solid part to set up a new sketch plane.

4. Construct two concentric circles of diameters 12 units and 30 units.

At this stage, do not add dimensions to position the sketch because including the position dimensions now will cause them to be included in the design element as well. (See Figure 4–99.)

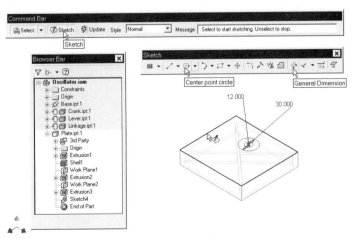

Figure 4–99 *Two concentric circles constructed*

366

5. Now extrude the sketch a distance of 10 units to join to the solid. (See Figure 4–100.)

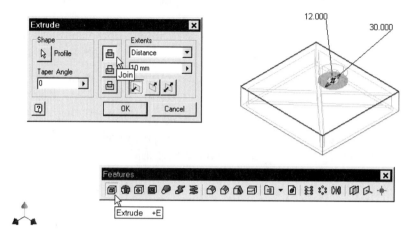

Figure 4–100   *Circles being extruded*

Create Design Element

6. Select Create Design Element from the Features toolbar or panel.

7. Select the extruded solid feature indicated in Figure 4–101.

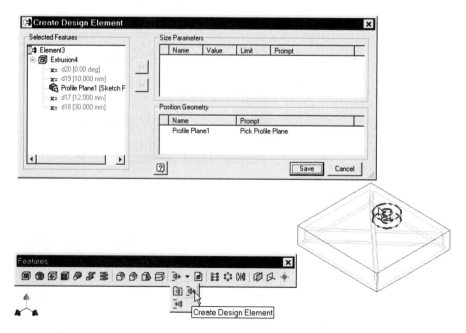

Figure 4–101   *Extruded feature selected*

8. In the Create Design Element dialog box, select the item having the value of 10 (extrude distance), and then select the >> button to put it in the Size Parameters box.

9. Repeat this for each of the values 12 (inside diameter) and 30 (outside diameter).

10. Select the Save button. (See Figure 4–102.)

11. Select the Catalog folder in your computer to save the design element file (file name: Join.ide). You can further categorize the elements and put them into various folders here. (See Figure 4–103.) A join design element is added.

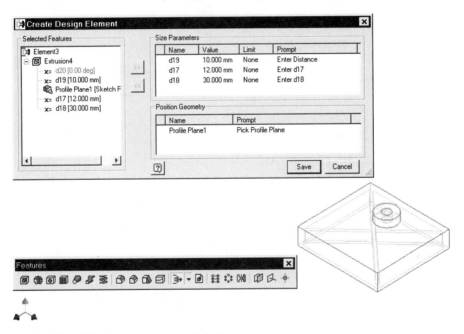

Figure 4–102 *Model parameters included*

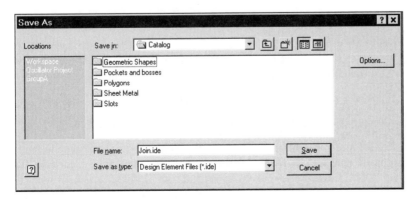

Figure 4–103 *Join design element file being saved*

12. Remember that you still need to position the extruded feature. Select the extruded solid feature in the browser bar, right-click, and select Edit Sketch.

13. Add two dimensions (length/2 and width/2) as shown in Figure 4–104.

14. Select Update from the Command Bar toolbar. The part is complete.

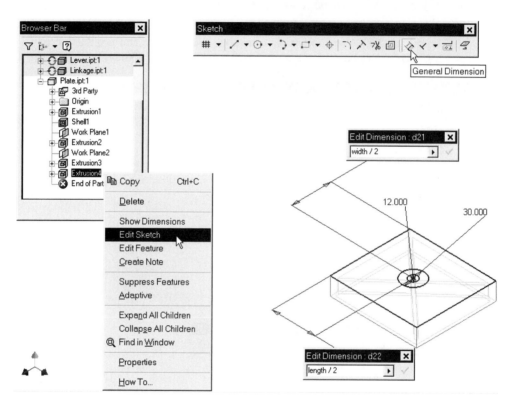

Figure 4–104 *Sketch modified*

Insert Design Element

15. Now you will insert the design element in the base of the assembly. Double-click the assembly in the browser bar to activate assembly mode.

16. Hide the plate and make the base visible.

17. Double-click the base to switch to part modeling mode.

18. Select Insert Design Element from the Features toolbar or panel and select the design element (Join.ide) that you constructed. (See Figure 4–105.)

19. Select the top face of the base to locate the design element. (See Figure 4–106.)

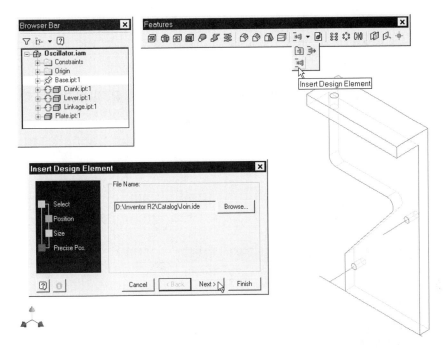

Figure 4–105   *Design element selected for insertion*

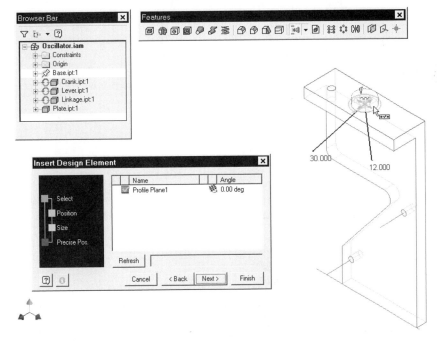

Figure 4–106   *Design element placed on the top face of the base*

20. Select the Next> button.

21. Set height to 20 units, the inside diameter to 12 units, and the outside diameter to 20 units. (See Figure 4–107.)

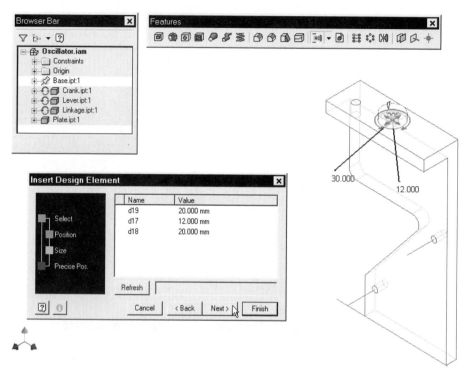

Figure 4–107 *Parameters set*

22. Select the Next> button.

23. Select Activate Sketch Edit Immediately.

24. Select the Finish button. (See Figure 4–108.)

25. To properly position the design element, select Concentric from the Sketch toolbar or panel and select the circles highlighted in Figure 4–109.

26. Select Update from the Command Bar toolbar. The design element is updated.

27. Now select the assembly in the browser bar.

28. Set all the parts to be visible. (See Figure 4–110.)

29. Save and close the assembly and the part files.

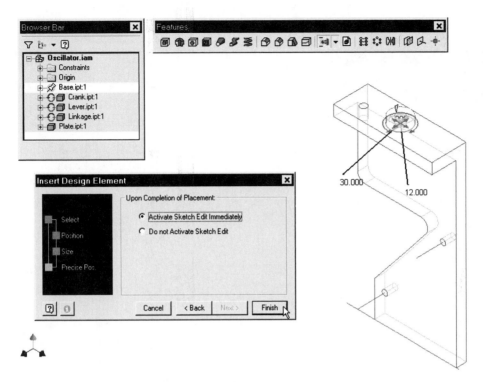

Figure 4–108 *Sketch Edit mode activated*

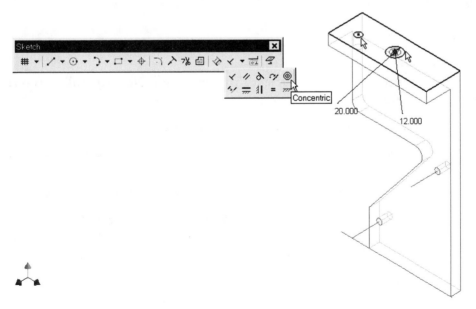

Figure 4–109 *Concentric constraint being applied*

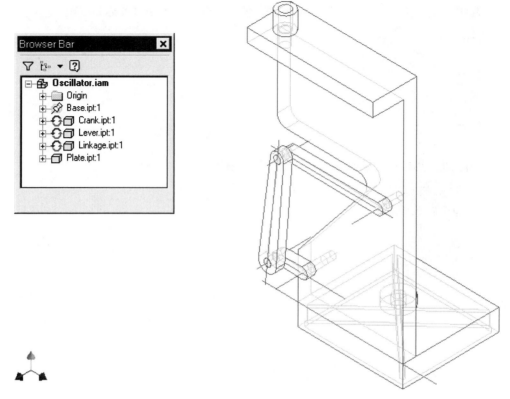

Figure 4–110 *Solid updated, and parts made visible*

CUT DESIGN ELEMENT

You have constructed a design element that joins a part. Now you will construct a design element that cuts a part.

1. Open the part file Crank.ipt and set the display to wireframe mode.

2. Select Sketch from the Command Bar toolbar and select the face indicated in Figure 4–111 to set up a new sketch plane.

3. Construct a rectangle and add four dimensions (2, 4, 6, and 16 units).

4. Extrude the sketch a distance of 5 units to cut the solid. (See Figure 4–112.)

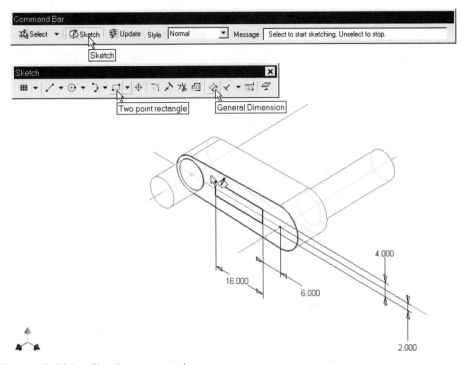

Figure 4–111   *Sketch constructed*

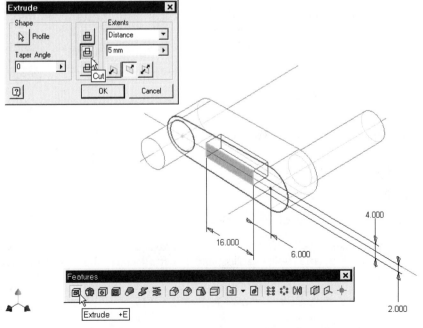

Figure 4–112   *Sketch being extruded*

Create Design Element

5. Select Create Design Element from the Features toolbar or panel.

6. Select the extruded cut feature and the parameters of the feature.

7. Save the design feature (file name: Cut.ide). (See Figure 4–113.)

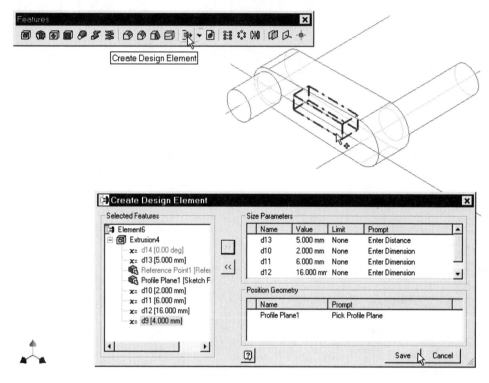

Figure 4–113   *Cut design element being created*

8. The part is complete. Save and close the file.

Insert Design Element

9. Now open the file Lever.ipt.

10. Set the display to wireframe mode.

11. Select Insert Design Element from the Features toolbar or panel, select Cut.ide from the design catalog, and select a face on the lever to position the design element. (See Figure 4–114.)

12. Change a dimension as shown in Figure 4–115.

13. Activate Sketch Edit mode.

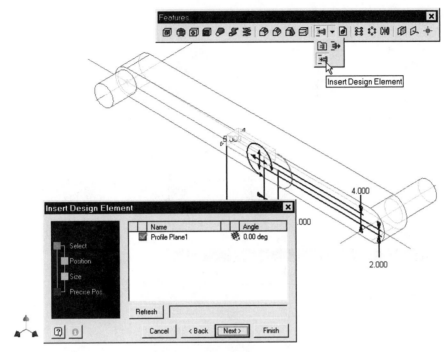

Figure 4–114   *Cut design element being inserted*

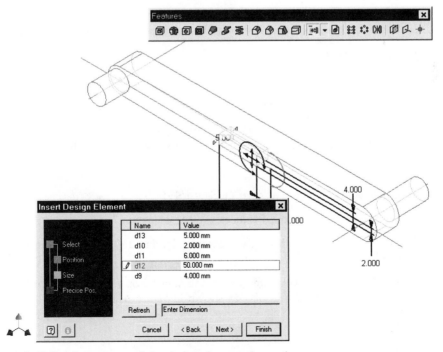

Figure 4–115   *Parameters of the design element changed*

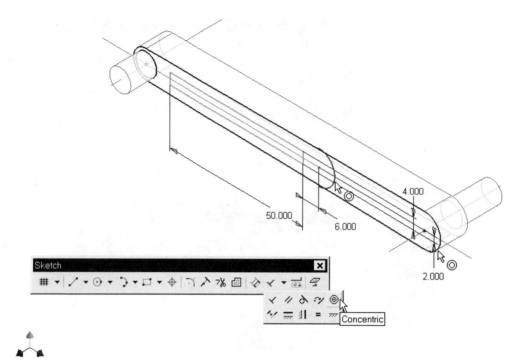

Figure 4–116 *Concentric constraint being applied to the inserted design element*

16. Select Update from the Command Bar toolbar. The lever is complete.

17. Save and close your file.

INTERSECT DESIGN ELEMENT

To learn how to construct an intersect element, you will construct two parts. You construct a design element from a part and insert the design element into another part.

1. Start a new part file.

2. Construct a sketch as shown in Figure 4–117.

3. Select Revolve from the Features toolbar or panel.

4. Revolve the sketch about the bottom edge for 360 degrees. (See Figure 4–118.)

5. Set the display to wireframe mode and select Rotate from the Standard toolbar to rotate the display as shown in Figure 4–119.

6. Select Sketch from the Command Bar toolbar to set up a new sketch plane.

7. Select Two point rectangle from the Sketch to construct a square and select General Dimension from the Sketch toolbar or panel to add four dimensions.

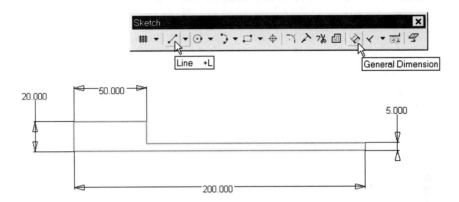

Figure 4–117   *Sketch constructed on new part file*

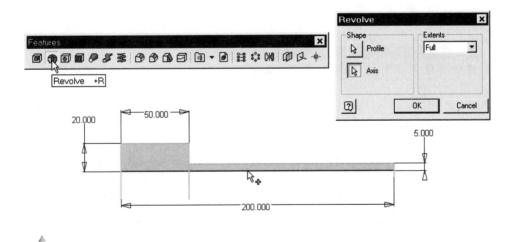

Figure 4–118   *Sketch being revolved*

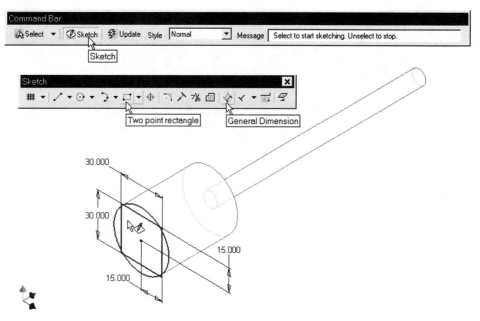

Figure 4–119   *Second sketch constructed*

8. Select Extrude from the Features toolbar or panel and extrude the sketch to intersect the solid. (See Figure 4–120.)

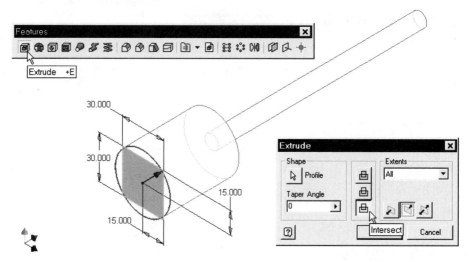

Figure 4–120   *Sketch being extruded*

9. The part is complete. Save your file (file name: Pin.ipt).

Create Design Element

10. Before you close the file, create a design element from the intersected extruded solid by selecting Create Design Element from the Features toolbar or panel (file name: Intersect.ide). (See Figure 4–121.)

11. The design element is complete. Close your file.

12. Now start another new part and construct a sketch as shown in Figure 4–122.

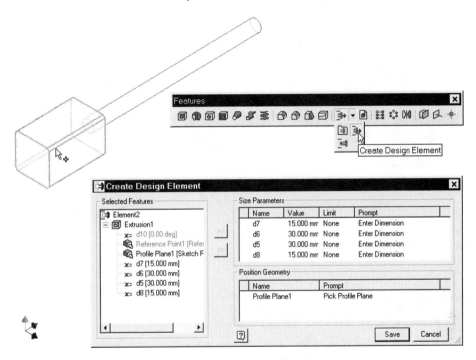

Figure 4–121   *Intersect design element being created*

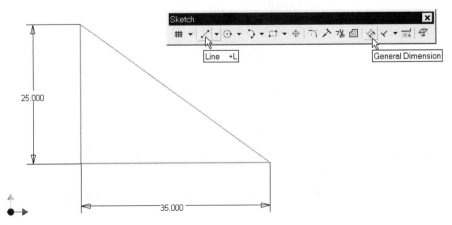

Figure 4–122   *Sketch on new solid part constructed*

13. Set the display to an isometric view.

14. Select Revolve from the Features toolbar or panel and revolve the sketch 360 degrees about the axis indicated in Figure 4–123.

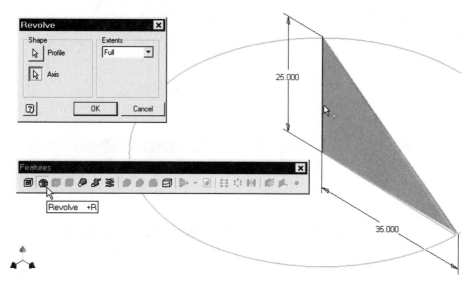

Figure 4–123   *Sketch being revolved*

Insert Design Element

15. Set the display to wireframe mode.

16. Select Insert Design Element from the Sketch toolbar or panel to insert the design element Intersect.ide.

17. Select the sketch plane indicated in Figure 4–124.

18. Select Next until you reach the Precise Pos. page of the Insert Design Element wizard, activate the sketch of the design element, and select Finish.

19. Select Concentric from the Sketch toolbar or panel to place a concentric constraint.

20. Select the circular edges highlighted in Figure 4–125.

21. Select Update from the Command Bar toolbar to update the part. (See Figure 4–126.)

22. The part is complete. Save and close your file (file name: Cone.ipt).

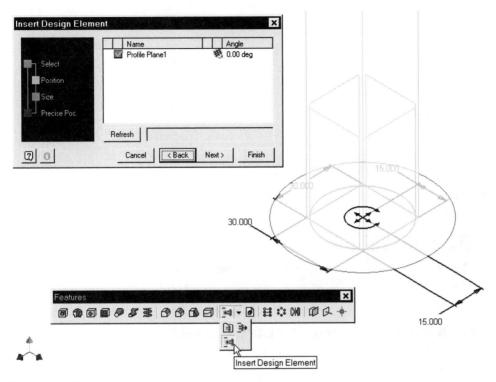

Figure 4–124   *Intersect design element being inserted*

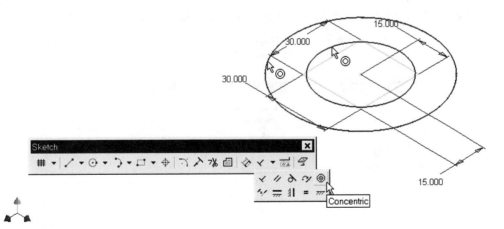

Figure 4–125   *Circular edges selected*

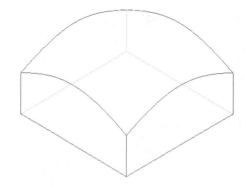

Figure 4–126 *Design element inserted and updated*

DESIGN NOTEBOOK

To record your design intention and other design information for later retrieval, you can maintain a notebook. In the notebook, you keep graphical and textual records of your design.

ADDING NOTES TO THE OSCILLATOR ASSEMBLY

The parts for the oscillator are modified. You will now work on the assembly to update the changes and to place a new component. While working on the assembly, you will use the design notebook to keep a record of all your work.

1. Open the file Oscillator.iam and set the display to wireframe mode.

2. Select Options from the Tools menu and select the Notebook tab.

3. Type your name in the User Information box (if it is not already present) and select the OK button. (See Figure 4–127.)

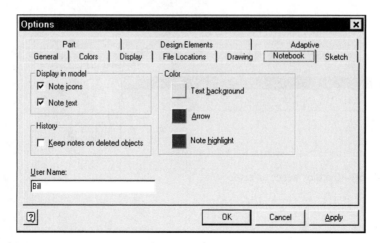

Figure 4–127 *Notebook tab of the Options dialog box*

4. Now select the assembly in the browser bar, right-click, and select Create Note. (See Figure 4–128.)

5. In your notebook, write down any design information that you want to keep for future use, for example, "This is a preliminary design." (See Figure 4–129.)

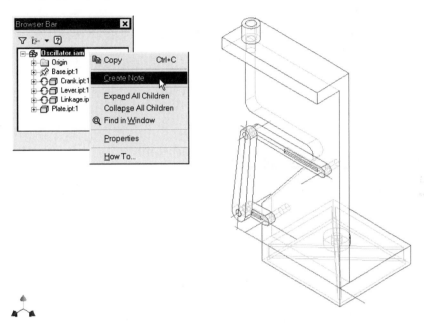

Figure 4–128   *Design notebook being activated*

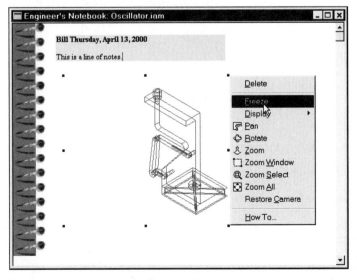

Figure 4–129   *Design diagram "frozen" in the notebook*

6. Because the assembly is going to be changed and you want to keep a record of the assembly before it is changed, select the figure in the notebook, right-click, and select Freeze. This way, the current status of your design is preserved in the notebook.

7. Now close the notebook, place a component in the assembly, and add appropriate assembly constraints.

8. Select Place Component from the Assembly toolbar or panel.

9. Place the part file Pin.ipt into the assembly. (See Figure 4–130.)

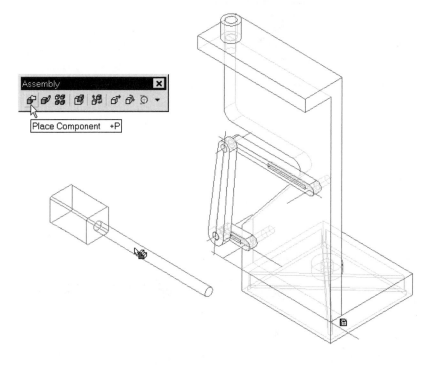

Figure 4–130   *Component placed in the assembly*

10. Select Place Constraint from the Assembly toolbar or panel.

11. Select the Mate button.

12. Select the faces highlighted in Figure 4–131 and select the Apply button.

13. Now select the axes highlighted in Figure 4–132 and select the Apply button.

14. Select the Angle button and select the axes indicated in Figure 4–133. Select the Flip buttons if the directions of the arrows are different.

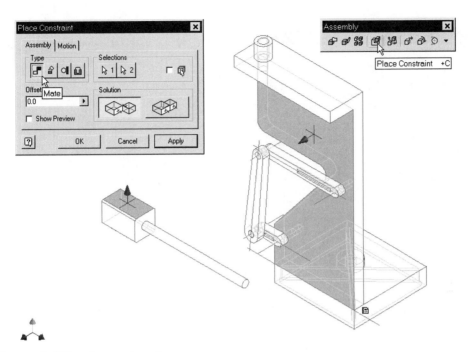

Figure 4–131 *Faces selected*

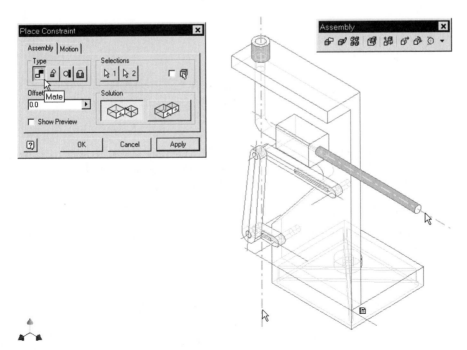

Figure 4–132 *Axes being mated*

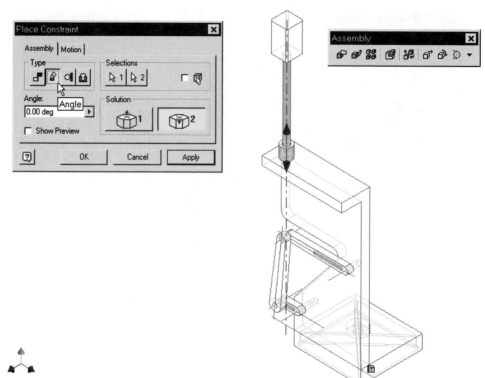

Figure 4–133 *Axes being aligned*

15. Select the Tangent button and select the faces highlighted in Figure 4–134 to apply a tangent constraint to the selected faces.

16. Select the OK button.

17. The constraints are complete. Select the assembly in the browser bar, right-click, and select Create Note to add a note in the design notebook, for example, "The mechanism is properly constrained." (See Figure 4–135.)

18. Close the design notebook.

19. Drive the angle constraint to test the mechanism as shown in Figure 4–136.

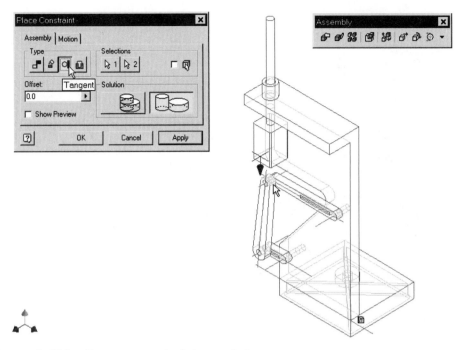

Figure 4–134   *Tangent constraint being applied*

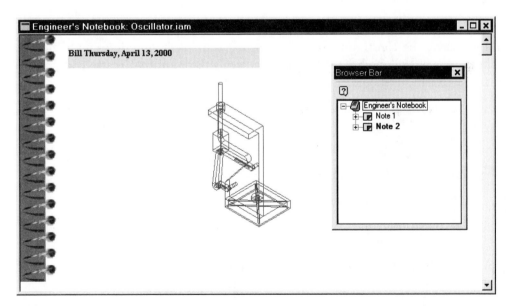

Figure 4–135   *Assembly completed*

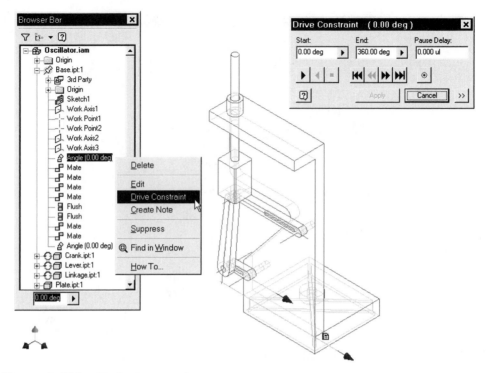

Figure 4–136   *Mechanism tested*

As you drive the constraint, you will find that the linkage interferes with the pin. Therefore, you will modify the linkage. To keep a record of this interference, record it in the design notebook.

20. Now open the file Linkage.ipt and edit the part. Select the sketch indicated in Figure 4–137 in the browser bar and select Edit Sketch.

21. Change the dimensions as shown in Figure 4–138.

22. Select the Update button from the Command Bar toolbar.

23. Save and close the file.

24. Now select the Update button from the Command Bar toolbar to update the assembly.

25. Drive the constraint again. (See Figure 4–139.)

26. If there is no more interference, save and close the file.

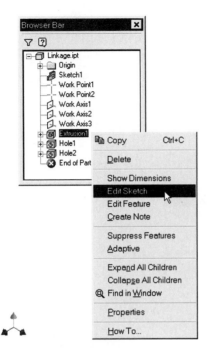

Figure 4–137   *Sketch being edited*

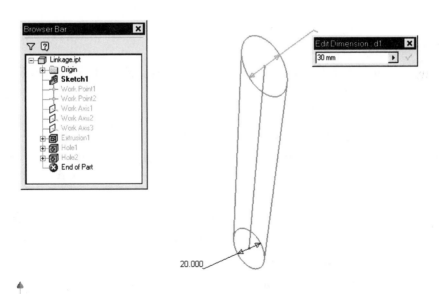

Figure 4–138   *Sketch edited*

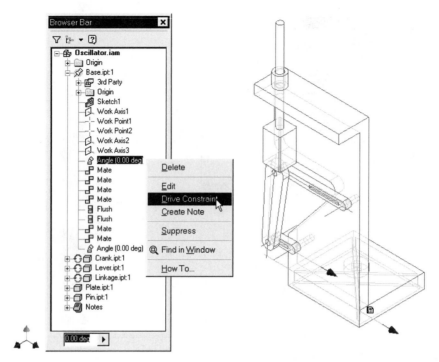

Figure 4–139 *Assembly completed*

MOTION SIMULATION

When you drive a constraint of a component in an assembly, the component moves. To cause a set of related components in an assembly to move simultaneously, you apply appropriate constraints or set relative motions to the components.

FRICTION DRIVE MECHANISM

Now you will construct a simple friction drive mechanism to learn how to apply relative motions in an assembly. In the mechanism, you will define relative motions among the drive members. Figure 4–140 shows the assembly.

Note that the standard engineering components such as bearings, seals, and fasteners are omitted for the sake of simplicity in illustration.

Construct Component Parts

The mechanism has five component parts: lower casing, upper casing, input shaft, intermediate shaft, and output shaft.

    1. Construct the part files as shown in Figures 4–141 through 4–145.

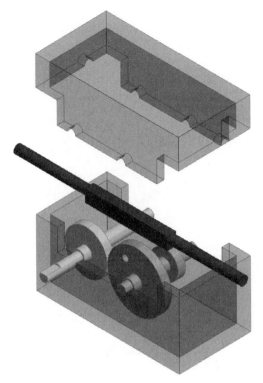

Figure 4–140 *Friction drive mechanism*

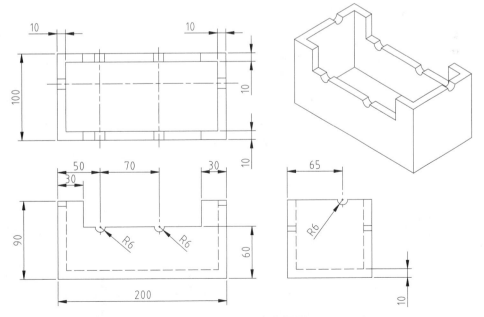

Figure 4–141 *Lower casing (file name: FrictionLower.ipt)*

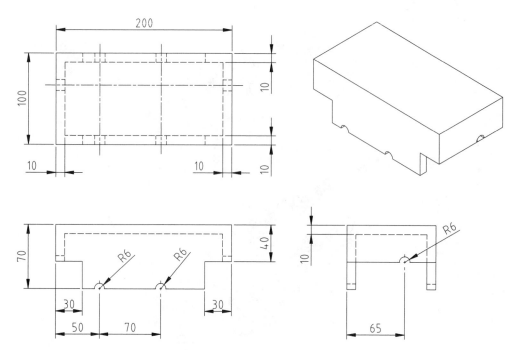

Figure 4–142 *Upper casing (file name: FrictionUpper.ipt)*

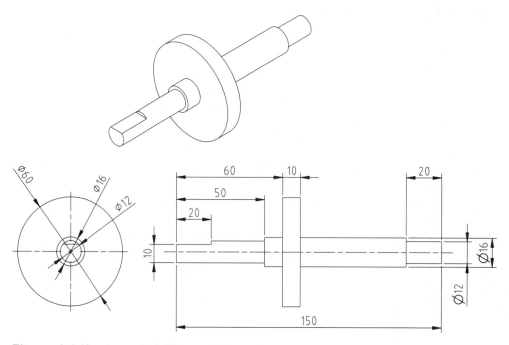

Figure 4–143 *Input shaft (file name: FrictionInput.ipt)*

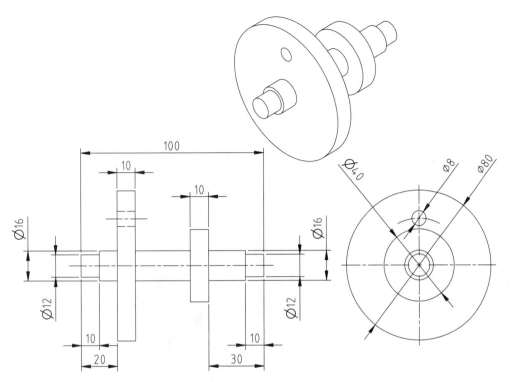

Figure 4–144 *Intermediate shaft (file name: FrictionIntermediate.ipt)*

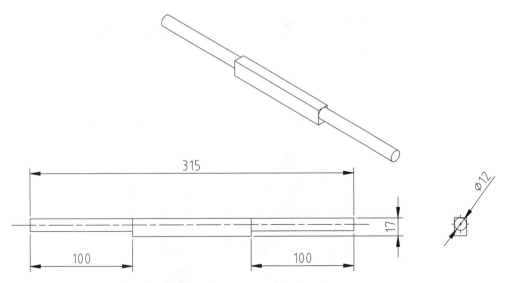

Figure 4–145 *Output shaft (file name: FrictionOutput.ipt)*

Assemble the Parts

2. Start an assembly file.

3. Place the part files FrictionLower.ipt, FrictionInput.ipt, and FrictionIntermediate.ipt in the assembly. (See Figure 4–146.)

4. Select Place Constraint from the Assembly toolbar or panel to apply an insert constraint to the circular edges highlighted in Figure 4–147. This constraint causes the input shaft to align with the lower casing.

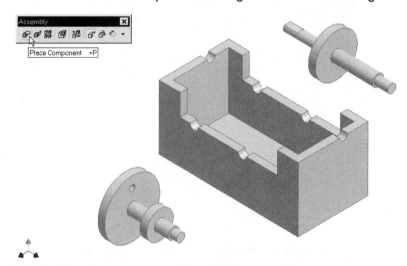

Figure 4–146 *Three component parts placed in the assembly*

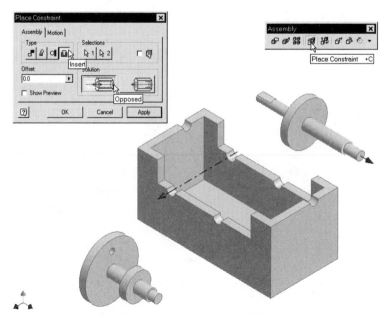

Figure 4–147 *Insert constraint being constructed*

5. To align the intermediate shaft with the lower casing, apply an insert constraint to the circular edges as shown in Figure 4–148.

6. Now place the component part Output.ipt in the assembly.

To assemble the output shaft properly, you will place three constraints.

7. First, apply an angle constraint to the faces highlighted in Figure 4–149.

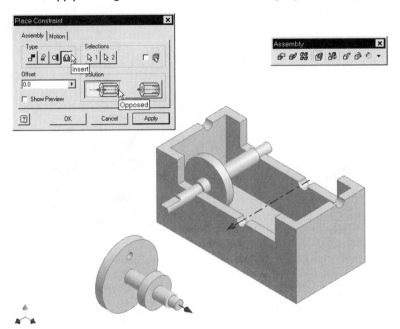

Figure 4–148   *Insert constraint being placed*

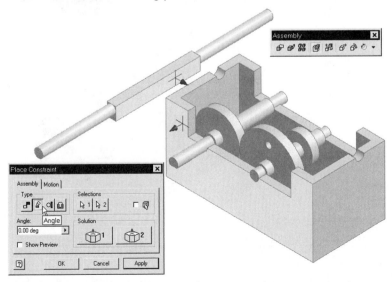

Figure 4–149   *Output shaft placed and angle constraint being constructed*

8. Construct a mate constraint to the selected axes highlighted in Figure 4–150.

9. Now construct a mate (opposed) constraint to the selected faces highlighted in Figure 4–151.

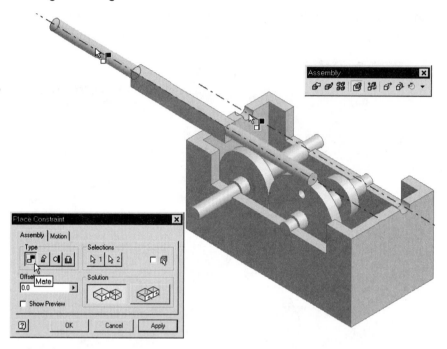

Figure 4–150   *Axes being mated*

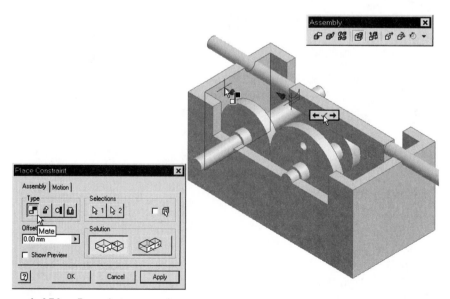

Figure 4–151   *Faces being mated*

10. Apply an angle constraint to the selected edges highlighted in Figure 4–152. The input shaft is fixed.

11. Now place the upper casing in the assembly.

12. Apply mate constraints to the selected faces highlighted in Figure 4–153.

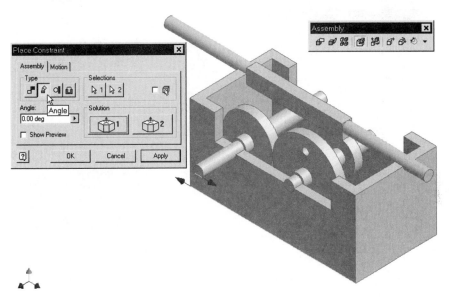

Figure 4–152   *Edges being constrained*

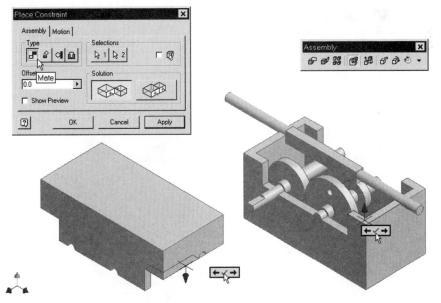

Figure 4–153   *Upper casing placed and faces being mated*

13. Apply a mate (flush) constraint to the selected faces highlighted in Figure 4–154.

14. Complete the assembly by constructing two more mate (flush) constraints to the paired faces highlighted in Figure 4–155.

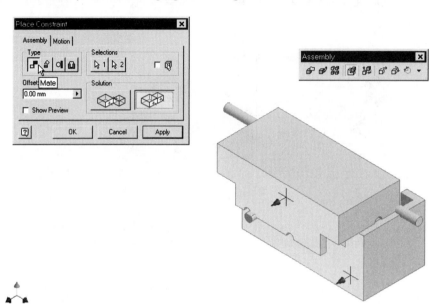

Figure 4–154 *Selected faces being mated*

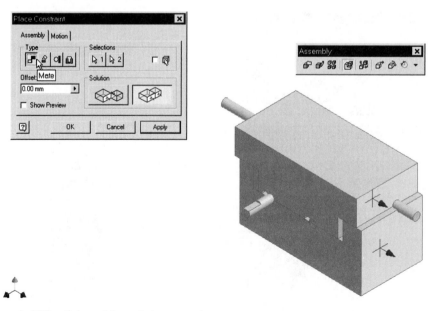

Figure 4–155 *Selected faces being mated*

15. Now the assembly is complete. In order to see the interior of the friction drive mechanism more clearly, select the upper casing in the browser bar, right-click, and deselect Visibility to hide it. (See Figure 4–156.)

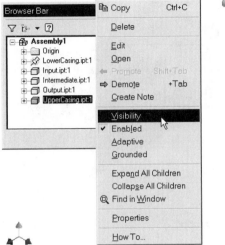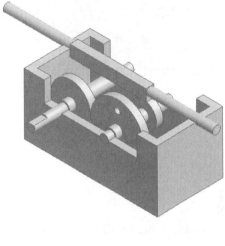

Figure 4–156 *Upper casing hidden*

Set Relative Motions

16. Now you will set relative motions among the components of the assembly. Select Place Constraint from the assembly toolbar or panel.

17. In the Place Constraint dialog box, select the Motion tab, select Rotation-Translation in the Type area, and select Forward in the Solution area.

18. Select the cylindrical face and the flat face highlighted in Figure 4–157.

19. Because the diameter of the friction disc of the intermediate shaft that is in contact with the output shaft is 40 mm, the distance that you should type in the Distance box should be 22/7*40. (π is 22/7.) Now select the Apply button.

20. Now select Rotation in the Type area and select Reverse in the Solution area of the Motion tab.

21. Select the input shaft and then the intermediate shaft highlighted in Figure 4–158.

22. Because the diameter of the friction disc of the input shaft is 60 mm and the diameter of the meshing friction disc of the intermediate shaft is 80 mm, the ratio that you should type in the Ratio box should be 30/40. Now select the OK button.

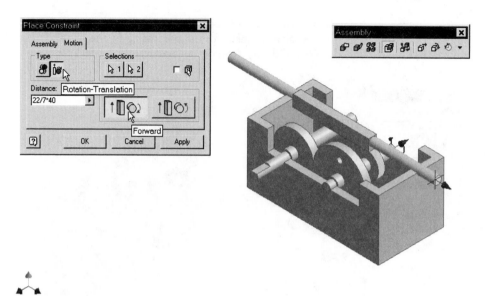

Figure 4–157 *Rotation-Translation motion being applied*

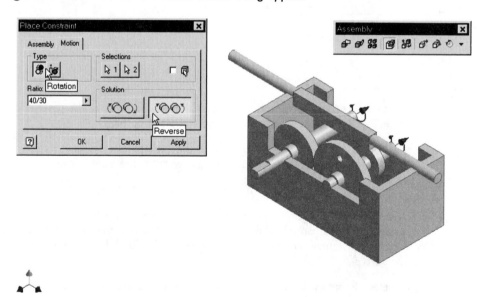

Figure 4–158 *Rotation motion being applied*

Now you will find "i" icons by two items in the hierarchy of the browser bar. They denote that the last motion that you set is not valid.

If you think carefully about the constraints that you applied to the input shaft and the output shaft, you will find that both of them are fixed. Therefore, a relative motion between them is not possible.

23. To make the motion feasible, select the mate constraint highlighted in Figure 4–159, right-click, and select Suppress. This way, the output shaft is free to translate.

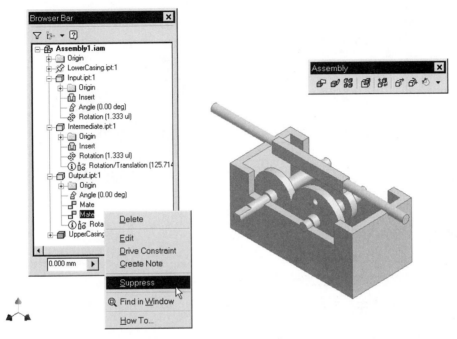

Figure 4–159 *Mate constraint being suppressed*

The motion settings among the components are complete.

Drive Constraint and Collision Detection

24. Now select the angle constraint highlighted in Figure 4–160. Right-click and select Drive Constraint.

25. In the Drive Constraint dialog box, set End to 360 degrees and select the >> button to expand the dialog box.

26. Set total # of steps to 10 ul (ul means unitless).

27. Select the Forward button. (See Figure 4–161.)

28. After you select the Forward button, the output shaft will translate to the right. You will find that the output shaft interferes with the casing. Select the Reverse button to reverse the mechanism.

29. To check whether there is any interference among the components, select Collision Detection in the Drive Constraint dialog box. (See Figure 4–162.)

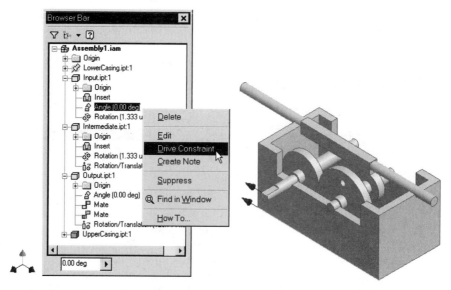

Figure 4–160 *Drive constraint selected*

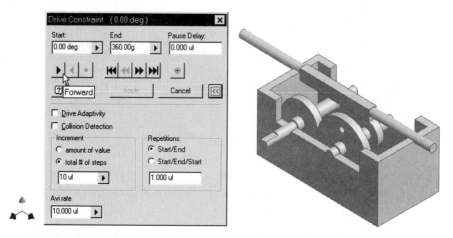

Figure 4–161 *Mechanism being set into motion*

30. Select the Forward button again. (See Figure 4–163.)

31. Now you discover that there is collision between the output shaft and the casing. To take remedial action, change the end position of the input shaft and try again.

32. The mechanism is complete. Save and close your file (file name: FrictionDrive.iam).

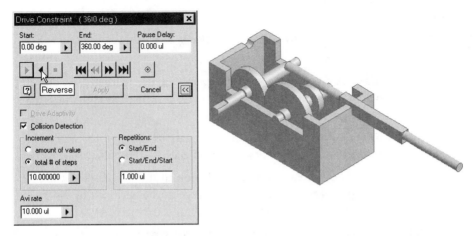

Figure 4–162 *Collision Detection selected*

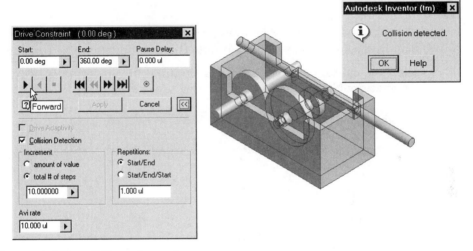

Figure 4–163 *Collision detected*

DESIGN COLLABORATION

To enable a team of designers to work collaboratively, you can set up a network so that files can be shared among the designers. You can establish a set of working directories to define file locations for each designer's computer and use Microsoft NetMeeting to facilitate communication among the designers.

NETWORK AND SHARE PERMISSIONS

A prior requirement for working collaboratively is to have a network connected among the designers.

Project Paths

To reiterate, you elect to create four sets of working directories that you will put in the project path file:

- **Workspace** A directory in each computer for use as each individual's workspace, the default location where each designer can construct new files on the local computer.

- **Local Search Paths** Search paths for locations assigned to each individual on their local computers, accessed only by the individual designer.

- **Workgroup** Search Paths Search paths for all the designers in the workgroup in designated computers, which can be accessed by all the designers working on the same project.

- **Library Search Paths** Search paths for the storage and retrieval of standard library parts in designated computers.

Share Rights

Having decided the directories and file locations, you assign share rights. To assign share rights, you must have administrator or administrator equivalent rights for the computer.

1. Select the directory, right-click, and select Sharing. Figure 4–164 shows the directory ModelCar on a C drive being selected and shared.

Figure 4–164 *Share rights being assigned*

2. In the Properties dialog box of the selected directory, select Shared As, specify the share name, decide the number of users allowed, and select the Permissions button. (See Figures 4–165 and 4–166.)

3. By default, the share right is assigned to everyone with full control. Select Everyone in the Access Through Share Permissions dialog box and select the Remove button.

4. Select the Add button. (See Figure 4–167.)

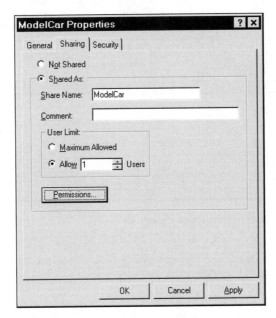

Figure 4–165   *Properties dialog box of the ModelCar directory*

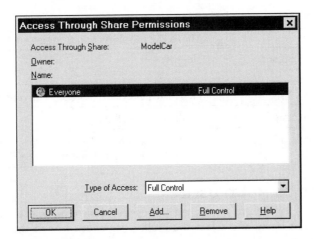

Figure 4–166   *Access Through Share Permissions dialog box*

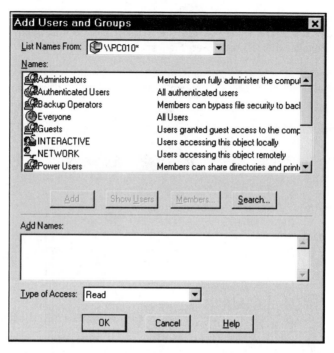

Figure 4–167 *Add Users and Groups dialog box*

5. In the Add Users and Groups dialog box, select the users and assign appropriate rights. (See table below.)

6. Select the OK button to exit.

7. Select the OK buttons in each dialog box to complete the share right assignment.

Table 4–1   Directory permissions

Directories	Share Permissions and Suggested Type of Access
Workspace in each designer's local computer	No share
Local Search Paths	Designated individual designer (Full control)
Workgroup Search Paths	All designers (Full control)
Library Search Paths	All designers who use the library (Read) Designated designers who help to build the library (Full control)

SETTING FILES TO MULTI USER

1. In order for files to be shared among the designers, select Options from the Tools menu and select the Multi User check box in the General tab. (See Figure 1–28).

File Reservation

2. After you enable multiple users to gain access to a file, select Open from the File menu. The Open dialog box has two additional buttons (Clear and Reserve) and a reservation note indicating who has reserved the file. (See Figure 4–168.)

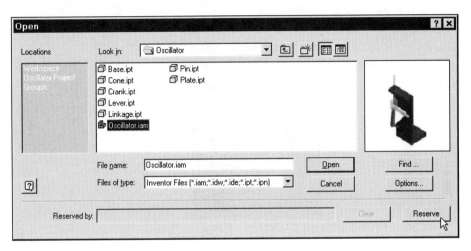

Figure 4–168   *File reservation*

Suppose a user named John has selected the Reserve button. Figure 4–169 shows another designer's Open dialog box. In the Reserved By box, John's name and the date when he reserved the file is displayed.

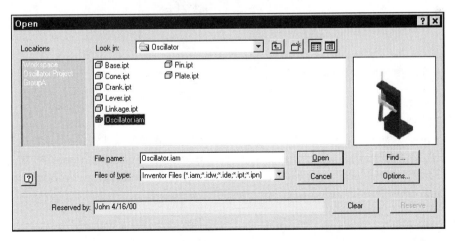

Figure 4–169   *File reserved*

3. To remove a reservation, select the Clear button in the dialog box.

WINDOWS NETMEETING

To communicate and share control rights in the process of designing, you can use NetMeeting. You need to install NetMeeting on each designer's computer in order for them to participate. Figure 4–170 shows the NetMeeting dialog box. To enable sharing, you need to select Enable Share from the Tools menu of the NetMeeting dialog box and specify sharing.

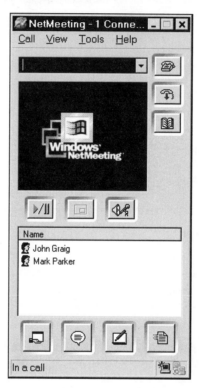

Figure 4–170 *NetMeeting dialog box*

Calling a NetMeeting

Now suppose you are John Graig and you are calling Mark Parker. In Inventor, select Add Participant from the Collaboration toolbar or select Tools ➤ Online Collaboration ➤ Meet Now. Figure 4–171 shows the assembly of the model car and participants are being added to collaborate in the design.

After you make a call, the NetMeeting dialog box displays a waiting message. (See Figure 4–172.)

Accepting a NetMeeting

In the other party's computer, an incoming message displays. (See Figure 4–173.)

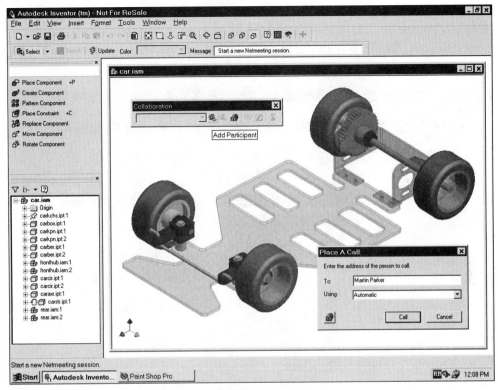

Figure 4–171   *Participant being added in the Netmeeting*

Figure 4–172   *Waiting message*

Figure 4–173   *Incoming message*

Requesting Control

After the participant selects the Accept button, Mark's computer shows John's screen display. Suppose Mark wants to take control: he selects Request Control from the Control menu. (See Figure 4–174.)

In John's computer, a request message displays. If John agrees, he selects the Accept button. (See Figure 4–175.)

Now Mark takes over the control and he continues the Inventor working session. In John's computer, Mark's work is displayed. After Mark finishes the work, he selects Release Control from the Control menu. John regains control over the working session.

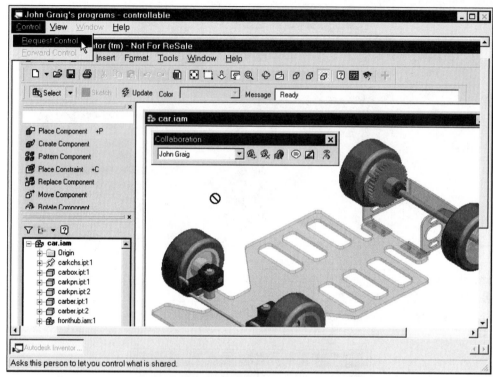

Figure 4–174 *John's Inventor screen displayed in Mark's computer*

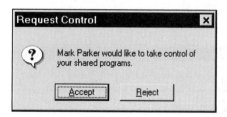

Figure 4–175 *Request control dialog box*

Chatting

Apart from sharing Inventor working session, you can chat with the other participants by selecting Chat in the Collaboration dialog box. Figure 4–176 shows the Chat dialog box.

Figure 4–176   *Chat dialog box*

Using a Whiteboard

To illustrate ideas, you can use the Whiteboard. (See Figure 4–177.)

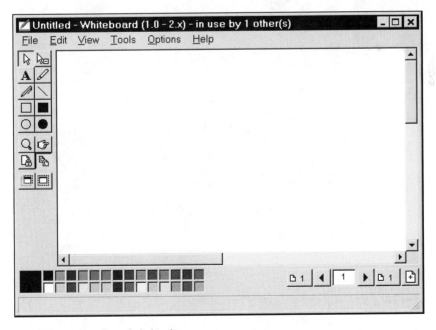

Figure 4–177   *Whiteboard dialog box*

Exercises

1. MODEL CAR STEERING BAR

In Chapter 3, you constructed an assembly of a scale model car. Now you will construct a steering bar by using a 2D layout drawing and adaptive technology. (See Figure 4–178.)

1. Create a new component (file name: carstr.ipt).

2. Construct a sketch in accordance with Figure 4–179. The two upper horizontal lines are collinear and the arcs are tangential to their adjacent lines.

3. Add two work points and a work axis in accordance with Figure 4–180.

4. Select the sketch, right-click, and select Adaptive.

5. Select the assembly in the browser bar to switch to assembly mode.

6. Select the new part in the browser bar, right-click, and select Adaptive. Now the new solid part and the sketch of the new solid part are adaptive.

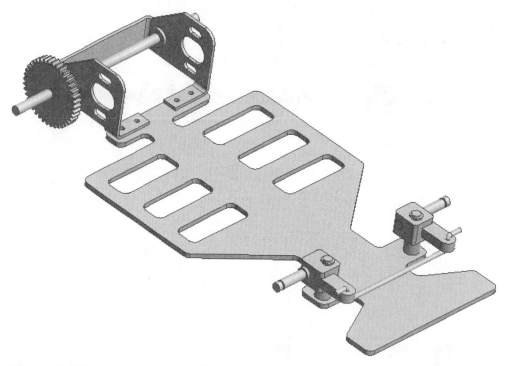

Figure 4–178 *Assembly with a steering bar added*

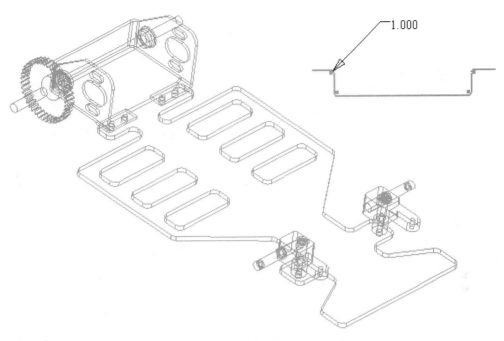

Figure 4–179   *New part created, and new sketch constructed*

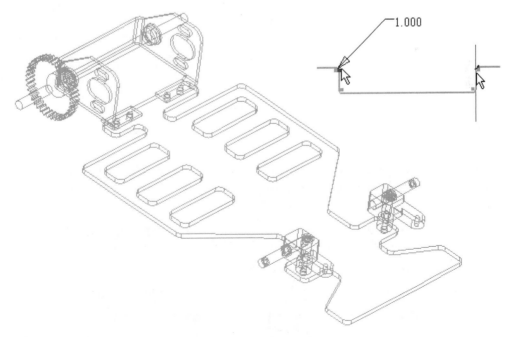

Figure 4–180   *Work points and work axis constructed*

7. Apply a mate constraint to the highlighted center point of the assembly and the highlighted work point of the steering bar in accordance with Figure 4–181.

8. Repeat the mate constraint on the highlighted center point and the highlighted work point in accordance with Figure 4–182.

9. Apply mate constraint on the axes highlighted in Figure 4–183.

10. The steering bar is properly assembled. (See Figure 4–184.) Select Save All from the File menu to save the assembly and the new part files.

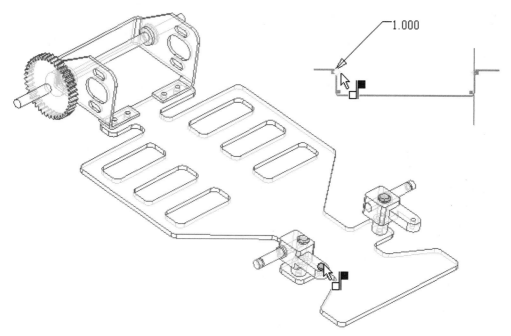

Figure 4–181 *Work point of the steering bar mated to the center point of the car*

In Chapter 3, you applied angle constraints to the front hub sub-assemblies. Because they are not movable and the steering bar is adaptive, the length of the steering bar adapts to the distance between the two sub-assemblies.

11. Now select the angle constraint of one of the front hub sub-assemblies in the browser bar and right-click. (See Figure 4–185.)

12. Select Delete. The angle constraint is deleted.

13. Now select the angle constraint of the other front hub sub-assembly.

14. Select Drive Constraint and set the start angle to –25 degrees and the end angle to 25 degrees. (See Figure 4–186.)

15. Select the Play button.

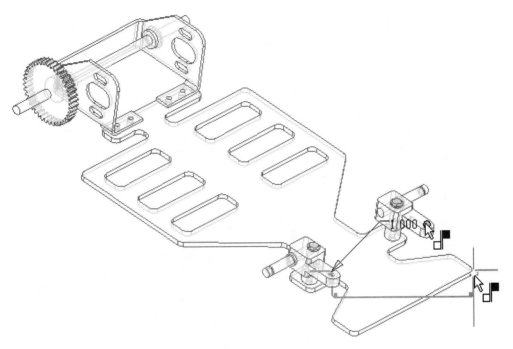

Figure 4–182 *Center point and work point being mated*

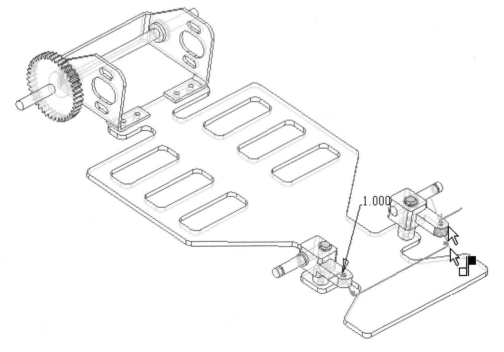

Figure 4–183 *Axes being mated*

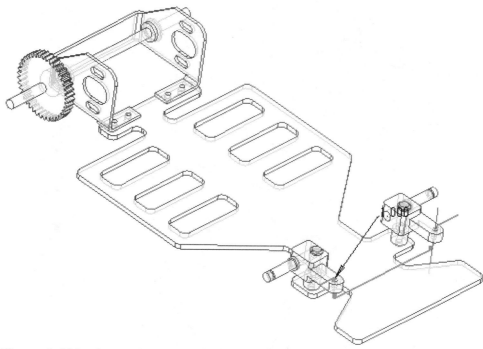

Figure 4–184 *Steering bar properly assembled*

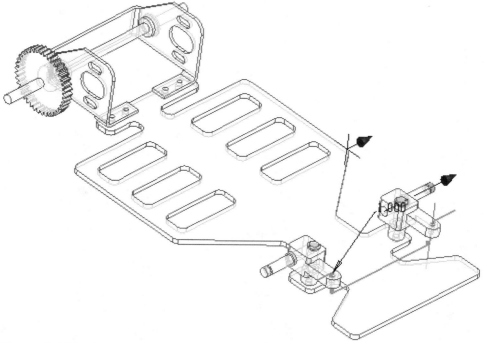

Figure 4–185 *Angle constraint selected and being deleted*

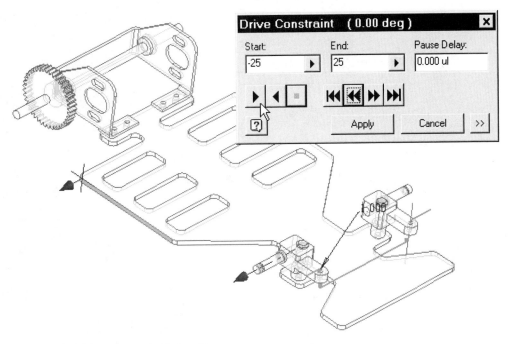

Figure 4–186 *Constraint being driven*

2. MODEL CAR ASSEMBLY

1. Open the file carstr.ipt.

2. Add dimensions in accordance with Figure 4–187.

3. Set the display to an isometric view.

4. Construct a work plane at the end point of the sketch.

5. Construct a sketch on the work plane and construct a circle. (See Figure 4–188.)

6. Construct a sweep solid.

7. Set the sketch plane, work points, work plane, and work axis to be invisible. (See Figure 4–189.) The steering bar is complete.

8. Now select the assembly file.

9. Update the assembly. (See Figure 4–190.)

10. Save and close all the files.

418

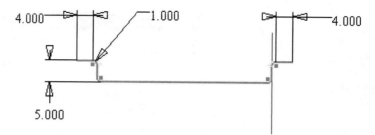

Figure 4–187   *Layout drawing of the steering bar*

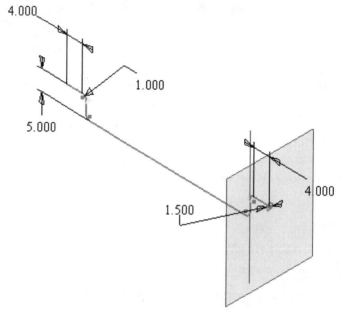

Figure 4–188   *Work plane, sketch plane, and circle constructed*

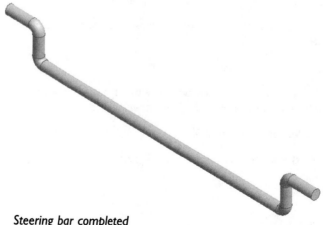

Figure 4–189   *Steering bar completed*

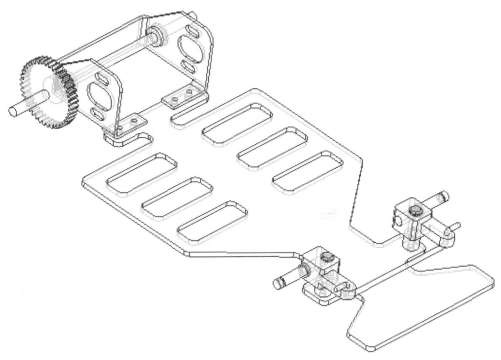

Figure 4–190 *Assembly updated*

3. MODEL CAR SPREADSHEET PARAMETERS

Figure 4–191 shows the front wheel and the rear wheel of the scale model car. Construct two part files and link the files to an Excel spreadsheet.

  1. Start a new part file.

  2. Construct a sketch in accordance with Figure 4–192.

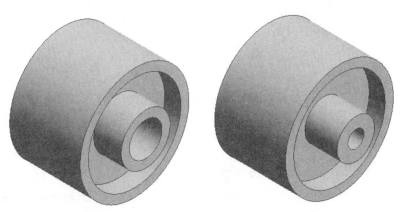

Figure 4–191 *Front and rear wheels*

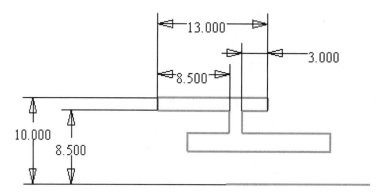

Figure 4–192 *Sketch*

3. Start a new spreadsheet and fill in the table in accordance with Figure 4–193.

4. Save and close the spreadsheet. (file name: car.xls)

5. Select Parameters from the Features toolbar and select the Link button in the Parameters dialog box.

6. Construct a link to an Excel Spreadsheet (file name: car.xls) (See Figure 4–194.)

7. Now add dimensions to complete the sketch by using the parameters defined in the spreadsheet. (See Figure 4–195.)

car.xls				
	A	B	C	D
1	holerad1	1.5		
2	bossrad1	3.5		
3	bosslgth1	14		
4	bossside1	6		
5	holerad2	3		
6	bossrad2	4.5		
7	bosslgth2	8		
8	bossside2	5		
9				
10				

Sheet1 / Sheet2 / Sh

Figure 4–193 *Excel spreadsheet*

8. Set the display to an isometric view.

9. Revolve the sketch, using the lower edge of the sketch as the centerline. (See Figure 4–196.)

10. Save the file (file name: carwhr.ipt).

11. Select Save As from the File menu to save to another file (file name: carwhf.ipt).

12. Now you have two files. Close carwhr.ipt and open carwhf.ipt.

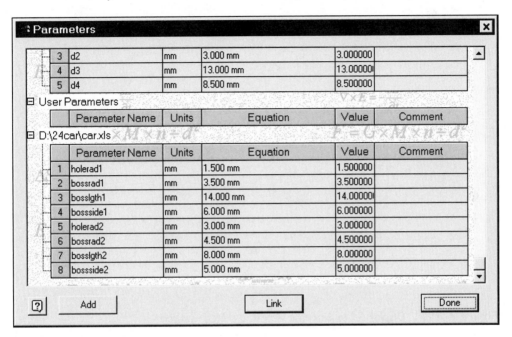

Figure 4–194   *Linked to a spreadsheet*

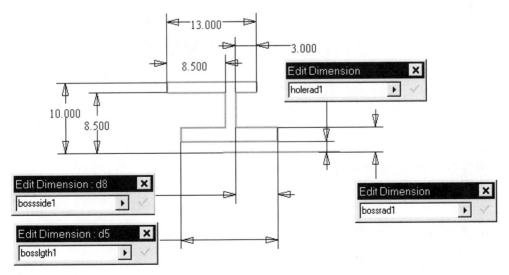

Figure 4–195   *Sketch completed*

13. Change the dimensions in accordance with Figure 4–197.

14. Update the part. (See Figure 4–198.)

15. Save and close the file.

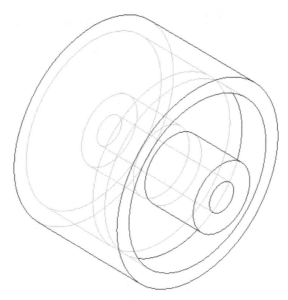

Figure 4–196   *Display set to isometric view, and sketch revolved*

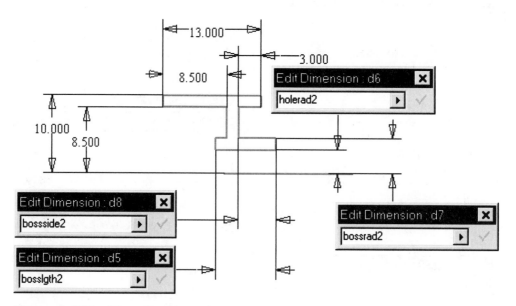

Figure 4–197   *Dimensions changed*

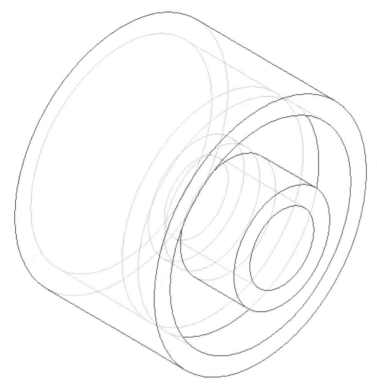

Figure 4–198 *Front wheel*

4. MODEL CAR TIRES AND COMPLETE ASSEMBLY

Figure 4–199 shows the complete assembly of the scale model car. You will construct a new part file (file name: cartir.ipt) for the tire of the model car. Then you will construct a sub-assembly of the rear wheel and the tire and add the front wheel, tire, and bearings to the front hub assembly. After that, you will complete the model car by placing the rear wheel assembly to the main assembly.

1. Start a new solid part file.

2. Construct a sketch in accordance with Figure 4–200.

3. Revolve the sketch about the lower horizontal line. (See Figure 4–201.)

4. Save and close the file (file name: cartir.ipt).

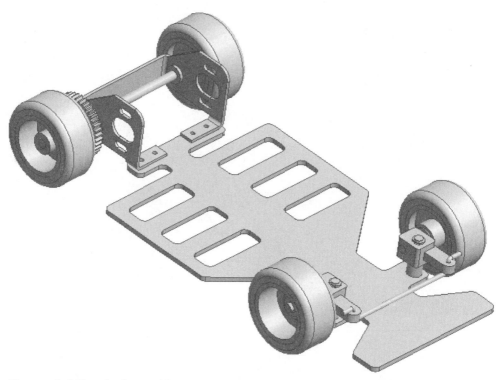

Figure 4–199   *Final assembly*

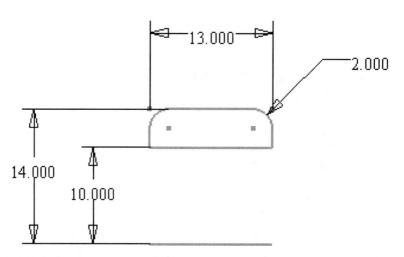

Figure 4–200   *Sketch constructed*

5. Start an assembly file.

6. Place the rear wheel and the tire. (See Figure 4–202.)

Figure 4–201   *Sketch revolved*

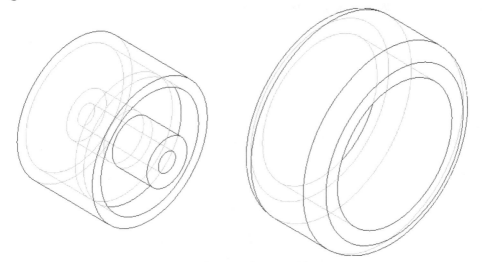

Figure 4–202   *Rear wheel and tire placed in the assembly*

7. Apply an insert constraint to the wheel and tire to assemble them together. (See Figure 4–203.)

8. Save and close the assembly file (file name: rear.iam).

9. Open the front hub assembly (file name: fronthub.iam).

10. Place a front wheel, a tire, a circlip, and two bearings. (See Figure 4–204.)

11. Assemble the components in accordance with Figure 4–205.

12. Save and close the file.

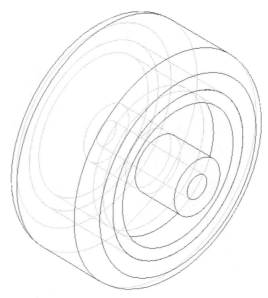

Figure 4–203   *Rear wheel assembly*

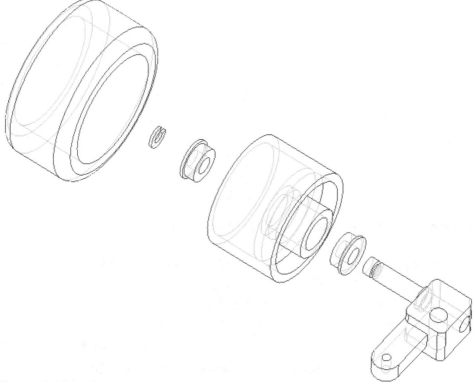

Figure 4–204   *Additional parts placed in the front hub assembly*

13. Open the main assembly file (file name: car.iam).

14. Place two copies of the rear wheel assembly (file name: rear.iam). (See Figure 4–206.)

15. Apply constraints to the rear wheel sub-assembly. (See Figure 4–207.)

16. The assembly is complete. Save and close your file.

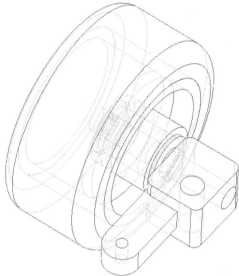

Figure 4–205   *Constraints applied*

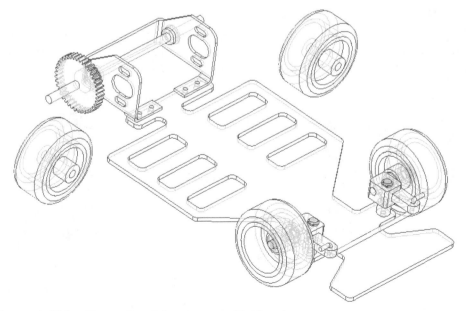

Figure 4–206   *Two copies of the rear assembly placed*

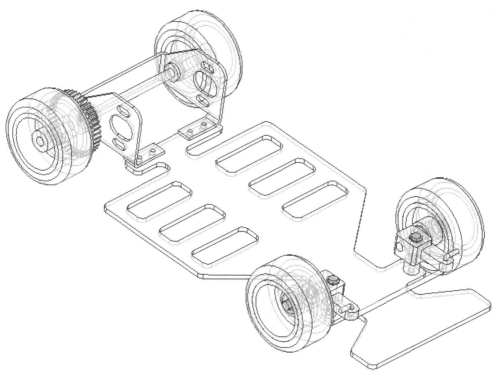

Figure 4–207 *Complete assembly*

5. GRIPPER

Design a gripper. Figure 4–208 shows two working positions of this gripper assembly, which has eight components: upper jaw, lower jaw, main body, cam unit, screw, push rod, hinge, and circlip. Figure 4–209 shows the exploded view.

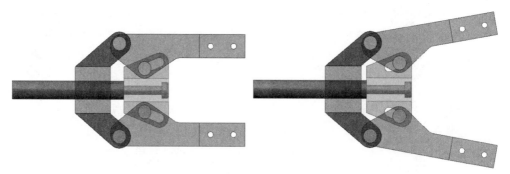

Figure 4–208 *Gripper assembly*

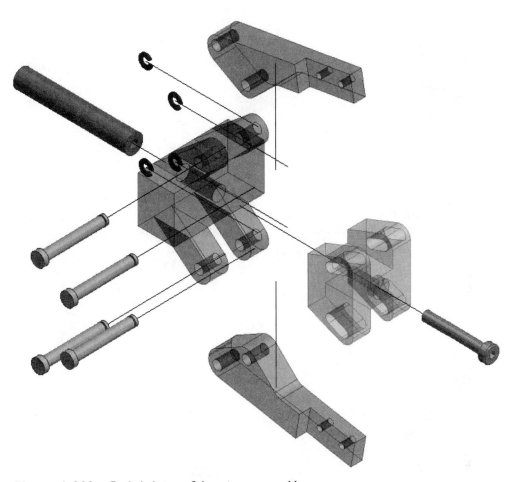

Figure 4–209 *Exploded view of the gripper assembly*

Referring to Figures 4–210 through 4–215 and the delineation below, design and construct the solid parts. Put them together in an assembly and construct a presentation file.

1. Figure 4–210 shows the upper jaw and the sketch for making the base solid feature. Construct the sketch and extrude the sketch a distance of 10 units.

2. Add two holes of 4 units diameter and two holes of M3 size. Do not fix the locations of the 4 units diameter holes and make its sketch adaptive.

3. After making the upper jaw, construct a derived solid from it to make the lower jaw.

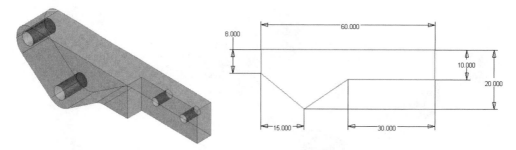

Figure 4–210 *Upper jaw*

4. Figure 4–211 shows the cam unit and the sketch for making the base solid feature. Construct the sketch and extrude it a distance of **26** units.

5. Design and construct the other details.

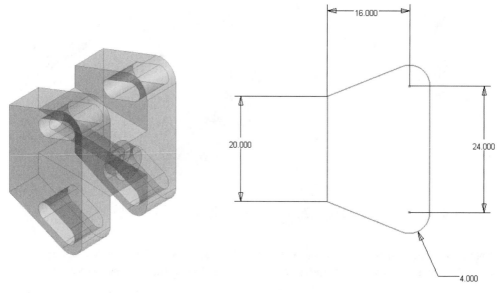

Figure 4–211 *Cam unit*

6. Figure 4–212 shows the main body and the sketch for its base solid feature. Extrude the sketch a distance of **40** units to make the base solid.

7. Design the other details, as shown in Figures 4–213 through 4–215.

Figure 4–213 shows the push rod. It is a cylindrical rod of 8 units diameter. At one end of the rod, there is a blind tapped hole of M4.

Figure 4–214 shows the screw.

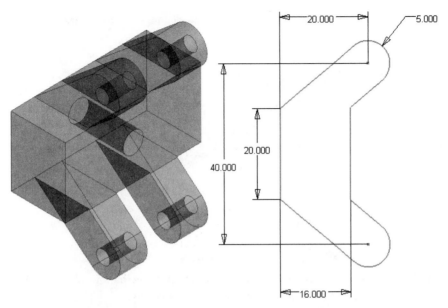

Figure 4–212 *Main body*

Figure 4–213 *Push rod*

Figure 4–214 *Screw*

Figure 4–215 shows the hinge and the circlip. The diameter of the hinge is 4 units.

Figure 4–215 *Hinge pin and circlip*

6. SKATE SCOOTER HANDLE ASSEMBLY

Figures 4–216 and 4–217 show the handle knob and the handle shaft of the skate scooter. Note that a diameter of the handle knob is adaptive.

    1. Construct the solid models of the components and assemble them together in accordance with Figure 4–218.

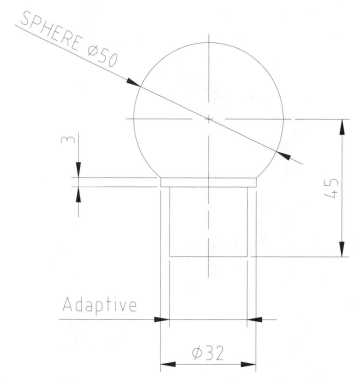

Figure 4–216 *Handle knob (file name: HandleKnob.ipt)*

Figure 4–217 *Handle shaft (file name: HandleShaft.ipt)*

Item	Description
1	Handle Shaft (File name: HandleShaft.ipt)
2	Handle Knob (File name: HandleKnob.ipt)

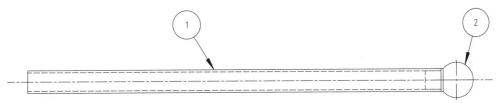

Figure 4–218 *Handle assembly (file name: Handle.iam)*

7. HINGE ASSEMBLY

Figures 4–219 through 4–221 shows the component parts of a hinge assembly (Figure 4–222). Construct the solid parts and the assembly in accordance with the dimensions given. Note that the upper hinge and the lower hinge are adaptive.

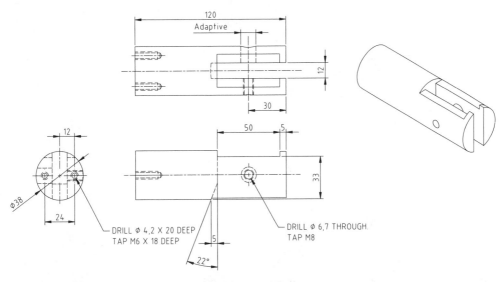

Figure 4–219 *Lower hinge (file name: HingeLower.ipt)*

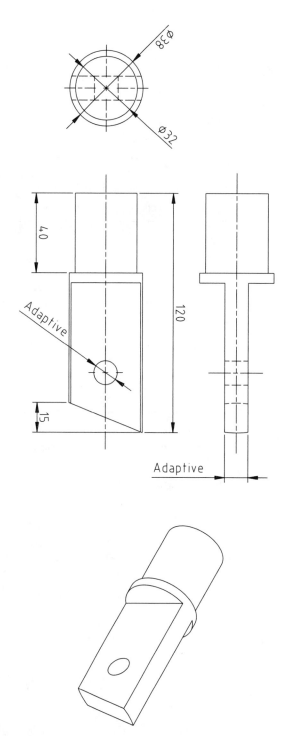

Figure 4–220   *Upper hinge (file name: HingeUpper.ipt)*

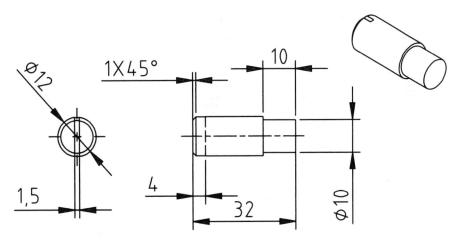

Figure 4–221   *Hinge pin (file name: HingePin.ipt)*

Item	Description
1	Lower Hinge (file name: HingeLower.ipt)
2	Upper Hinge (file name: HingeUpper.ipt)
3	Hinge Pin (file name: HingePin.ipt)

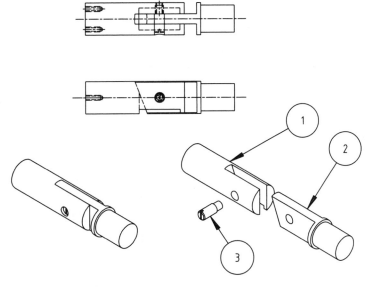

Figure 4–222   *Hinge assembly (file name: Hinge.iam)*

8. SKATE SCOOTER SHANK ASSEMBLY

The skate scooter has seven kinds of screws. See Figure 4–223 and Table 4–2:

Table 4–2 Skate scooter screws

	Screw A	Screw B	Screw C	Screw D	Screw E	Screw F	Screw G
Length	92	75	34	105	20	60	40
Shank Length	25	25	9	53	6	40	5
Shank Diameter	12	12	7	12	5	5	5
Screw Length	55	38	18	30	9	16	30
Screw Diameter	10	10	5	10	5	5	5
Hex Head Thickness	10	10	5	10	3.75	3.75	3.75
Across Corner	18	18	10.5	18	10	10	10

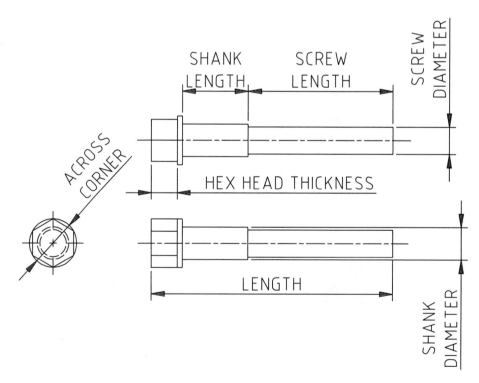

Figure 4–223 *Screw*

In addition to seven kinds of screws, the skate scooter has two kinds of nuts. See Figure 4–224 and Table 4–3.

Table 4–3   Skate scooter nuts

	M10 Nut	**M5 Nut**
Thread Size	M10	M5
Hex Head Thickness	10	5
Thickness	12	6
Across Corner	20	10

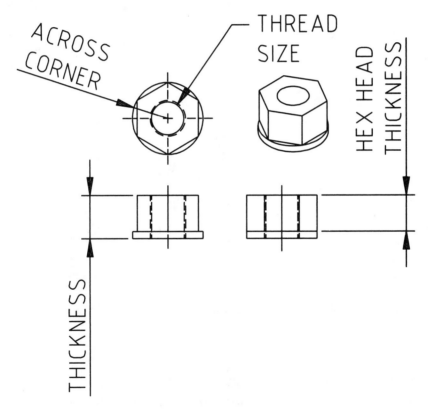

Figure 4–224   *Nut*

Open the shank assembly that you constructed in Chapter 3 and place the hinge assembly, a screw F, and a M5 nut. See Figures 4–225 and 4–226.

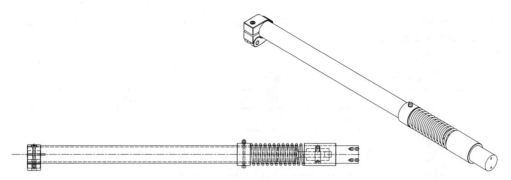

Figure 4–225   *Shank assembly (file name: Shank.iam)*

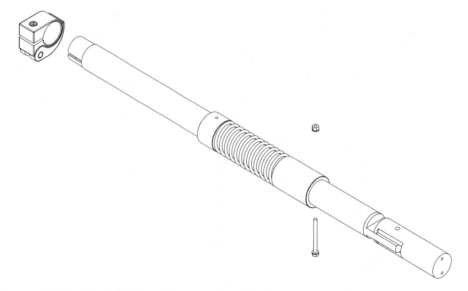

Figure 4–226   *Exploded view (file name: Shank.ipn)*

9.   SKATE SCOOTER FRONT END ASSEMBLY

1. Open the lower front end assembly that you constructed in Chapter 3.

2. Place two screw As, two screw Bs, two screw Cs, four M10 nuts, and two M5 nuts.

3. Complete the assembly in accordance with Figures 4–227 and 4–228.

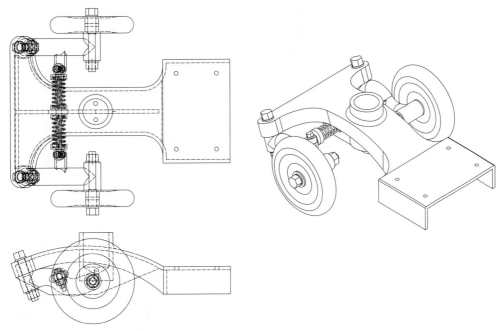

Figure 4–227   *Lower front end assembly*

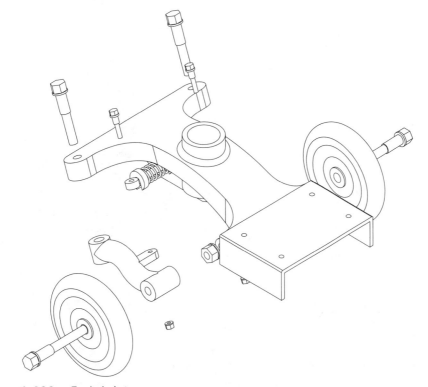

Figure 4–228   *Exploded view*

10. SKATE SCOOTER ASSEMBLY

1. Start a new assembly file.

2. Place the lower front end assembly, the shank assembly, the handle assembly, and two Screw Gs.

3. Assemble the components in accordance with Figures 4–229 and 4–230.

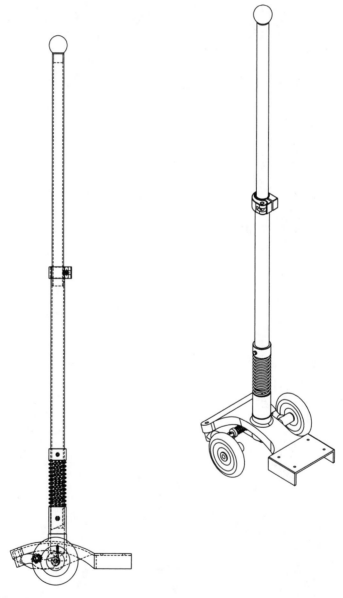

Figure 4–229 *Front end assembly (file name: FrontEnd.iam)*

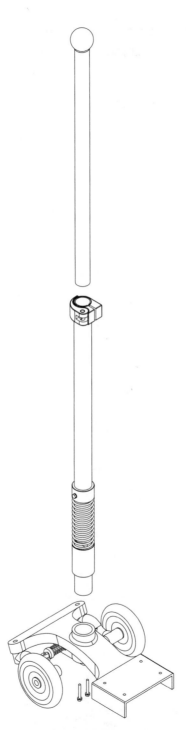

Figure 4–230 *Exploded view (file name: FrontEnd.ipn)*

11. SKATE SCOOTER REAR END ASSEMBLY

Open the rear end assembly file and place a D screw and a M10 nut. See Figures 4–231 and 4–232.

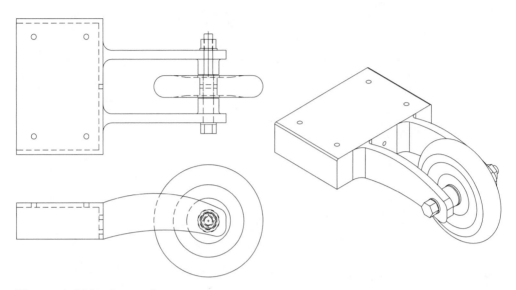

Figure 4–231 *Rear end*

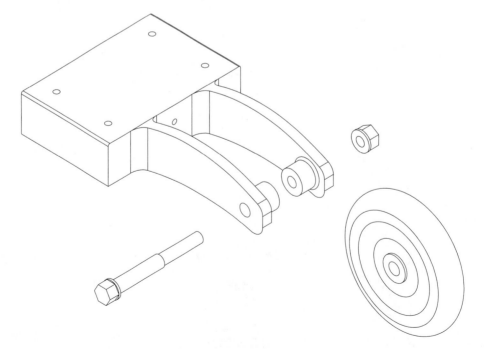

Figure 4–232 *Exploded view*

SUMMARY

You can use 2D layout drawings to design and validate a working mechanism. You construct simple 2D layout drawings of the linkages of a mechanism, construct work features on the linkages, assemble the linkages, and validate the design. After refining the design, you build 3D solid parts from the 2D layouts.

To ensure that corresponding 3D solid parts of an assembly match in size and dimension, you can use the adaptive technique by setting a selected solid feature of a solid part to be adaptive in the solid part and the solid part to be adaptive in the assembly. After you apply assembly constraints to adaptive solid parts, the adaptive solid feature will adapt to changes of the mating solid feature.

While constructing a 3D solid part, you can set the display to either shaded or wireframe mode. In wireframe mode, all the edges are displayed. In shaded mode, the interiors are hidden. When a sketch is constructed on a plane inside an existing solid part, you can set the display to shaded mode and slice the graphics along the sketch plane to clip away the portion of the solid part in front of the sketch plane. This way, you see the sketch plane more clearly.

There are four kinds of design parameters—model, user, linked, and embedded. Model parameters are parameters that are assigned by Inventor to a dimension of the solid part. User, linked, and embedded parameters are parameters that you add to the parameter table. User parameters are parameters that you add individually. Linked and embedded parameters are sets of parameters that you add collectively by using an Excel spreadsheet. You can use these parameters to control the model parameter. By linking a set of solid parts to a single Excel spreadsheet, you control the dimensions of the solid parts collectively.

To enhance design efficacy and efficiency, you can set up a design catalog by exporting selected solid features as design elements and then you can insert design elements in new solid parts. Design elements are solid features that you export from a solid part to the design catalog. An element that is joined to, cut from, or intersected with the solid part in the source solid part file will join, cut, or intersect the new solid part.

While designing, you record what you want to be retrieved later in a design notebook. The design notebook gives clues and information to you and other designers.

By setting a relative motion among the components in an assembly, you simulate complex mechanical motion.

To work collaborative among a group of designers, you can set up a network, establish appropriate search paths, and share files. You need to install NetMeeting properly in all the participants' computers, to enable sharing, and to activate NetMeeting from within the Inventor environment.

REVIEW QUESTIONS

1. What are the key features that you need to construct on a 2D layout drawing for them to be assembled? Use an example to illustrate your answer.

2. Briefly describe the meaning of adaptive technology. State the advantages of using adaptive technology in design. Use an example to illustrate the use of adaptive technology in design.

3. Compare wireframe display with shaded display. How can a shaded display be sliced along the current sketch plane?

4. Explain the four kinds of design parameters in a solid part file. State the difference between linking and embedding a parameter spreadsheet.

5. State the three kinds of design element and depict how a design catalog can be constructed.

6. What object can you include in your file to store textual design information?

7. How can you set up your computers to work collaboratively among a team of designers? Outline the steps for enabling a file to be shared in a NetMeeting.

CHAPTER 5

Sheet Metal Modeling

OBJECTIVES

This chapter explains the key concepts of sheet metal modeling, the ways to construct a sheet metal part, and how to convert a 3D solid part to a 3D sheet metal component. After studying this chapter, you should be able to

- Describe the key concepts of sheet metal modeling
- Construct 3D sheet metal parts
- Develop a 2D flat pattern from a 3D sheet metal part
- Convert 3D solid parts to 3D sheet metal parts

OVERVIEW

A sheet metal component is a special kind of 3D object. You make a 3D object from a piece of 2D sheet metal of uniform thickness by cutting out a flat pattern and folding it into the final shape of the component. When you design a sheet metal component, you think about how you will manufacture the 3D component from a flat sheet, round off the joints between the faces of a sheet metal part, and provide recesses at the joints of the faces or the bend corners. Furthermore, you must think about how you will unfold the 3D component into a 2D flat pattern for manufacturing purposes. To cope with these requirements, you need a special set of modeling tools.

In this chapter, you will learn how to use Autodesk Inventor to construct sheet metal components and construct flat patterns of the sheet metal parts. Figures 5–1 shows a sheet metal part and Figure 5–2 shows the flat pattern of the part.

445

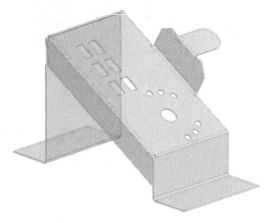

Figure 5–1 *Sheet metal component*

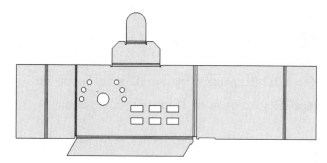

Figure 5–2 *Flat pattern of the sheet metal component*

SHEET METAL CONCEPTS

To manufacture a 3D object, you use material cutting or material forming processes. Material cutting processes involve the removal of material: You start with a piece of material that is larger than the overall size of the final shape of the 3D object. Then you use various machining processes to remove unwanted material from it. Typically, you use turning, milling, etc. Material forming processes involve the deformation of material to change its shape to the final appearance of the product. Sheet metal work is a kind of material forming process. You cut out a 2D flat pattern from a large sheet of material that has a uniform thickness. Then you bend the 2D flat pattern into a 3D complex object.

To design a 3D sheet metal component, you should, in addition to thinking about the functional requirement of the component, think about the bend radii of the faces and relieves at the bend of the faces. Most importantly, you should think about how to unfold the 3D component to make a 2D flat pattern.

SKETCHED SOLID FEATURES AND PLACED SOLID FEATURES

Basically, constructing a 3D sheet metal component is similar to constructing the 3D solid part that you learned about in earlier chapters. You start your design by making a sketch, which you extrude to form a face of the sheet metal. Then you elaborate your design by adding more faces to it or by cutting holes on it.

Like other 3D solid parts, a sheet metal part has three kinds of features: sketched solid features, placed solid features, and work features. Because of the unique characteristics of the sheet metal part, the sheet metal modeling tool set is slightly different from the set of tools that you learned about earlier. Figure 5–3 shows the sheet metal toolbar.

Figure 5–3 *Sheet metal toolbar*

The sheet metal toolbar and panel have 17 buttons, the first seven of which are sheet metal tools. Table 5–1 describes the choices.

Table 5–1   Sheet metal toolbar and panel options

Option	Description	Tool
Settings	Enables you to set sheet metal thickness and bend parameters.	Sheet metal tool
Flat Pattern	Enables you to flatten a 3D sheet metal component into a 2D flat pattern.	Sheet metal tool
Face	Enables you to extrude a sketch to form a sheet metal face (sketched feature).	Sheet metal tool
Cut	Enables you to extrude a sketch to cut an opening in a sheet metal face (sketched feature).	Sheet metal tool
Flange	Enables you to construct a sheet metal flange at the edge of a sheet metal face (placed feature).	Sheet metal tool
Corner Seam	Enables you to treat the construct between two disjoint sheet metal faces (placed feature).	Sheet metal tool
Bend	Enables you to construct a sheet metal bend at the intersection of two non-parallel sheet metal faces (placed feature).	Sheet metal tool
Hole	Enables you to construct a hole on a sheet metal face (placed feature).	General tool

Option	Description	Tool
Corner Round	Enables you to place a fillet at the corner of a sheet metal face (placed feature).	General tool
Corner Chamfer	Enables you to place a chamfer at the corner of a sheet metal face (placed feature).	General tool
View Catalog/ Create Design Element/ Insert Design Element	View Catalog displays the design element catalog. Create Design Element enables you to create a design element. Insert Design Element enables you to insert a design element.	General tools
Work Plane	Enables you to construct a work plane (work feature).	General tool
Work Axis	Enables you to construct a work axis (work feature).	General tool
Work Points	Enables you to construct a work point (work feature).	General tool
Rectangular Pattern	Enables you to place a rectangular pattern of features (placed feature).	General tool
Circular Pattern	Enables you to place a circular pattern of features (placed feature).	General tool
Mirror Feature	Enables you to place a mirror feature (placed feature).	General tool

Face

You start a sheet metal part by making a sketch. Then you determine the thickness of the sheet metal and other settings. After that, you make a face of a sheet metal component from the sketch. (See Figure 5–4.) Making a face from a sketch is similar to constructing an extruded solid. The extrusion height is determined by the thickness of the sheet metal.

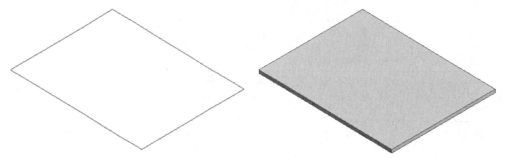

Figure 5–4 *Face of a sheet metal part*

The first face of a sheet metal part is the base part. To add a face to the sheet metal component, you set up a sketch plane and construct a sketch. When a face is made from the sketch, bends and notches are created automatically. (See Figure 5–5.)

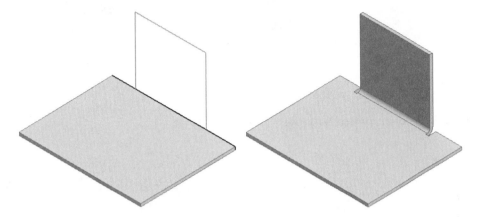

Figure 5–5 *Rounded bends and notches added automatically at the joint of new faces*

Flange

To construct a face along the entire edge of a face, you do not have to make a sketch; you specify a flange. (See Figure 5–6.)

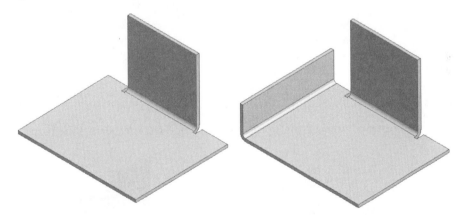

Figure 5–6 *Rounded bends added automatically at the joint of the flange*

Cutting

In contrast to adding faces and flanges, you remove unwanted portions of a sheet metal by cutting. You construct a sketch to depict the shape of the cut. (See Figure 5–7.)

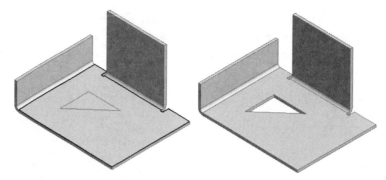

Figure 5–7 *Cutting made on a face*

Hole

If you want to make a circular cutting on a face, you place a hole feature. (See Figure 5–8.)

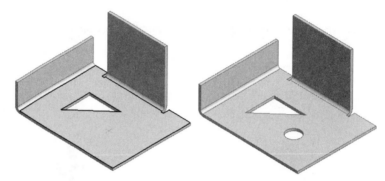

Figure 5–8 Hole feature placed

Corner Seam

To refine the corner of a sheet metal component, you add a corner seam. (See Figure 5–9.)

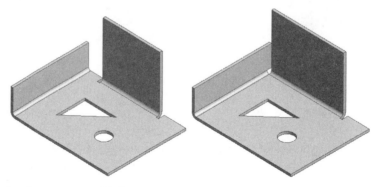

Figure 5–9 *Corner seam added*

Bend

To lengthen two non-parallel faces of a sheet metal part and construct a round edge at the intersection, you add a bend. (See Figure 5–10.)

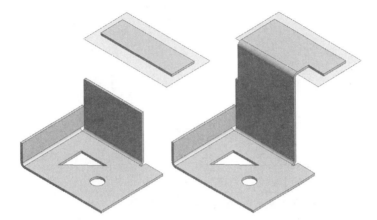

Figure 5–10 *Bend added*

Corner Round and Corner Chamfer

To modify the sharp corners of a face, you add corner rounds and corner chamfers. (See Figure 5–11.)

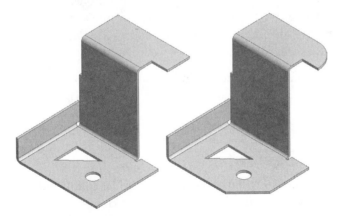

Figure 5–11 *Corner round and corner chamfer*

Flat Pattern

Perhaps the most important aspect regarding the design of a sheet metal component is to make a flat pattern (development) of the component so that it can be fabricated. (See Figure 5–12.)

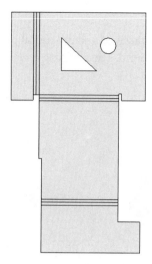

Figure 5–12 *Flat pattern of a sheet metal component*

To make a sheet metal component, you need special tools as well as common solid modeling tools. The following tools are also applicable to sheet metal design: design elements, work features, rectangular pattern, circular pattern, and mirror features.

CONVERSION FROM A SOLID PART

There are two ways to design a 3D sheet metal component. Apart from using the sheet metal modeling tool to construct the faces of the sheet metal component one by one and unfolding the 3D component to obtain a 2D flat pattern, you can convert a solid model to a sheet metal object.

To convert a 3D solid part, select Sheet Metal from the Applications menu. Then select a face of the 3D solid part and select Flat Pattern from the Sheet Metal toolbar or menu.

SHEET METAL DESIGN

Now you will construct a sheet metal component. The completed component and its flat pattern are shown in Figure 5–1 and 5–2. In making this sheet metal component, you will learn how to use various sheet metal modeling tools.

SHEET METAL PART FILE

1. Start a new file by selecting New from the File menu.

2. In the New dialog box, select the Default tab and choose Sheet Metal.ipt as the template. (See Figure 5–12.)

3. Select the OK button.

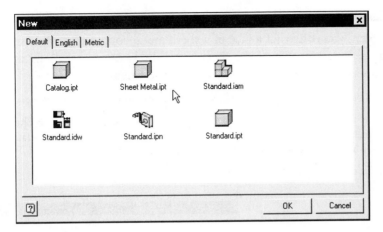

Figure 5–13 *Sheet Metal.ipt template being selected*

Sketch

Construction of a sheet metal part begins with making a face, which is a sketched feature. You construct a sketch and then construct a sheet metal face from it.

4. Select Two point rectangle from the Sketch toolbar or panel to construct a rectangle as shown in Figure 5–14.

5. Select General Dimension from the Sketch toolbar or panel to add two dimensions.

6. Exit sketch mode.

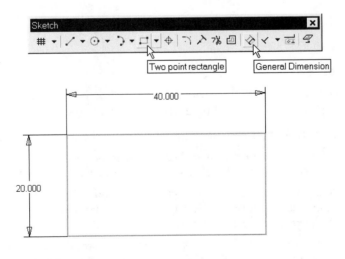

Figure 5–14 *Sketch constructed*

Set Thickness

With a sketch, you construct a face of a sheet metal component. Before you make a face from a sketch, you specify the sheet metal thickness. Sheet metal thickness and other parameters are controlled by sheet metal settings.

7. Select Settings from the Sheet Metal toolbar or panel. (See Figure 5–15.)

8. In the Sheet Metal Settings dialog box, there are three tabs: Sheet, Bend, and Corner. In the Sheet tab, specify Steel, Mild as the material, millimeter as the measurement unit, and 1unit thickness.

9. Select the OK button.

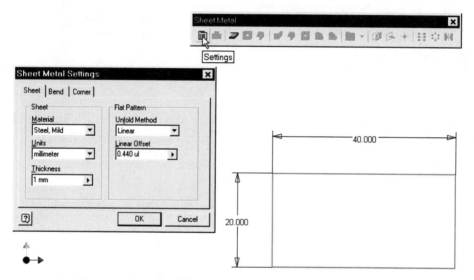

Figure 5–15 *Sheet tab of the Sheet Metal Settings dialog box*

Construct the Face

A sheet metal face is a sketched feature; you extrude a sketch to a sheet metal face. After specifying the sheet metal parameters, you will construct a sheet metal face.

10. Right-click and select Isometric View to set the display to an isometric view.

11. Select Face from the Sheet Metal toolbar or panel to extrude the sketch to become a face of the sheet metal component. (See Figure 5–16.)

12. A sheet metal face is constructed. Now you will construct another face. Select Sketch from the Command Bar toolbar and select the face indicated in Figure 5–17 to set up a new sketch plane.

13. Select Two point rectangle from the Sketch toolbar or panel to construct a sketch and select General Dimension to add dimensions to the sketch.

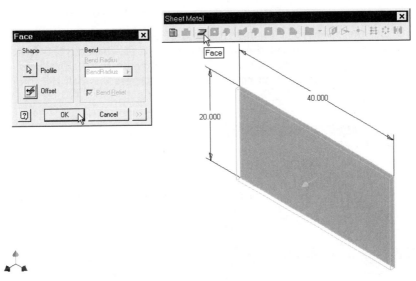

Figure 5–16 *Sketch being extruded to become a face*

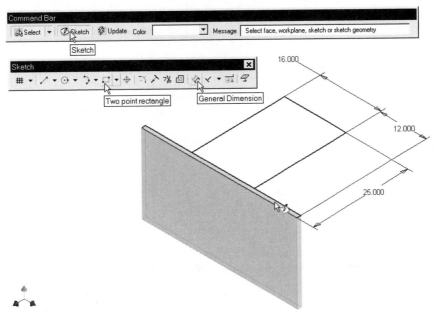

Figure 5–17 *Sketch constructed*

Set Bend Parameters

14. As we explained earlier, a bend will be placed at the joint automatically. To determine the parameter of the bend, select Settings from the Sheet Metal toolbar or panel. (See Figure 5–18.)

15. In the Sheet Metal Settings dialog box, select the Bend tab.

16. Set the bend radius to the Thickness of the metal, use Default Straight bend relief shape, set the minimum remnant to Thickness*2.0 (twice the thickness of the metal), set the bend relief width to the Thickness, and accept the default bend relief depth of Thickness*0.5 (half the thickness).

17. Select the OK button.

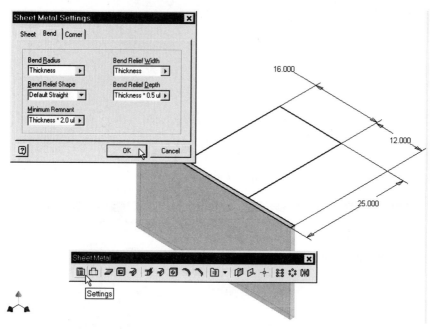

Figure 5–18 *Bend settings*

Construct Second Face

Now select Face from the Sheet Metal toolbar or panel. In the Face dialog box, select the Profile button and select the area highlighted in Figure 5–19.

18. Examine the direction of extrusion. If the direction differs from that shown in Figure 5–19, select the Offset button to flip the direction.

19. Select the OK button. The sketch is extruded into a face and joined to the sheet metal component. At their joint, you will find a bend and two relief notches. (See Figure 5–20.)

CORNER ROUND AND CORNER CHAMFER

In a solid part, you place fillets and chamfers. Here in a sheet metal, you place corner rounds and chamfers at the corners of the sheet metal faces. Corner Round and Corner Chamfer are placed features.

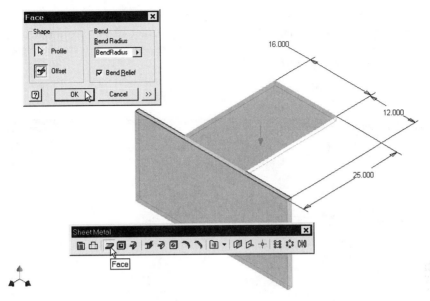

Figure 5–19 *A face being constructed*

Place Corner Round

20. Select Corner Round from the Sheet Metal toolbar or panel, set the radius to 8 units, and select two edges to round off the corners of the sheet metal component.

21. Select the OK button. (See Figure 5–20.)

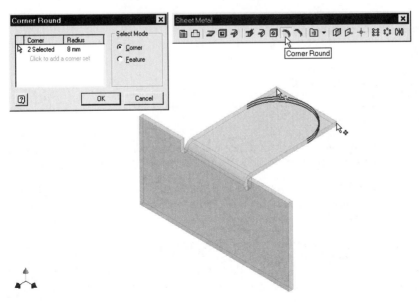

Figure 5–20 *Bend and bend relief placed, and corners being rounded*

Place Corner Chamfer

22. To bevel the corners of a face, select Corner Chamfer from the Sheet
Metal toolbar or panel, set the chamfer distance to 6 units, and select two
edges indicated in Figure 5–21.

23. Select the OK button.

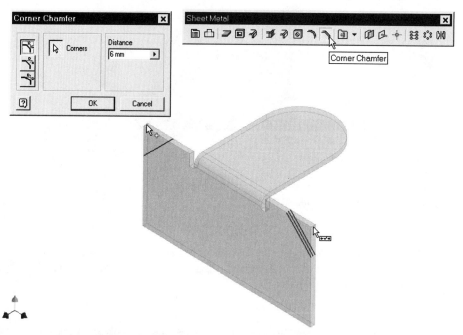

Figure 5–21 *Corners being chamfered*

SHEET METAL FLANGES

A sheet metal flange is a placed feature. You select an edge and specify the length
and angle of the flange.

24. Now you will construct a face and then two flanges. Select Sketch from the
Command Bar toolbar and select the lower face indicated in Figure 5–22.

25. Select Two point rectangle from the Sketch toolbar or panel to construct
a rectangle and select General Dimension from the Sketch toolbar or
panel to add four dimensions.

26. Exit sketch mode, and then using the sketch as the profile, construct a face
by selecting Face from the Sheet Metal toolbar or panel. (See Figure 5–23.)

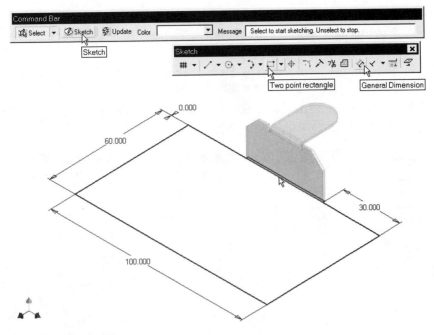

Figure 5–22 *New sketch constructed*

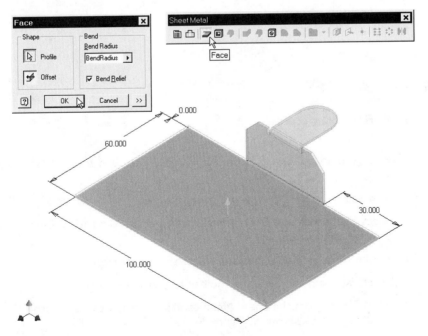

Figure 5–23 *Sketch being extruded to form a face*

Place Flanges

27. Now select Zoom Window from the Standard toolbar and zoom the display as shown in Figure 5–24.

28. Select Flange from the Sheet Metal toolbar or panel and select an edge indicated in the figure to construct a flange.

29. In the Flange dialog box, set the angle to 120 degrees and select the OK button.

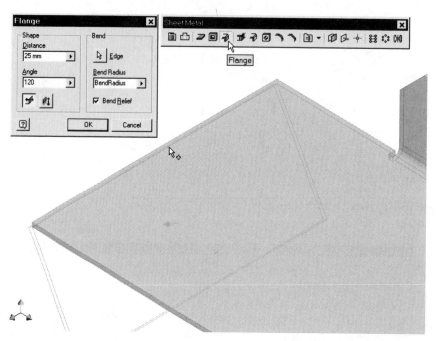

Figure 5–24 *Flange being added*

30. A flange is added. Select Zoom All from the Standard toolbar. (See Figure 5–25.)

31. Now select Zoom Window and zoom the display as shown in Figure 5–26.

32. Select Flange from the Sheet Metal toolbar or panel to add a flange at the edge indicated in the figure.

33. Set the angle to 60 degrees and select the OK button.

34. Select Pan from the Standard toolbar to pan the display.

35. Select Flange from the Sheet Metal toolbar or panel to add another flange at 90 degrees as shown in Figure 5–27.

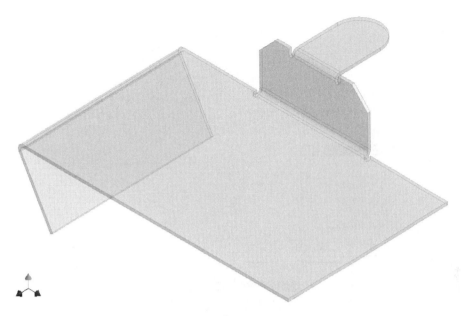

Figure 5–25 *Display zoomed to display all*

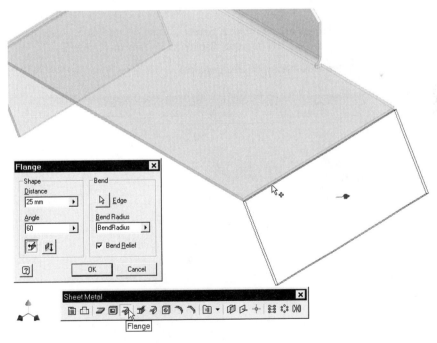

Figure 5–26 *A flange being added*

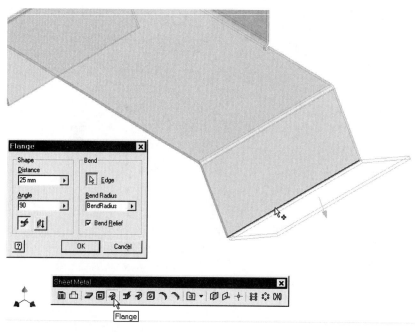

Figure 5–27 *Flange being added*

36. Now three flanges are complete. Select Zoom All and then Rotate from the Standard toolbar to set the display as shown in Figure 5–28.

37. Save your file (file name: Sheetmetal.ipt).

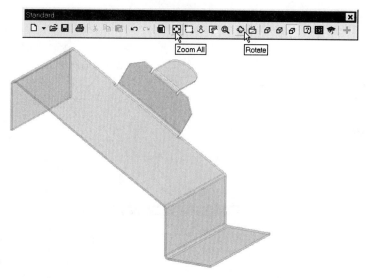

Figure 5–28 *Display set*

SHEET METAL BEND

A sheet metal bend is a placed feature. You lengthen two non-parallel faces of a sheet metal component in a direction so that they meet each other and construct a bend at the intersection. Working with the same file, Sheetmetal.ipt, you will construct a face and then form a bend between two faces.

1. Select Sketch from the Command Bar toolbar and select the face indicated in Figure 5–29 to set up a new sketch plane.

2. Select Look At from the Standard toolbar and select the plane to set the display to the top view of the new sketch plane.

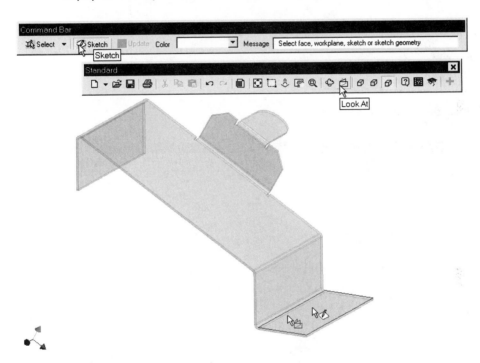

Figure 5–29 *Sketch plane constructed*

3. Select Two point rectangle and then General Dimension from the Standard toolbar to construct a rectangle and add dimensions as shown in Figure 5–30.

4. Right-click and select Previous View to set the display to the previous view.

5. Select Face from the Sheet Metal toolbar or panel to construct a sheet metal face. (See Figure 5–31.)

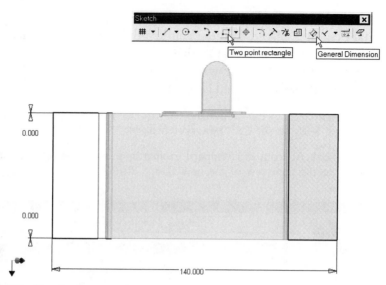

Figure 5–30 *Sketch constructed*

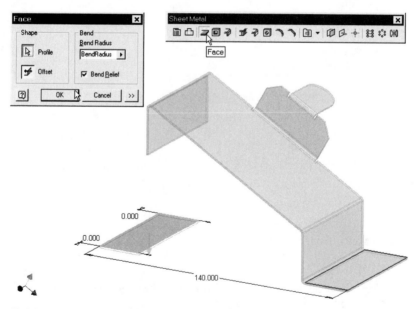

Figure 5–31 *Display set and face being constructed*

Bend

6. To construct the bend between the new sheet metal face and the main body of the sheet metal component, select Bend from the Sheet Metal toolbar or panel and select the edges indicated in Figure 5–32, and then select the OK button.

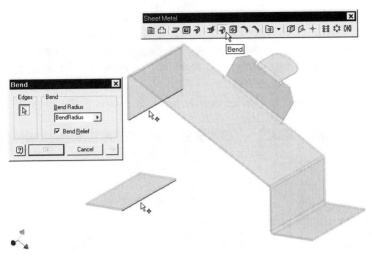

Figure 5–32 *Edges selected*

SHEET METAL CORNER SEAMS

A corner seam is a placed feature. You treat the seam between two faces by removing or adding materials to the faces.

7. Now you will make a flange and then construct a corner seam at the joint between two faces. Zoom the display and select Flange from the Sheet Metal toolbar or panel to add another flange as shown in Figure 5–33.

8. In the Flange dialog box, set the distance to 15 mm and the angle to 90 degrees.

9. Select the OK button.

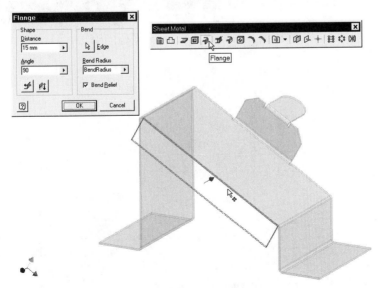

Figure 5–33 *Flange being added*

Construct Corner Seams

10. To refine the corners of the flanges, select Corner Seam from the Sheet Metal toolbar or panel and select the edges indicated in Figure 5–34. In the Corner Seam dialog box, there are three kinds of seams; select the Reverse Overlap button.

11. Now construct the other corner seam as shown in Figure 5–35.

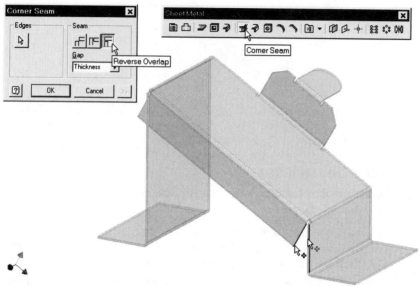

Figure 5–34 *Corner seam being constructed*

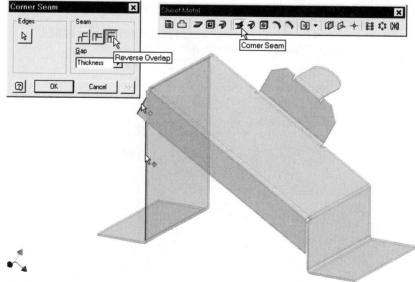

Figure 5–35 *Second corner seam being constructed*

SHEET METAL CUT

A sheet metal cut is a sketched feature. You construct a sketch to depict the shape of the cut.

Sketch

12. Set the display to an isometric view.

13. Select Sketch from the Standard toolbar and select the face highlighted in Figure 5–36 to set up a sketch plane.

14. Select Two point rectangle and then General Dimension from the Sketch toolbar or panel to construct a rectangle and add four dimensions.

Cut

15. Select Cut from the Sheet Metal toolbar or panel and select the area highlighted in Figure 5–37.

16. Cut the rectangle a distance of the sheet metal thickness. (See Figure 5–37.)

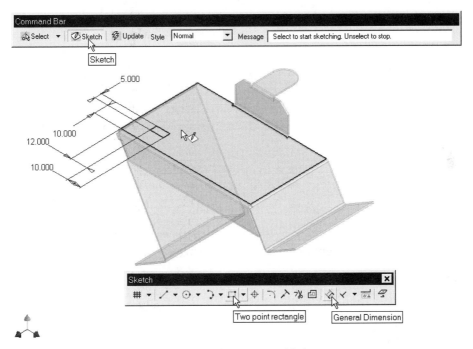

Figure 5–36 *Rectangle constructed and dimensions added*

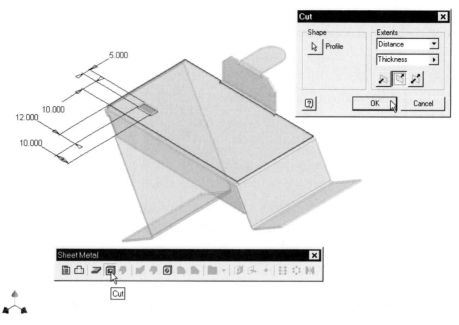

Figure 5–37 *Rectangle cut on the sheet metal component*

RECTANGULAR PATTERN

Rectangular pattern, hole, circular pattern, and mirror features are placed features. Making them on a sheet metal part is the same as making similar features on a 3D solid part.

17. Select Rectangular Pattern from the Sheet Metal toolbar or panel to place a rectangular pattern of the cut on the sheet metal.

18. Select the edges and specify the directions as indicated.

19. Specify the parameters of the pattern as shown in Figure 5–38, and then select the OK button.

HOLE

Now you will place two holes in the sheet metal part.

20. Select Sketch from the Command Bar toolbar and select the face indicated in Figure 5–39 to set up a new sketch plane.

21. Select Point, Hole Center from the Sketch toolbar or panel to specify a hole center.

22. Select General Dimension from the Sketch toolbar or panel to place two dimensions. (See Figure 5–39.)

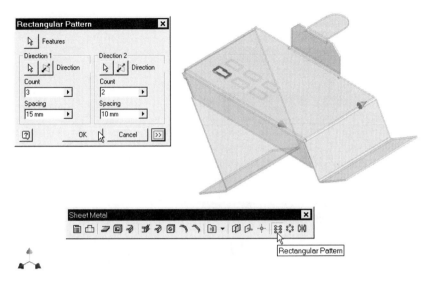

Figure 5–38 *Rectangular pattern being constructed*

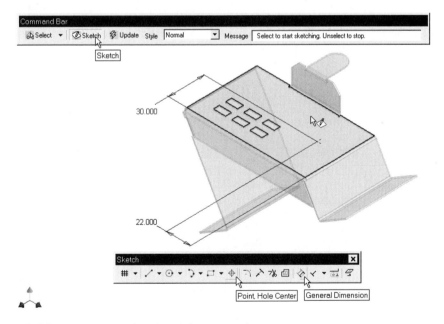

Figure 5–39 *Hole center placed and dimensioned*

23. Select Hole from the Sheet Metal toolbar or panel. In the Holes dialog box, specify Through All and set the diameter to 10 units. (See Figure 5–40.)

24. Select Sketch from the Command Bar toolbar and select the same face again to set up another sketch plane.

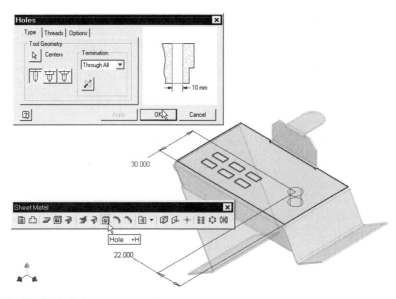

Figure 5–40 *Hole being constructed*

25. Select Point, Hole Center from the Sketch toolbar or panel to place a hole center.

26. Select General Dimension from the Sketch toolbar or panel to add two dimensions as shown in Figure 5–41.

27. Select Hole from the Sheet Metal toolbar or panel to place a through all hole of 6 units diameter. (See Figure 5–42.)

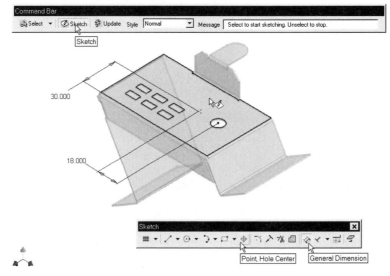

Figure 5–41 *Hole center placed*

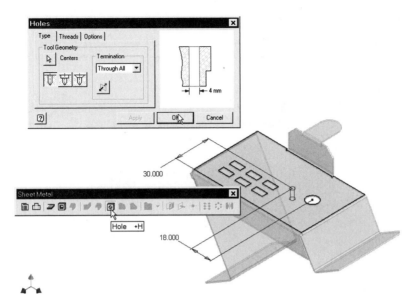

Figure 5–42 *Hole placed*

CIRCULAR PATTERN

Now you will place a circular pattern on the sheet metal part.

28. Select Circular Pattern from the Sheet Metal toolbar or panel.

29. Select the small hole as the feature to form a pattern and select the large hole to use its axis as the rotation axis.

30. Set the number of placement count to 3 and the incremental angular distance to 25 degrees. (See Figure 5–43.)

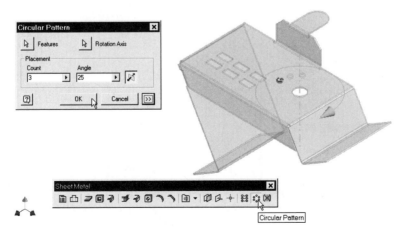

Figure 5–43 *Circular pattern being placed*

MIRROR

Now you will place a mirror pattern of features. Because a mirror pattern requires a mirror plane, you will construct a work axis and a work plane.

Construct Work Axis

31. Select Work Axis from the Sheet Metal toolbar or panel and select the hole indicated in Figure 5–44 to construct a work axis.

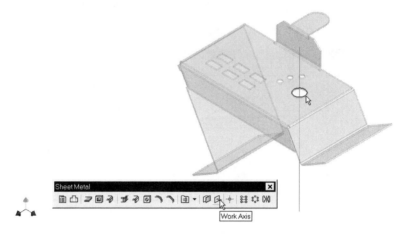

Figure 5–44 *Work axis constructed*

Construct Work Plane

32. Select Work Plane from the Sheet Metal toolbar or panel and select the face and work axis indicated in Figure 5–45 to construct a work plane. Set the angle of the work plane to 90 degrees.

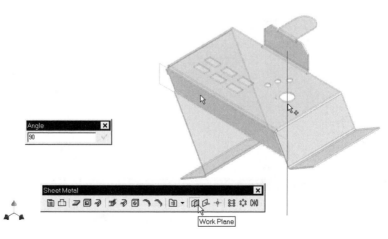

Figure 5–45 *Work plane being constructed*

Mirror

33. Now select Mirror from the Sheet Metal toolbar or panel.

34. Select the features highlighted in Figure 5–46 as the mirror feature and select the work plane as the mirror plane.

35. Select the OK button.

36. Save your file.

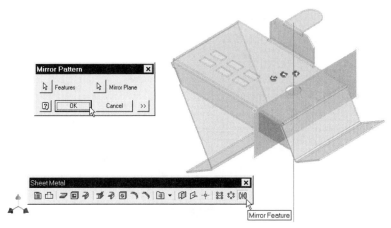

Figure 5–46 *Mirror feature being constructed*

SHEET METAL DESIGN ELEMENT

In Chapter 4, you learned how to construct a design catalog and incorporate design elements in your new design. Here in a sheet metal model, you can use a design element as well. Figure 5–47 shows the Design Element buttons on the Sheet Metal toolbar.

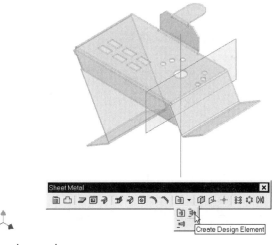

Figure 5–47 *Design element buttons*

SHEET METAL FLAT PATTERN

A 2D flat pattern is a development of the 3D sheet metal component.

1. The main body of the sheet metal part is complete. Hide the work axis and the work plane.

2. Select Flat Pattern from the Sheet Metal toolbar or panel to construct a flat pattern of the component in a separate window. (See Figure 5–48.)

3. To return to the sheet metal part, select Flat Pattern in the browser bar, right-click, and select Close Window. (See Figure 5–49.)

4. Save and close your file (file name: Sheetmetal.ipt).

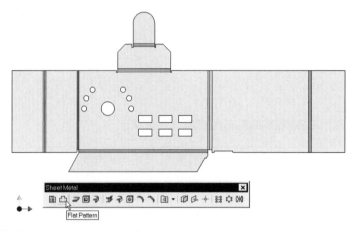

Figure 5–48 *Flat pattern constructed*

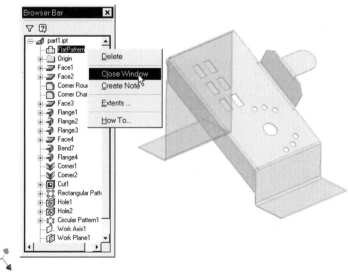

Figure 5–49 *Completed sheet metal component*

CONVERSION

Apart from using the set of sheet metal tools to construct a 3D sheet metal component, you can construct a 3D solid part and convert it to a sheet metal part.

PART FILE

1. Start a new part file.

2. Select Line from the Sketch toolbar or panel to construct four line segments (among the lines, two lines are horizontal and one line is vertical) and select General Dimension from the Sketch toolbar or panel to construct three dimensions as shown in Figure 5–50.

3. Exit sketch mode.

4. Select Revolve from the Features toolbar or panel to revolve the sketch 90 degrees as shown in Figure 5–51.

5. Rotate the view as shown in Figure 5–52.

6. Select Shell from the Features toolbar or panel to place a shell feature and select the four faces indicated in Figure 5–52.

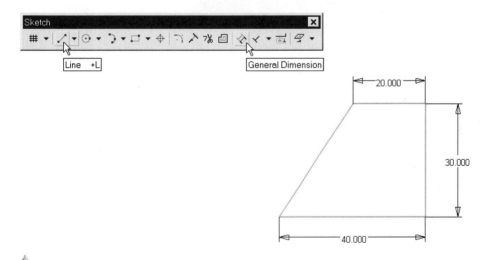

Figure 5–50 *Sketch constructed*

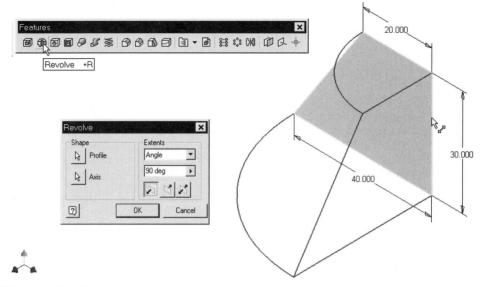

Figure 5–51 *Sketch being revolved*

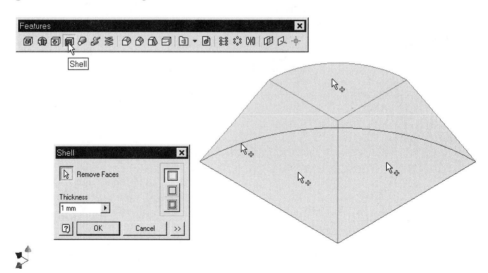

Figure 5–52 *Solid part being hollowed*

CONVERTED SHEET METAL PART

7. The solid part is complete. Now select Sheet Metal from the Applications menu.

8. Select the face indicated in Figure 5–53.

9. Select Flat Pattern from the Sheet Metal toolbar or panel.

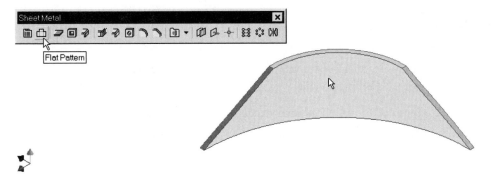

Figure 5–53 *Solid part being converted*

The solid part is converted to a sheet metal part. Figure 5–54 shows the flat pattern.

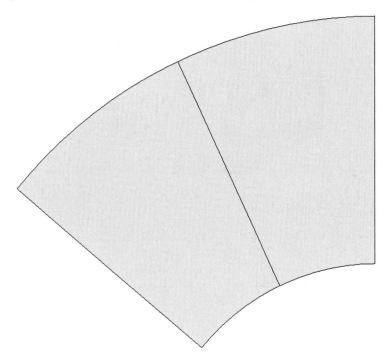

Figure 5–54 *Flat pattern*

10. The part is complete. Save and close your file (file name: Convert.ipt).

Exercises

I. SHEET METAL GEAR BOX

Figure 5–55 shows a sheet metal component. Follow the steps to construct a sheet metal solid part of this component.

1. Construct a sketch in accordance with Figure 5–56.

2. Set the display to isometric and set the sheet metal thickness to 1 unit.

3. Construct a sheet metal face from the sketch. (See Figure 5–57.)

4. Construct a flange at a distance of 25 units and an angle of 90 degrees in accordance with Figure 5–58.

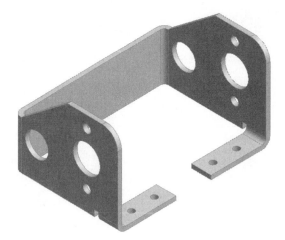

Figure 5–55 *Sheet metal gear box*

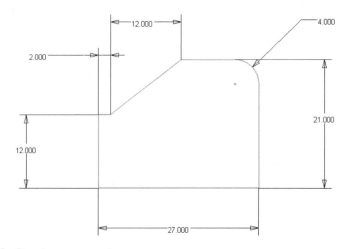

Figure 5–56 *Sketch constructed*

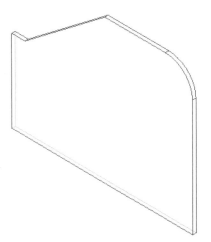

Figure 5–57 *Face extruded to the thickness of the sheet metal component*

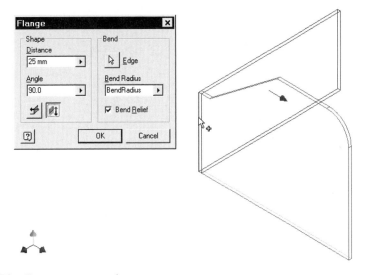

Figure 5–58 *Flange constructed*

5. Construct a work plane that is **39** units offset from the face indicated in Figure 5–59.

6. Set up a new sketch plane on the new work plane.

7. Copy the sketch of the base feature to the new sketch plane. (See Figure 5–60.)

8. Add two dimensions to the new sketch. (See Figure 5–61.)

9. Construct a face from the sketch. (See Figure 5–62.)

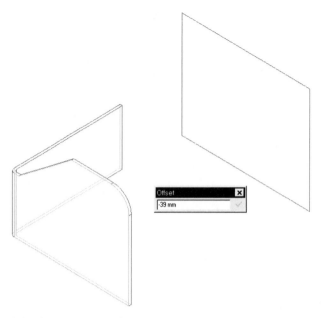

Figure 5–59 *Work plane constructed*

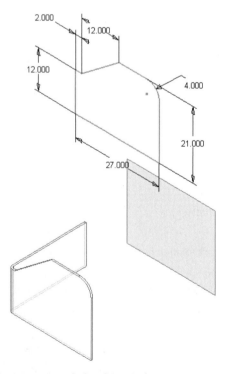

Figure 5–60 *Sketch plane set up and sketch copied*

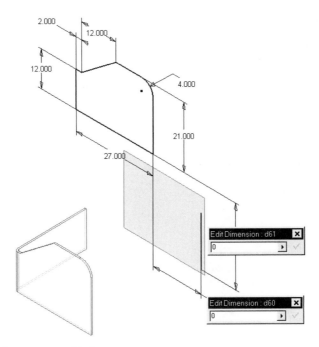

Figure 5–61 *Dimensions added*

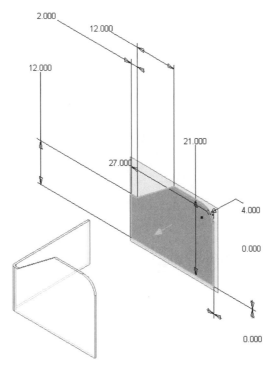

Figure 5–62 *Face constructed*

10. Construct a bend at the edges highlighted in Figure 5–63.

11. Hide the work plane, select the lower edge of the part, and set up a sketch plane.

12. Construct a rectangle and add four dimensions in accordance with Figure 5–64.

13. Construct a face in accordance with Figure 5–65.

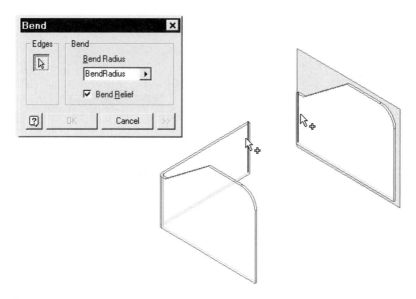

Figure 5–63 *Bend being constructed*

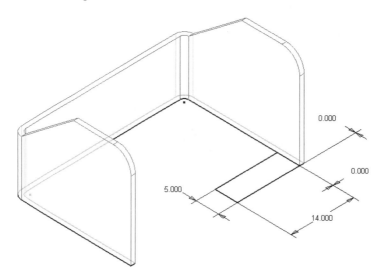

Figure 5–64 *Sketch plane set up and rectangle constructed*

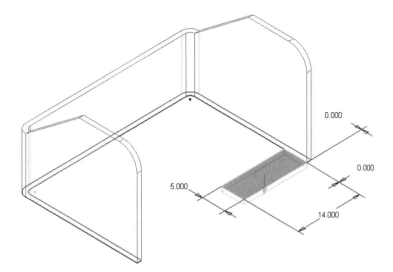

Figure 5–65 *Face being constructed*

14. Set up another sketch plane and construct a rectangle in accordance with Figure 5–66.

15. Construct a face from the sketch. (See Figure 5–67.)

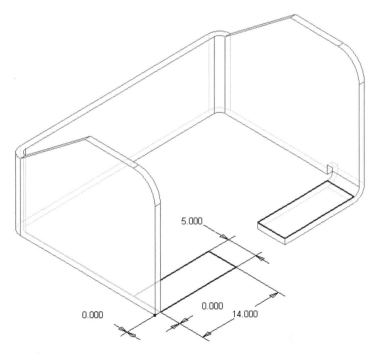

Figure 5–66 *Sketch constructed*

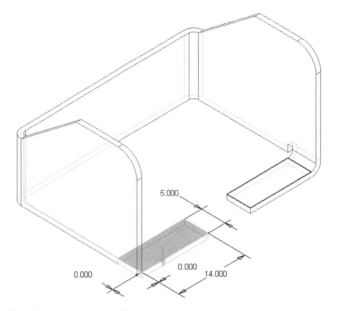

Figure 5–67 *Face being constructed*

16. Set the display in accordance with Figure 5–68.

17. Construct a sketch and add dimensions.

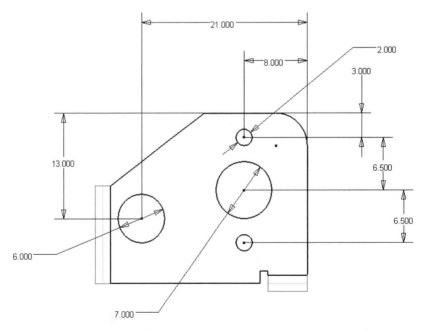

Figure 5–68 *Sketch constructed*

18. Set the display to an isometric view and cut through the sheet metal using the sketch as the profile. (See Figure 5–69.)

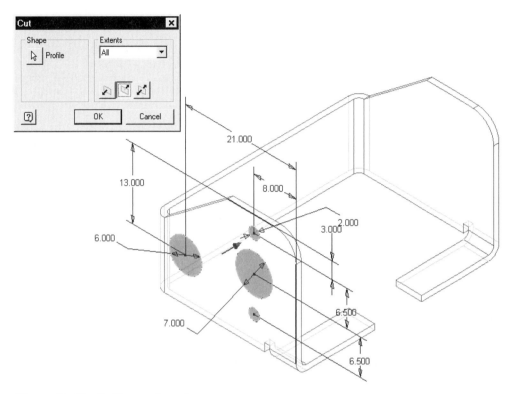

Figure 5–69 *Profile cut through*

19. Now set up a sketch plane and construct four center points in accordance with Figure 5–70.

20. Construct four holes of M2 size in accordance with Figure 5–71.

21. Now construct a flat pattern. (See Figure 5–72.)

22. The component is complete. Save and close your file (file name: SheetBox.ipt).

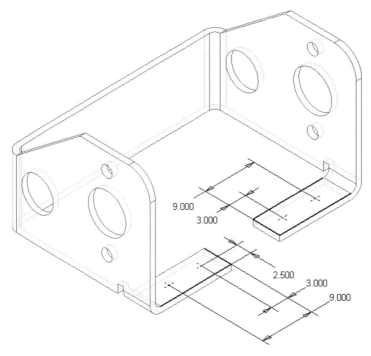

Figure 5–70 *Center points constructed*

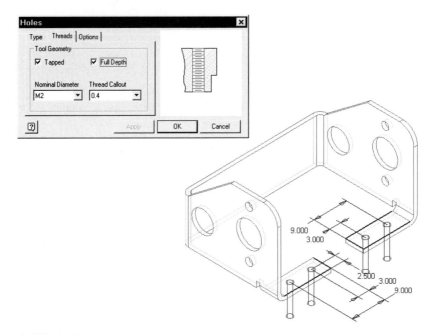

Figure 5–71 *Hole cut*

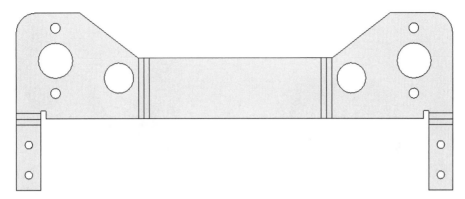

Figure 5–72 *Flat pattern*

2. MODEL CAR GEAR BOX

In Exercise 9 of Chapter 2, you constructed a solid part of the gear box of the model car. You constructed this component by using the 3D solid modeling tool. Now open the file (Box.ipt), convert it to a sheet metal component, and construct a flat pattern. (See Figure 5–73.)

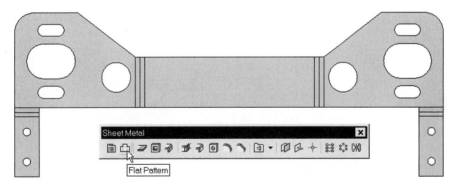

Figure 5–73 *Flat pattern of the converted solid part*

3. TOGGLE CLAMP

Design a sheet metal toggle clamp. Figure 5–74 shows two operating positions of the assembly of this clamp assembly. In essence, the assembly is a four-bar linkage. It has six component parts: base, jaw, handle, linkage, pin, and circlip. Among them, the base, jaw, and handle are sheet metal parts. Figure 5–75 shows the exploded view of the assembly.

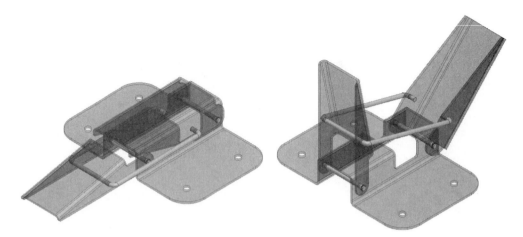

Figure 5–74 *Sheet metal clamp*

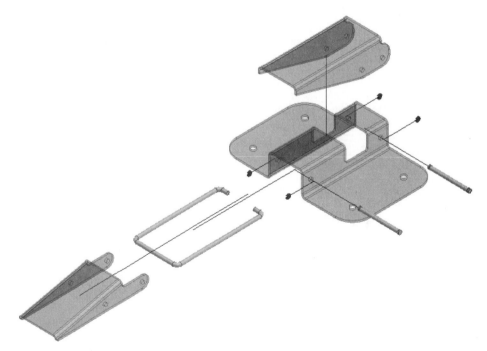

Figure 5–75 *Exploded view of the sheet metal clamp*

Use the top-down approach, 2D layout drawings, and adaptive techniques to design the parts and put them together into an assembly. You may refer to Figures 5–76 through 5–78 below for the major dimensions. The sheet metal thickness is 1 unit.

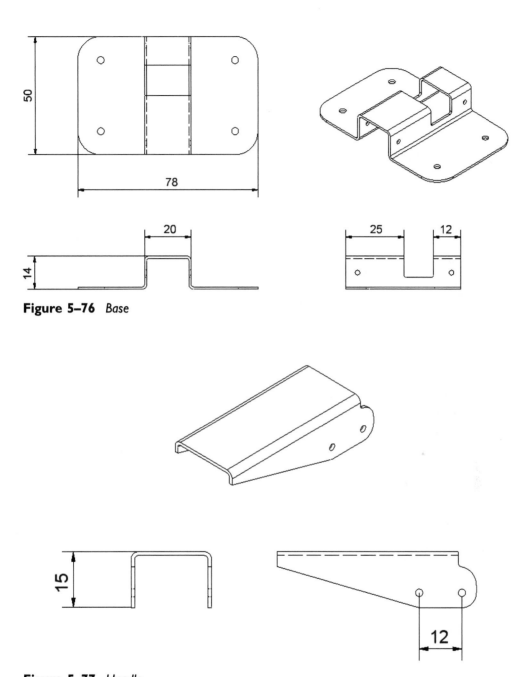

Figure 5–76 *Base*

Figure 5–77 *Handle*

490

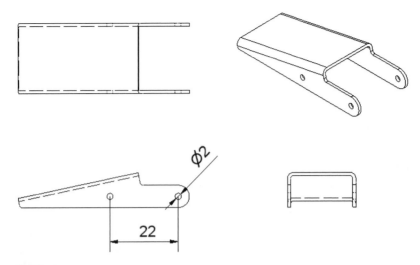

Figure 5–78 Jaw

4. SKATE SCOOTER REAR END ASSEMBLY

1. Construct a sheet metal component in accordance with the dimensions shown in Figure 5–79.

2. Open the rear end assembly of the skate scooter and place the sheet metal component, two screw Es, and two M5 nuts.

3. Assemble the components together in accordance with Figures 5–80 and 5–81.

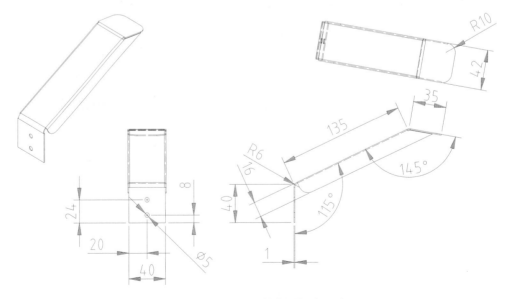

Figure 5–79 Rear wheel brake (file name: RearWheelBrake.ipt)

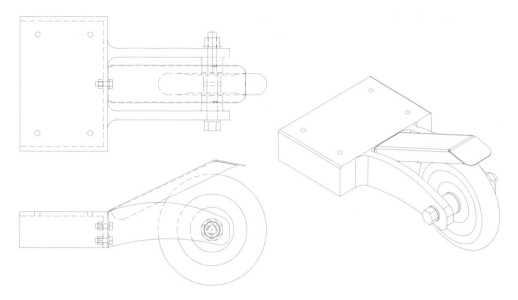

Figure 5–80   *Rear end assembly*

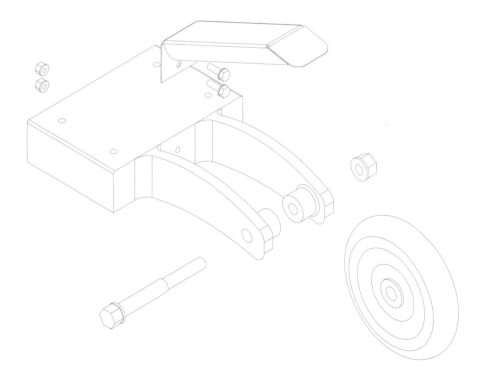

Figure 5–81   *Exploded view*

5. SKATE SCOOTER SHEET METAL PLANK

1. Construct a sheet metal plank of the skate scooter in accordance with Figure 5–82.

2. Start a new assembly file.

3. Place the front end assembly, rear end assembly, the plank, eight screw Es, and eight M5 nuts in the assembly file and assemble the components in accordance with Figures 5–83 and 5–84.

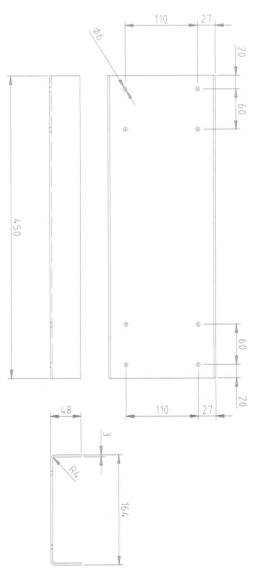

Figure 5–82 *Plank (file name: Plank.ipt)*

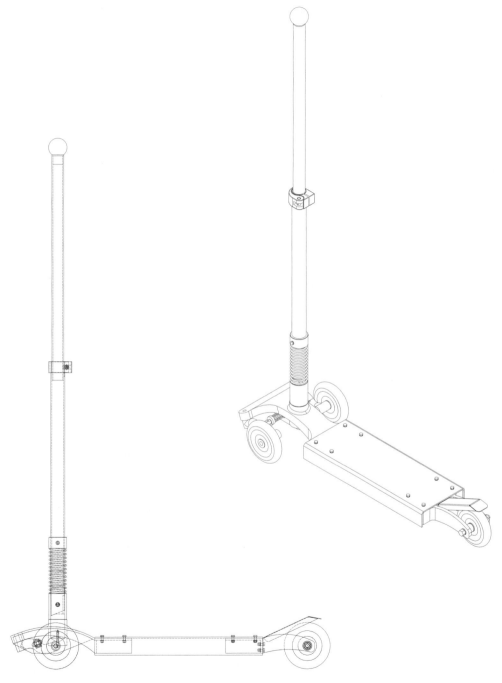

Figure 5–83 *Skate scooter assembly (file name: SkateScooter.iam)*

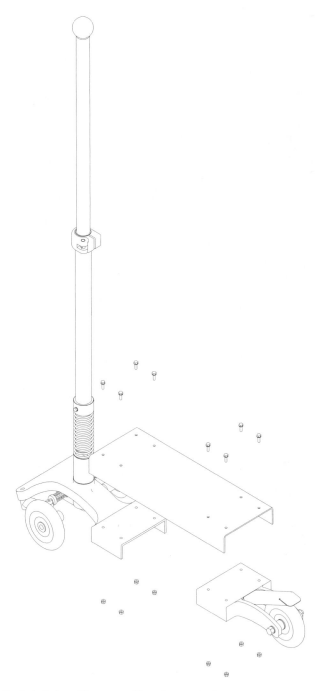

Figure 5–84 *Exploded view (file name: SkateScooter.ipn)*

SUMMARY

Sheet metal components are a special kind of 3D solid object. A sheet metal part is composed of a number of flat faces of uniform thickness that are joined together at rounded bends. To cope with practical manufacturing needs, you place relief cuts at the joints.

The design tools that are specific to sheet metal part construction are face, cut, flange, corner seam, and bend. To develop the sheet metal part for manufacturing, you use the flat pattern tool. Before making the bends in a sheet metal part, you determine the sheet metal settings. While you make a model, bends and relieves are placed automatically on the sheet metal component.

In addition to the specific tools, you can use corner round, corner chamfer, special design elements, work planes, work axes, work points, rectangular patterns, circular patterns, and mirror.

If you already have a shelled solid part or solid part that can be unfolded into a flat pattern, you can convert the solid to a sheet metal part.

REVIEW QUESTIONS

1. What are the design tools that are specific to sheet metal design?

2. State the similarities and differences between the 3D part modeling tool set and the 3D sheet metal modeling tool set.

CHAPTER 6

Engineering Drafting

OBJECTIVES

The aims of this chapter are to explain the concepts of engineering drafting and computer-generated engineering drafting, to illustrate how to prepare drawing sheets for constructing engineering drawings, to delineate the steps to construct engineering drawing views from parts and assemblies, flat pattern views from sheet metal parts, and presentation views of assemblies, and to outline the ways to add annotations to a drawing. After studying this chapter, you should be able to

- Describe the key concepts of engineering drafting and computer-generated engineering drafting

- Prepare drawing sheets for the construction of engineering drawing views

- Construct engineering drawing views of 3D parts, sheet metal parts, and assemblies

- Add annotations to a drawing

OVERVIEW

In a modern factory, you use the digital data about 3D parts and assemblies for downstream manufacturing operations. The need for and importance of 2D engineering drawings are diminishing. However, there are still many occasions when you need to produce 2D orthographic engineering drawings. Therefore, you will learn how to output engineering drawings from the 3D solid parts and assemblies in this chapter.

With 3D parts and assemblies, making the 2D engineering drawing is done by the computer and is semi-automatic. You start a drawing file, select a 3D part file, assembly file, or presentation file and let the computer generate the orthographic views. Then you can add annotations to the drawing.

ENGINEERING DRAFTING CONCEPTS

The traditional way to construct an engineering drawing is to think about how a 3D object would look when you project it orthogonally on a 2D plane and to construct the orthographic views in accordance with your perception of the object's 2D appearance.

ORTHOGRAPHIC PROJECTION

To depict a 3D object on a piece of 2D drawing paper, you use orthographic projection. The word "ortho" is a Greek word that means right or true. Orthographic projection is an engineering communication tool to represent 3D objects on 2D drawing sheets by using multiple-view drawings. You project the 3D object perpendicularly onto a projection plane with parallel projectors. (See Figure 6–1.)

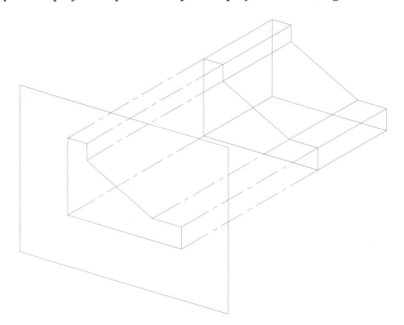

Figure 6–1   *An 3D object projected onto a plane*

Basically, you can use six projection planes that are mutually perpendicular to each other to construct six drawing views showing the front, right side, left side, rear side, top, and bottom of the 3D object. The six projection planes form a box. (See Figure 6–2.)

Now you can imagine the 3D object inside the box and project views orthogonally onto the six walls of the box. Because it is inconvenient to carry the box around, you cut and spread the box onto a common plane to obtain a drawing showing the six basic views. (See Figure 6–3.)

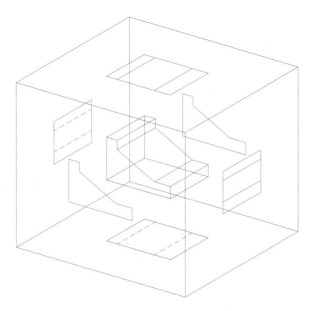

Figure 6–2 *Six projection planes forming a box with the 3D object placed inside*

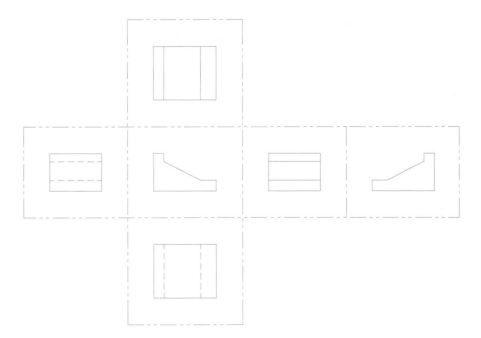

Figure 6–3 *Cutting and spreading the box*

Projection Systems

There are two kinds of orthographic projection systems. In one system, you put the projection plane in front of the 3D object. In the other system, you place the projection plane at the far side of the 3D object. (See Figure 6–4 and compare it with Figure 6–1.)

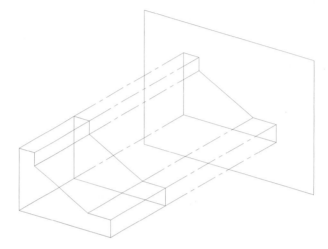

Figure 6–4   *Projection plane placed at the far side of the 3D object*

Similarly, there are six basic orthogonal views that form a box. (See Figure 6–5 and compare it with Figure 6–2.)

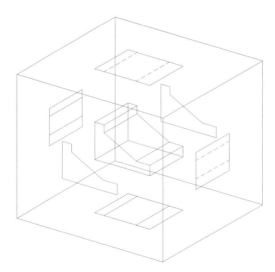

Figure 6–5   *Six projection planes*

Again, you will cut and spread the box onto a common plane to get six basic views on a drawing sheet. (See Figure 6–6 and compare it with Figure 6–3.)

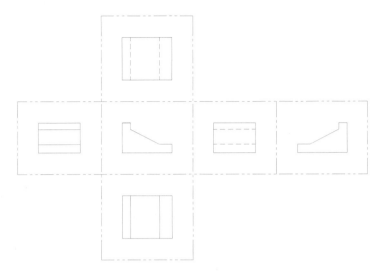

Figure 6–6 *Six basic views*

First and Third Angle Projection

In Figures 6–3 and 6–6, you can see that the front view and the rear view, the left side view and the right side view, and the top view and the bottom view are quite similar. To describe this 3D object, three drawing views (front, side, and top) are sufficient. (See Figure 6–7.)

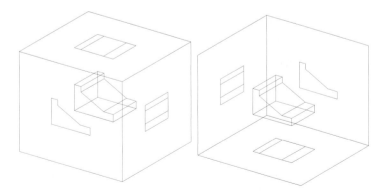

Figure 6–7 *Three views of the 3D object in two projection systems*

If you put the two projection systems together in 3D space, you will find a very interesting picture. (See Figure 6–8.)

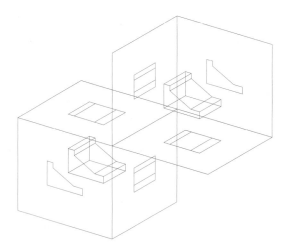

Figure 6–8 *Two systems put together*

In Figure 6–8, one system falls neatly in the first quadrant and the other in the third quadrant of the 3D space. Because we have to give the two projection systems names to identify which system we are using, we call one system the first angle projection system and the another one the third angle projection system.

Projection Symbols

To indicate the system of projection that you are using, you place a symbol on your drawing sheet. The projection symbol is the engineering drawing of the front and side views of a conical object. (See Figure 6–9.)

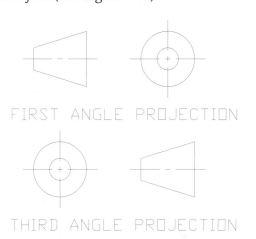

Figure 6–9 *Projection symbols*

COMPUTER-GENERATED ENGINEERING DRAWING VIEWS

If you have already constructed 3D solids and assembly of 3D solids in the computer, the construction of an engineering drawing is very simple. You start a drawing file, select a 3D solid or an assembly, construct a drawing sheet, let the computer project orthographic views from the solid or assembly, and add annotations to the views.

Starting an Engineering Drawing File

To construct an engineering drawing, you use the drawing template from the New dialog box.

1. Select New from the File menu.

2. Select the Standard.idw template. (See Figure 6–10.)

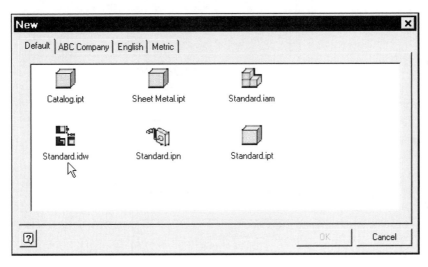

Figure 6–10   *Standard.idw template selected*

After starting a drawing file, you will find a default drawing sheet in the graphics area and two objects in the browser bar: Drawing Resources and Sheet1. (See Figure 6–11.)

3. Expand the browser bar. You will find four kinds of drawing resources (Sheet Formats, Borders, Title Blocks, and Sketched Symbols) and the drawing sheet with a default border and title block. (See Figure 6–12.)

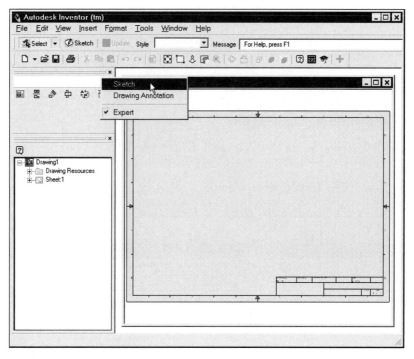

Figure 6–11 *A drawing file started*

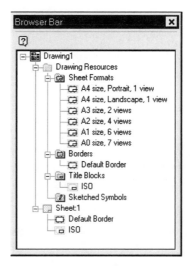

Figure 6–12 *Expanded browser bar of a drawing file*

To make a drawing, you use three toolbars/panels: Drawing Management, Sketch, and Drawing Annotation. Figure 6–13 shows the Drawing Management toolbar.

Figure 6–13 *Drawing Management toolbar*

The Drawing Management toolbar has six buttons. Table 6–1 describes the choices:

Table 6–1   Drawing Management toolbar and panel options

Option	Description
Create View	Enables you to construct an engineering drawing view from a selected part, assembly, or presentation.
Project Views	Enables you to construct orthogonally projected engineering drawing views and isometric views.
Auxiliary View	Enables you to construct an auxiliary view.
Section View	Enables you to construct a section view.
Detail View	Enables you to construct a detail view.
New Sheet	Enables you to place a drawing sheet on which you construct engineering drawing views.

Figure 6–14 shows the Drawing Annotation toolbar.

Figure 6–14 *Drawing Annotation toolbar*

The Drawing Annotation toolbar has fourteen button areas. Table 6–2 describes the choices:

Table 6–2   Drawing Annotation toolbar and panel options

Option	Description
General Dimension	Enables you to construct general dimensions on the drawing sheet.
Ordinate Dimension	Enables you to construct ordinate dimensions on the drawing sheet.
Hole/Thread Notes	Enables you to construct a hole or thread note on the drawing sheet.
Center Mark/ Center Line/ Center Line Bisector/ Centered Pattern	Enable you to construct various kinds of centerlines on the drawing sheet.

Option	Description
Surface Texture Symbol	Enables you to construct a surface texture symbol on the drawing sheet to specify surface roughness requirement.
Weld Symbol	Enables you to construct a weld symbol on the drawing sheet to specify welding requirement.
Feature Control Frame	Enables you to construct a feature control frame to specify geometric tolerance requirement.
Feature Identifier Symbol	Enables you to construct a feature identifier symbol to specify an identifier for a particular feature of the part.
Datum Identifier Symbol	Enables you to construct a datum identifier symbol to specify geometric tolerance data reference.
Datum Target- Leader/ Datum Target-Circle/ Datum Target-Line/ Datum Target-Point/ Datum Target-Rectangle	Enable you to construct various kinds of datum target symbols on the drawing sheet in order to specify geometric tolerance requirement.
Text	Enables you to construct text on the drawing sheet.
Leader Text	Enables you to construct a leader together with lines of text on the drawing sheet.
Balloon/ Balloon All	Enable you to construct a balloon or a set of balloons in an assembly drawing sheet.
Parts List	Enables you to place a part list in an assembly drawing sheet.

Figure 6–15 shows the Sketch toolbar.

Figure 6–15 *Sketch toolbar*

Sheet

To construct engineering drawing views of 3D objects, you need one or more drawing sheets. On a drawing sheet, you include a border, a title block, and appropriate symbols (sketched symbols) to comply with appropriate engineering drawing standards.

When you start a drawing file, it already displays a default drawing sheet with default title block and borders. If you find these objects appropriate, you can start placing engineering drawing views on the drawing sheet.

Drawing Resources

If you want to configure a drawing sheet, use the drawing resources provided. Drawing Resources contains various kinds of sheet formats, borders, title blocks, and sketched symbols.

To use a different kind of drawing sheet, select one from Sheet Formats in Drawing Resources or specify a new sheet by selecting Sheet from the Insert menu.

To use a different kind of border, select one from Borders in Drawing Resources or define a new border by selecting Define New Border from the Format menu.

To use a different kind of title block, select one from Title Blocks in Drawing Resources or define a new title block by selecting Define New Title Block from the Format menu.

To include a symbol on the drawing sheet, define a new symbol by selecting Define New Symbol from the Format menu and insert it in the drawing sheet.

Drawing Views

After you have prepared a drawing sheet, you construct drawing views and add annotations. To construct drawing views, you use the Drawing Management toolbar or panel. To add annotations to the drawing, you use the Drawing Annotation toolbar or panel.

Shortcut Keys

Apart from the appropriate toolbars, Table 6–3 shows the shortcut keys available:

Table 6–3   Engineering drawing shortcut keys

Shortcut Key	Function
O	Drawing ordinate dimensions
B	Balloon
F	Feature control frame

Procedure

Construction of engineering drawing through Autodesk Inventor involves three major tasks:

1. Prepare a 2D drawing sheet.

2. Construct engineering drawing views on the drawing sheets.

3. Add annotations to the drawing.

Figure 6–16 shows the completed engineering drawing for the main body of the food grinder.

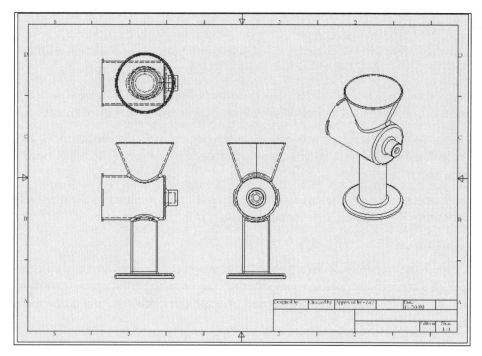

Figure 6–16 *Engineering drawing of the main body of the food grinder*

In the next sequence of steps, you will learn how to set up a drawing sheet for making an engineering drawing.

DRAWING SHEET PREPARATION

To reiterate, a drawing file has two major kinds of objects: drawing resources and drawing sheet. Drawing resources are objects that are needed by the drawing sheet. There are four kinds of resources: sheet formats, borders, title blocks, and sketched symbols. To construct engineering drawing views, you need a drawing sheet. When you start a new drawing file, a default sheet is given, and on it a set of default borders and a default title block.

To comply with appropriate engineering standards, you need to select a proper sheet size, construct standardized borders and title blocks, and make sketched symbols on the title block.

ENGINEERING DRAWING STANDARDS

Before you start making a drawing, you set appropriate engineering drawing standards.

1. Select Standards from the Format menu. (See Figure 6–17.)

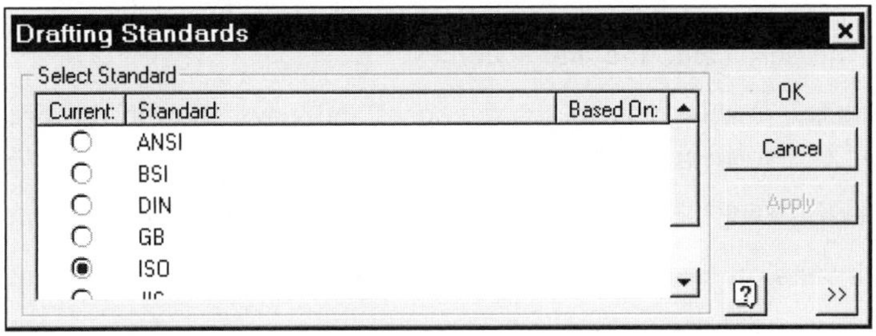

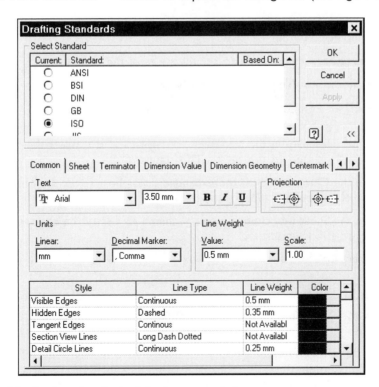

Figure 6–17 *Drafting Standards dialog box*

There are six kinds of drawing standards: ANSI, BSI, DIN, GB, ISO, and JIS.

In addition to the six kinds of standards, you can create your own standard. In the Drafting Standards dialog box, select the Click to add a new standard choice at the bottom of the standards list. Then specify a name for your new standard.

2. Now select the >> button to expand the dialog box. (See Figure 6–18.)

Figure 6–18 *Drafting Standards dialog box expanded*

In the expanded dialog box, there are thirteen tabs:

- **Common** Enables you to set the text style, projection direction, units of measurement, and line style.
- **Sheet** Enables you to set the sheet and view labels in the browser bar and the color scheme of the drawing sheet.
- **Terminator** Enables you to set the style of arrows and datum references.
- **Dimension Value** Enables you to set the style of the dimension values.
- **Dimension Geometry** Enables you to set the geometry of the dimension line and extension line and the location of dimension text.
- **Centermark** Enables you to set the proportion of the dashes of a center mark.
- **Welding Symbol** Enables you to set the weld symbols.
- **Surface Texture** Enables you to set the surface texture symbols.
- **Control Frame** Enables you to set the control frame of geometric tolerances.
- **Datum Target** Enables you to set the datum target of geometric tolerances.
- **Parts list** Enables you to set the styles of the parts list.
- **Balloon** Enables you to set the styles of the balloons.
- **Hatch** Enables you to set the styles of the hatch patterns.

3. Now select ISO in the standards list, right-click, and select the Set ISO Defaults button to use default ISO standard.
4. On the Common tab of the expanded dialog box, select the Third angle of projection button. (See Figure 6–19.)
5. On the Sheet tab, choose a color scheme to display the engineering drawing in your screen.
6. Select OK to close the dialog box.

Now you will learn how to prepare a drawing sheet by selecting objects from Drawing Resources and defining your own sheet, borders, title block, and sketched symbols.

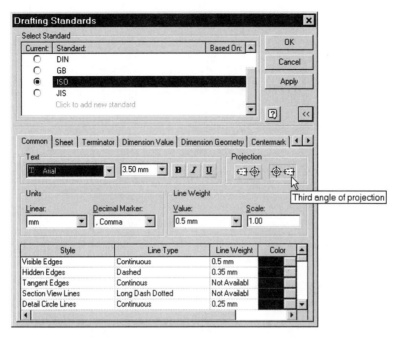

Figure 6–19 *Third Angle of Projection selected*

SHEET

A drawing sheet is analogous to a piece of drawing paper that you use to construct an engineering drawing. In a drawing file, you can insert more than one drawing sheet by adding new drawing sheets. You can delete drawing sheets, but you must have at least one drawing sheet.

There are two ways to insert a new drawing sheet to the drawing file: You can select and double-click a sheet from Sheet Formats in the browser bar and you can insert a sheet by selecting Sheet from the Insert menu.

Add a Sheet

7. Now add a drawing sheet from the resources provided. In the browser bar, select A0, 7 views, right-click, and select New Sheet. (See Figure 6–20.)

Because the sheet already specifies seven default drawing views, you have to select a solid part file or an assembly file.

8. In the Select Component dialog box, select the Explore directories button. (See Figure 6–21.)

9. Select the solid part file Cap.ipt that you constructed in Chapter 2.

512

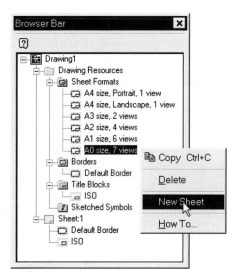

Figure 6–20 *Insertion of new drawing sheet from the browser bar*

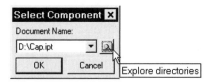

Figure 6–21 *Select Component dialog box*

A new drawing sheet with seven drawing views is inserted to the drawing file. (See Figure 6–22.)

10. Now insert another drawing sheet. Select Sheet from the Insert menu.

11. Select A3 from the Size box in the Format area and select Landscape in the Orientation area.

12. Select the OK button. (See Figure 6–23.)

Because the drawing sheet that you inserted does not have any drawing view, you have a blank drawing sheet. In total, you have three drawing sheets: a default sheet with borders and title, an A0 sheet with borders, title, and seven drawing views, and a blank drawing sheet.

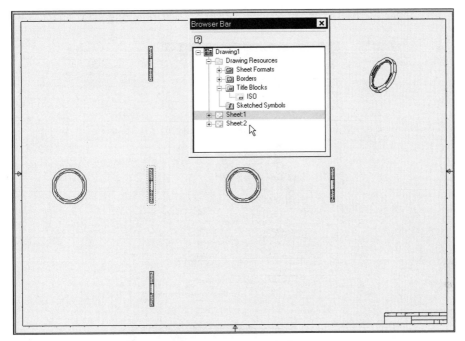

Figure 6–22 *A0 drawing sheet with seven views inserted*

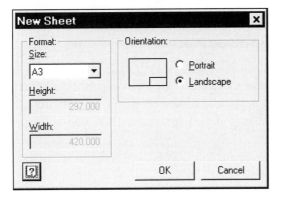

Figure 6–23 *New Sheet dialog box*

Delete a Sheet

Now you will learn how to delete drawing sheets.

13. Select the default drawing sheet in the browser bar, right-click, and select Delete to delete this sheet. (See Figure 6–24.)

14. Similarly, delete the A0, 7 views sheet so that only one blank sheet is left.

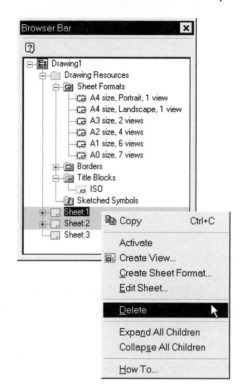

Figure 6–24 *Deleting a drawing sheet*

A drawing file can have more than one drawing sheet. Theoretically, you can construct different solid parts/assemblies on different drawing sheets. However, it is impractical and confusing if you use more than one solid part or assembly in a drawing file.

BORDERS

A drawing sheet should have four border lines around the edges. To add borders to a blank drawing sheet, you can insert the default borders from the drawing resources of the browser bar or construct them by sketching.

Insert Borders

15. To insert a set of standard borders in the blank sheet, select Default Border in the browser bar, right-click, and select Insert Drawing Border to display the Default Drawing Borders Parameters dialog box. (See Figure 6–25.)

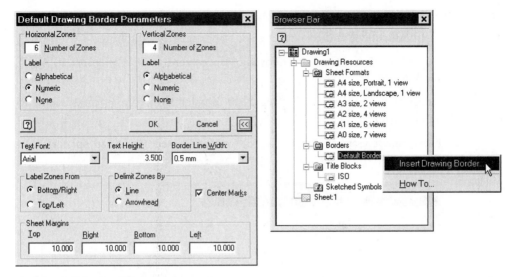

Figure 6–25 *Borders being inserted*

16. In the dialog box, set the number of horizontal zones to 6 and the number of vertical zones to 4 and select the OK button. A set of borders is inserted in the drawing sheet. (See Figure 6–26.)

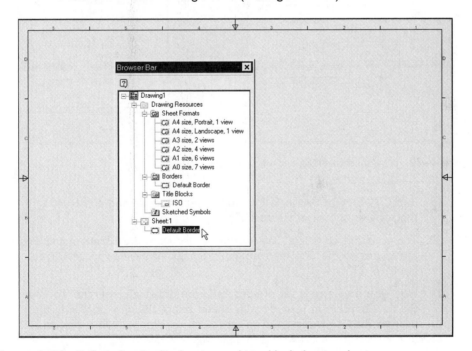

Figure 6–26 *Default drawing borders inserted in a blank drawing sheet*

Define New Borders

You will add a blank sheet and construct a set of borders by sketching.

17. Select New Sheet from the Drawing Management toolbar to insert an A3 blank sheet.

Before you start sketching, select Define New Border from the Format menu to initialize a new border.

18. Now select Grid from the Sketch toolbar to set the snap spacing to 2 mm x 2 mm as shown in Figure 6–27.

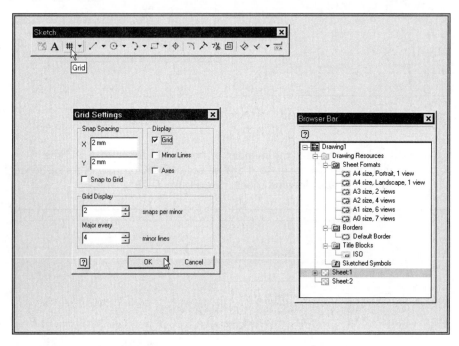

Figure 6–27 *New sheet inserted, and snap settings made*

19. Select Two point rectangle from the Sketch toolbar and construct a rectangle as shown in Figure 6–28.

20. The border is complete. Select Save Border from the Format menu to save the format to the drawing resources (border name: newborder). (See Figure 6–29.)

21. Now you have a new set of borders in the drawing resources. To insert the new borders in your drawing sheet, select the new border in the browser bar, right-click, and select Insert, or simply double-click on newborder in the browser bar. (See Figure 6–30.)

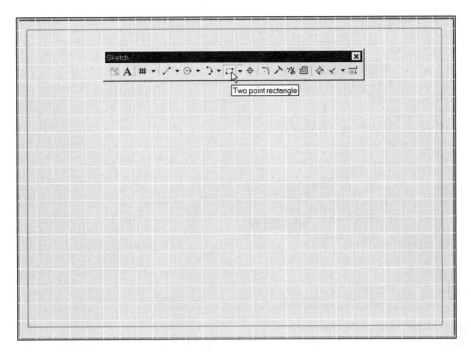

Figure 6–28 *Rectangle constructed*

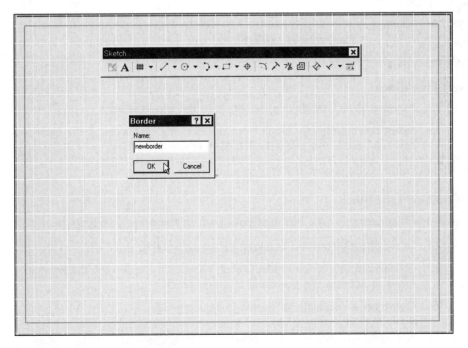

Figure 6–29 *New border saved*

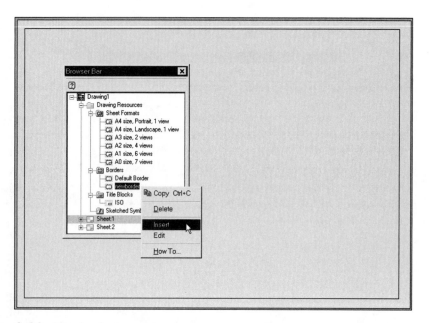

Figure 6–30 *New border inserted*

TITLE BLOCKS

Now you will learn how to add title blocks to the drawing sheets in your drawing file. Again, there are two ways to use a title block: you insert a default title block or you construct a title block and insert it in your drawing sheet.

Insert Title Block

22. Activate Sheet1 by double-clicking it in the browser bar.

23. Select the default title block from the drawing resources of the browser bar, right-click, and select Insert. (See Figure 6–31.)

The default title block is now inserted in one of the drawing sheets.

Define New Title Block

Now you will construct a title block of your own in the other drawing sheet.

24. Select Sheet2 from the browser bar to activate it.

25. To define a title block, select Define New Title Block from the Format menu to activate the sketching of a title block.

26. Construct a rectangle as shown in Figure 6–32.

A title block should have textual information such as the name of the designer, the title of the project etc. Include these objects as property fields in the title block.

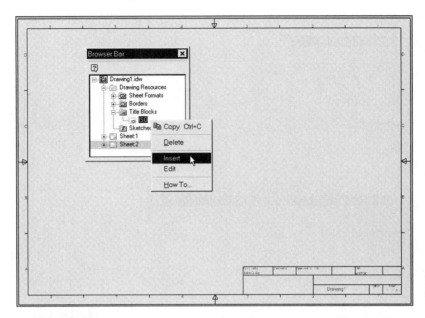

Figure 6–31 *Title block inserted*

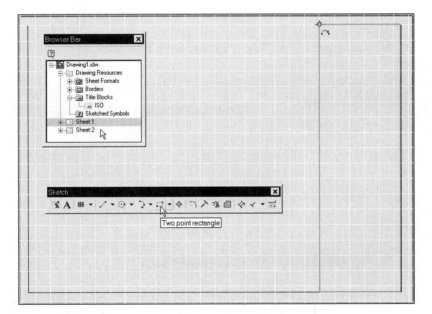

Figure 6–32 *Title block sketching activated, and rectangle constructed*

27. Select Property Field from the Sketch toolbar.
28. Select a point as shown in Figure 6–33.

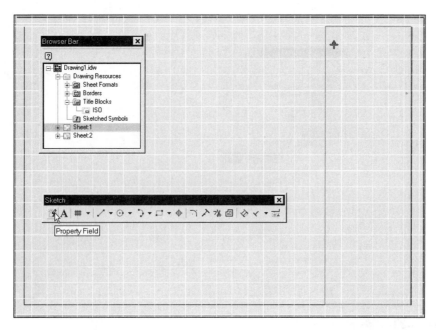

Figure 6–33 *Property field being added to the title block*

29. In the Format Field Text dialog box, select Description. (See Figure 6–34.) This way, a description field is added to the title block.

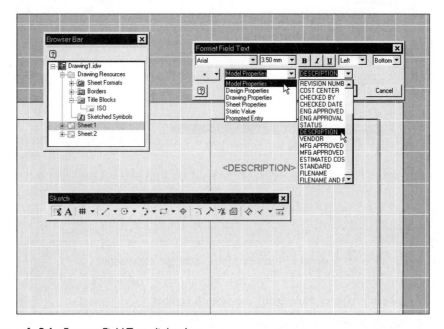

Figure 6–34 *Format Field Text dialog box*

30. In the Format Field Text dialog box, there are six sets of properties that you can add to the title block. Now add property fields and add some lines to the title block as shown in Figure 6–35.

31. Now the title block is complete. Select Save Title Block from the Format menu (title block name: newtitle). (See Figure 6–36.)

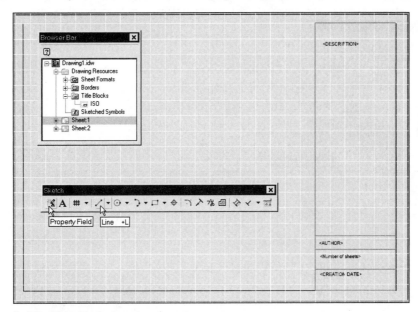

Figure 6–35 *Title block completed*

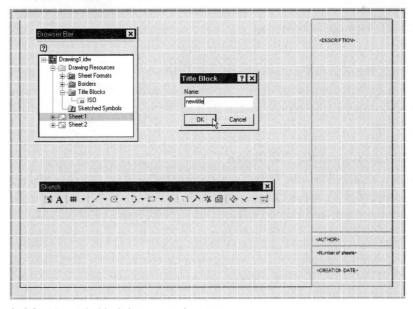

Figure 6–36 *New title block being saved*

32. Now the new title block is saved as a drawing resource. To insert the new title block to your drawing sheet, select the new title block from the drawing resources of the browser bar, right-click, and select Insert. (See Figure 6–37.)

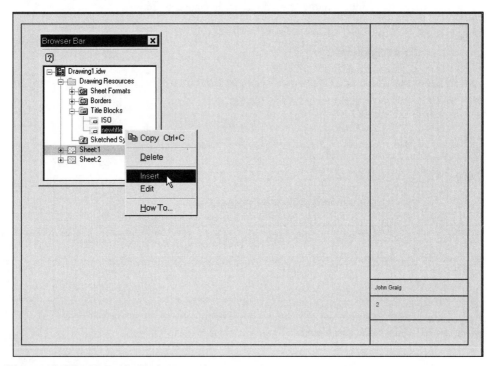

Figure 6–37 *New title block inserted*

To reiterate, there are two ways to add a title block: You can select a default title block from drawing resources of the browser bar and insert it to the drawing, and you can construct a title block, save it as a drawing resource, and insert it in the drawing sheet.

Specify Information in the Title Block

Information that will be displayed in the property field of the title block is called drawing properties To specify these properties, select Properties from the File menu and fill in your name and related information. (See Figure 6–38.)

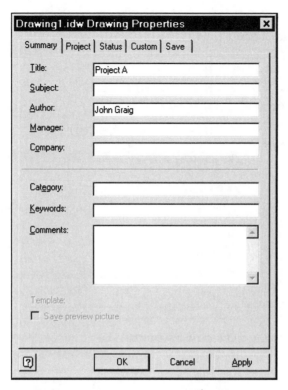

Figure 6–38 *Drawing Properties dialog box*

SKETCHED SYMBOLS

In addition to the borders and title block, you can construct sketched symbols for insertion in the drawing sheet.

33. Double-click Sheet1 in the browser bar to activate Sheet1.

34. Select Define New Symbol from the Format menu.

35. Select Two point rectangle and then Center point circle from the Sketch toolbar or panel to construct a rectangle and a circle as shown in **Figure 6–39**.

36. The sketched symbol is complete. Select Save Sketched Symbol from the Format menu to save the sketched symbol to the drawing resources (sketched symbol name: Logo). (See Figure 6–40.)

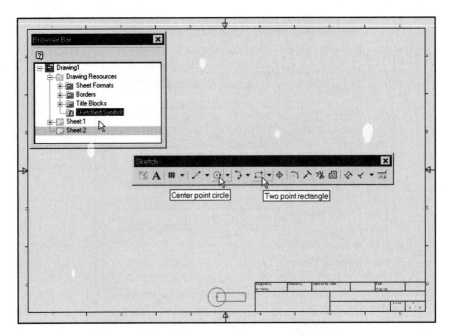

Figure 6–39 *A rectangle and a circle constructed*

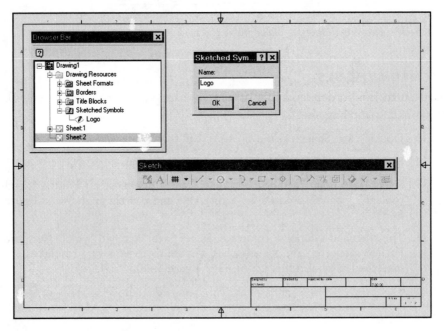

Figure 6–40 *New sketched symbol saved*

37. Now you have a new sketched symbol in the drawing resources. To insert the symbol in the drawing sheet, select the symbol in the browser bar, right-click, and select Insert. (See Figure 6–41.)

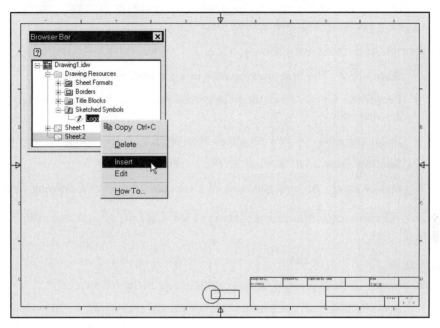

Figure 6–41 *Sketched symbol inserted*

TEMPLATE

38. To make the borders, title block, and sketched symbol available to other drawing files, select Save Copy As from the File menu.

39. Save the file to the Inventor Template directory (file name: newsheet1.idw). (See Figure 6–42.)

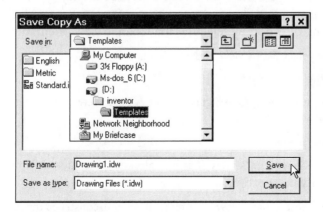

Figure 6–42 *Save as a template*

ENGINEERING DRAWING VIEWS

After you have prepared a drawing sheet, you select a 3D solid part, an assembly, or a presentation of an assembly and let the computer generate the orthographic views on the drawing sheets of a drawing file.

There are five kinds of drawing views:

- **Base view**   The first drawing view of a set of projected views.

- **Projected view**   An orthographic view generated from an existing drawing view.

- **Auxiliary view**   A view projected at an angle from an existing drawing view.

- **Section view**   The section across a cutting plane.

- **Detail view**   An enlarged view of a selected portion of a drawing view.

Now you will construct engineering drawings for a solid part, an assembly, and a sheet metal part.

BASE VIEW

The first drawing view of a set of orthographic views is the base view.

1. Activate Sheet1 in the browser bar by double-clicking it.

2. To construct a drawing view, select Create View from the Drawing Management toolbar.

Because this is the first drawing view of the drawing sheet, you have to select a part file, assembly file, or a presentation file.

3. Select the Explore directories button and select the file Base.ipt of the oscillator project. (See Figure 6–43.)

After selecting a file, you select a viewing direction, display type, and display scale. There are eleven standard drawing views to choose from. In addition, you can set your own viewing direction by selecting the Change view orientation button.

4. Now select Front view and Hidden line style.

5. Set the scale to 0.5.

6. Select a location. (See Figure 6–44.) A base view of the selected solid part is constructed.

PROJECTED VIEWS

Projected views derive from a view that you already constructed. You select a view to project and specify the locations.

7. Now use the base view as a foundation and construct three more views. Select Projected View from the Drawing Management toolbar.

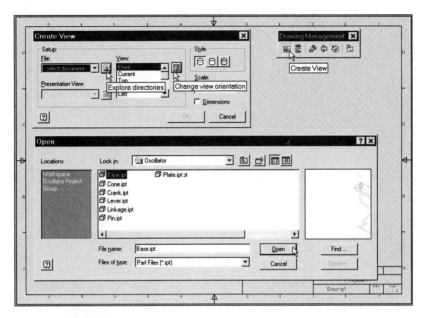

Figure 6–43 *File selected*

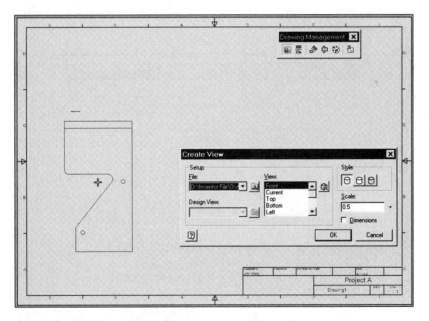

Figure 6–44 *Front view constructed*

8. Select the front view and select a location to the right side of the front view. (See Figure 6–45.) A side view is specified.

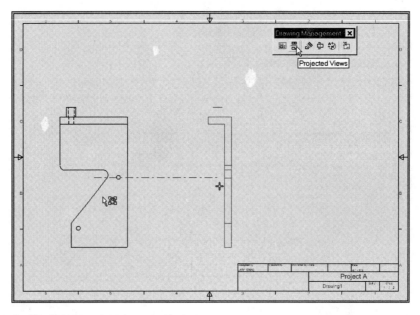

Figure 6–45 *Side view location specified*

9. Select a location above the front view. (See Figure 6–46.) A top view is specified.

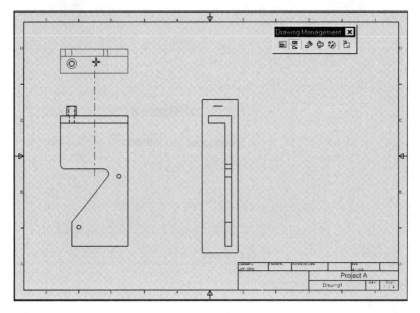

Figure 6–46 *Top view location specified*

10. Select a location at the upper right corner of the drawing sheet to specify an isometric view. (See Figure 6–47.)

11. After specifying the location of the drawing views, right-click and select Create. The specified views are constructed.

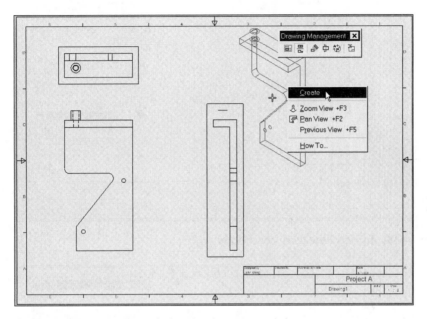

Figure 6–47 *Isometric view specified and views constructed*

TEMPLATE

Now you have four engineering drawing views in the drawing sheet. If you save this file as a template and use this template in a new drawing file, you will have four views once you specify a 3D part or an assembly.

12. To save a file as a template, select Save Copy As from the File menu and save the file in the template directory of Inventor (file name: newsheet2.idw).

AUXILIARY VIEW

An auxiliary view is an orthographic view that you project from a specified direction.

13. Select Auxiliary View from the Drawing Management toolbar.

14. Select the front view.

15. Select an edge as shown in Figure 6–48.

16. Select a location as shown in Figure 6–49 to construct an auxiliary view.

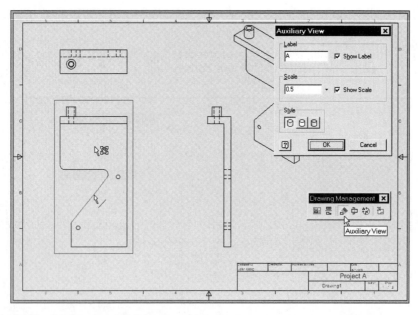

Figure 6–48 *Auxiliary view being constructed*

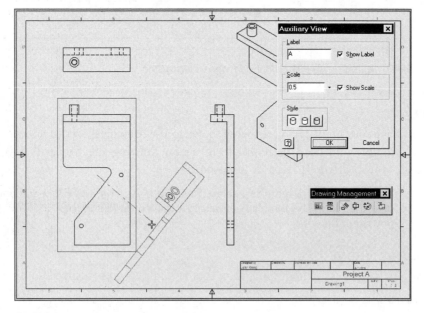

Figure 6–49 *Auxiliary view location specified*

SECTION VIEW

A section view depicts a plane seen on a cutting plane. You select a view to locate a cutting plane and specify a location of the section view. Because you will place hatching lines in the section view, you need to set the hatching style prior making a section view.

17. Select Standards from the Format menu. In the expanded Drafting Standards dialog box, select the Hatch tab.

18. Select a hatch pattern type and set the line weight, hatch pattern angle, and scale.

19. Select OK to have changes applied.

20. You will construct a base view and a section view on another drawing sheet. Double-click Sheet2 in the browser bar to activate it.

21. Select Create View from the Drawing Management toolbar.

22. Select the solid part file Base.ipt and select a location. (See Figure 6–50.)

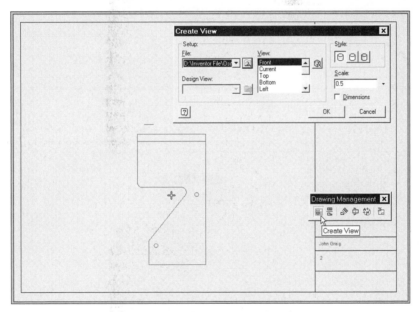

Figure 6–50 *Front view placed on the second drawing sheet*

Now you have a base view on the second drawing sheet of the drawing file. Before you construct a section view, you will construct a centerline to precisely define a cutting plane that passes through the centerline of a cylindrical hole.

23. Select Center line bisector from the Drawing Annotation toolbar.

24. Select two vertical lines highlighted in Figure 6–51 to construct a centerline.

25. Select Section View from the Drawing Management toolbar.

26. Select the upper end point of the centerline and a point near the lower horizontal line to define a section plane. (See Figure 6–52.)

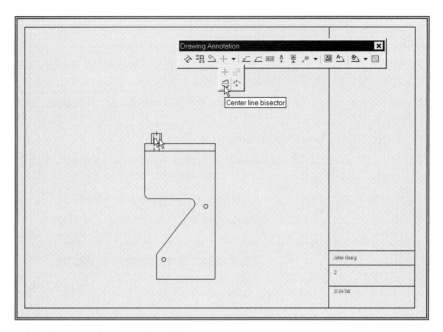

Figure 6–51 *Center line constructed*

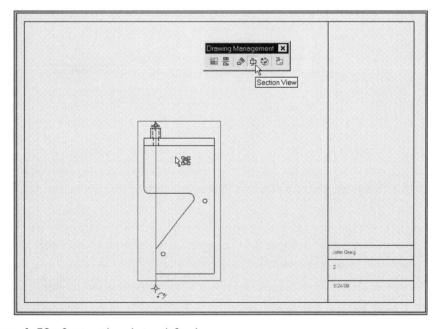

Figure 6–52 *Section plane being defined*

27. Right-click and select Continue. (See Figure 6–53.)

28. Select a location at the left side of the front view as shown in Figure 6–54.

29. Select the OK button. A section view is constructed.

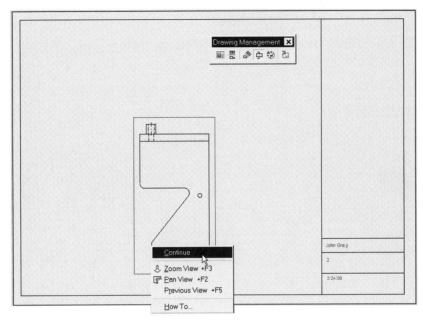

Figure 6–53 *Section plane defined*

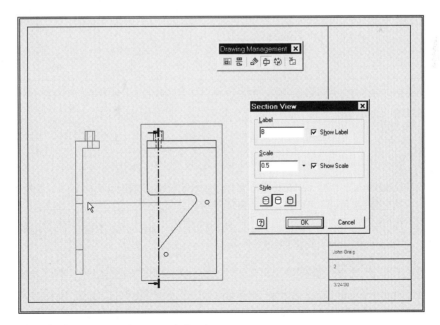

Figure 6–54 *Section view location defined*

DETAIL VIEW

A detail view is an enlarged view of a small portion of an existing drawing view. To make a detail view, you specify an enlargement scale, define a small portion of a view by describing a circle, and specify a location.

30. Select Detail View from the Drawing Management toolbar and select the front view as shown in Figure 6–55.

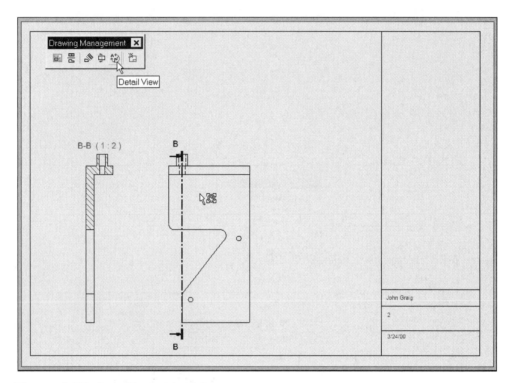

Figure 6–55 *Detail View selected*

31. Select the upper left corner of the front view.

32. Drag the cursor to describe a circle as shown in Figure 6–56 to define a circular area for making a detail view.

33. Select the lower right corner of the drawing sheet to define a location for the detail view. (See Figure 6–57.)

34. Select the OK button. The detail view is constructed. (See Figure 6–58.)

35. The drawing is complete. Save and close your file (file name: Base.idw).

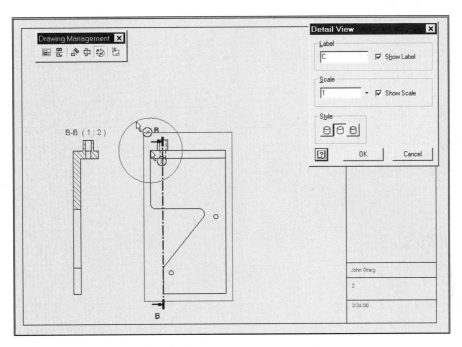

Figure 6–56 *Circular area described*

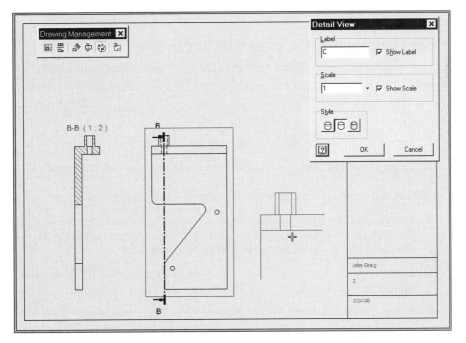

Figure 6–57 *Location for the detail view defined*

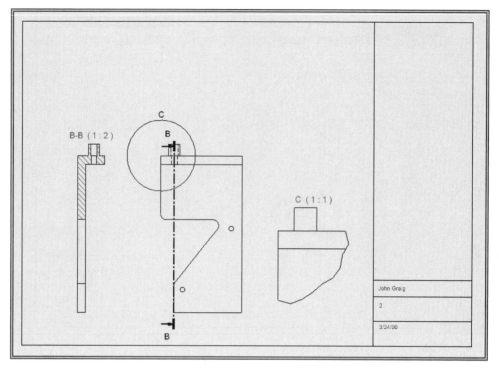

Figure 6–58 *Detail view constructed*

FLAT PATTERN VIEW OF A SHEET METAL PART

For a sheet metal part, you place the flat pattern of the component as well as the orthographic views in an engineering drawing. The orthographic views display the form, size, and shape of the 3D sheet metal component and the flat pattern view displays the 2D development of the component.

1. Start a new drawing file. Use the default drawing sheet.

2. Select the sheet metal component that you constructed in Chapter 5.

After selecting a part file, you will see a preview of the default view.

3. Because the view orientation of the part does not match the top view of the component, you will change the view orientation by selecting the Change View Orientation button. (See Figure 6–59.)

4. Select Look At from the Custom View toolbar and select the face highlighted in Figure 6–60.

5. A top view is displayed in the Custom View dialog box. (See Figure 6–61.) Select Exit Custom View from the Custom View toolbar to return to the Create View dialog box.

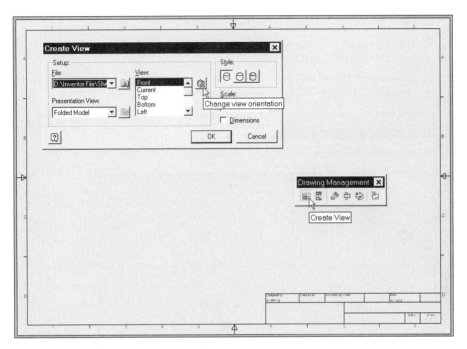

Figure 6–59 *Sheet metal component selected*

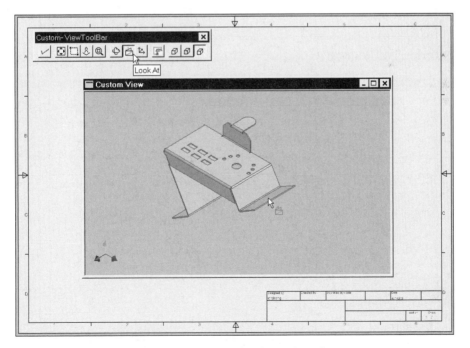

Figure 6–60 *Display changed to isometric and a face selected*

6. Now select a location for the custom view.

7. Set the scale to 0.5 and select the OK button. (See Figure 6–62.) A custom view is constructed.

8. Now select Projected View from the Drawing Management toolbar to construct three drawing views as shown in Figure 6–63.

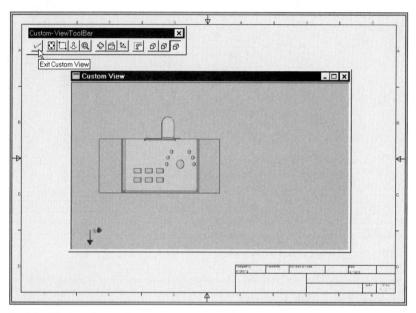

Figure 6–61 *Display changed*

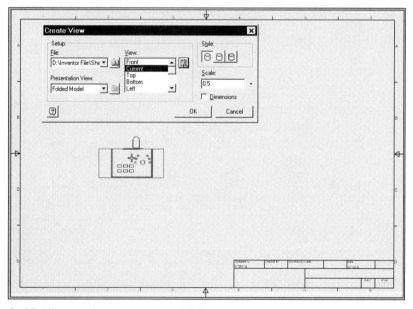

Figure 6–62 *Custom drawing view constructed*

9. To construct a flat pattern view, select Create View from the Drawing Management toolbar.

10. Select Flat Pattern in the Create View dialog box.

11. Select a location and select the OK button. (See Figure 6–64.)

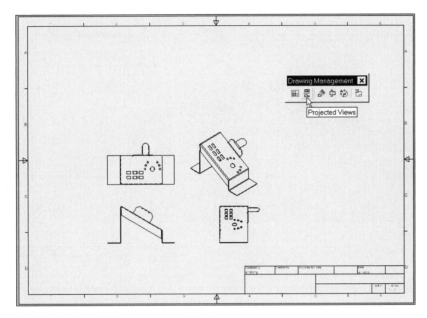

Figure 6–63 *Projected views constructed*

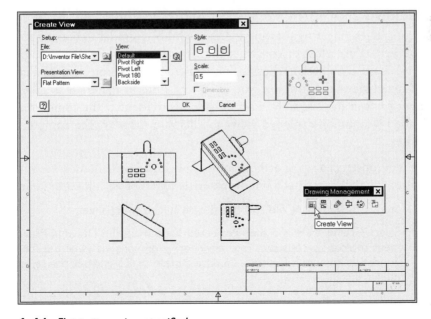

Figure 6–64 *Flat pattern view specified*

12. The drawing is complete. (See Figure 6–65.) Save and close your file (file name: Sheetframe.idw).

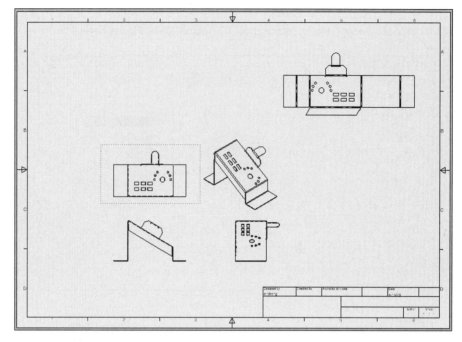

Figure 6–65 *Completed drawing*

ENGINEERING DRAWING VIEWS OF AN ASSEMBLY

Constructing an engineering drawing for an assembly is similar to making a drawing for a solid part. You select or construct drawing sheets and construct engineering drawing views on the drawing sheets. You construct base, projected, auxiliary, section, and detail views. In addition, you construct a bill of materials to tabulate the component parts in the assembly, a set of balloons to identify the component parts, and include presentation exploded views to illustrate the ways the components are put together.

Now you will construct a set of orthographic views and a presentation view. Later in this chapter, you will construct a bill of materials together with a set of balloons.

1. Start a new drawing file and use the default drawing sheet.

2. Select Create View and then Projected View from the Drawing Management toolbar to construct four engineering drawing views of the food grinder as shown in Figure 6–66. Use a scale of 0.5 and set the style to hidden line.

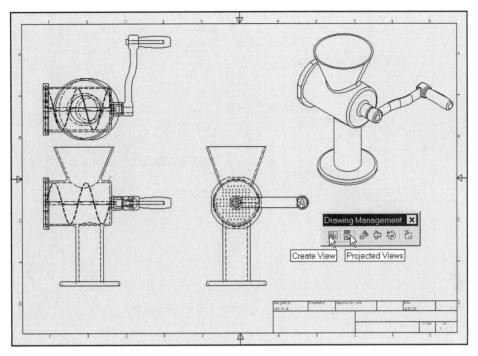

Figure 6–66 *Four engineering drawing views of the food grinder*

PRESENTATION VIEW OF AN ASSEMBLY

A presentation view is an exploded view of an assembly. You will use a new drawing sheet to construct a presentation view in the assembly drawing.

3. Select New Sheet from the Drawing Management toolbar to insert a new drawing sheet.

4. Select Default Border in the browser bar, right-click, and select Insert Drawing Border to insert a set of borders.

5. Select the title block in the drawing resources of the browser bar, right-click, and select Insert to insert a title block. (See Figure 6–67.)

6. A drawing sheet is constructed. Now select Create View from the Drawing Management dialog box and select Explore directories in the Create View dialog box.

7. Select Presentation Files in the Files of Type box of the Open dialog box.

8. Select the presentation file of the food grinder as shown in Figure 6–68.

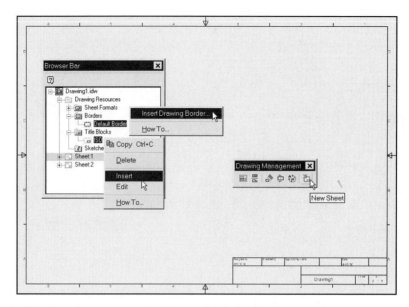

Figure 6–67 *New drawing sheet, borders, and title block constructed*

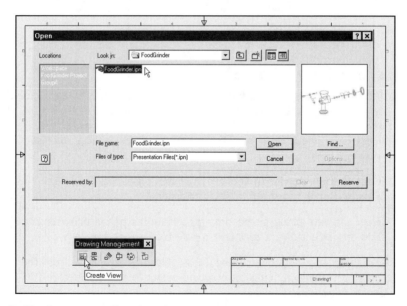

Figure 6–68 *Presentation file selected*

9. To set the orientation of the exploded view, select the Change View Orientation button in the Create View dialog box and rotate the view as shown in Figure 6–69

10. After setting the view orientation, select Exit Custom View to exit.

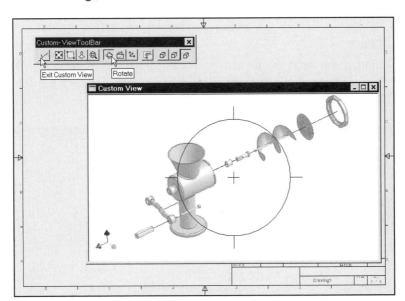

Figure 6–69 *Custom view being constructed*

11. Now select a location for the presentation view and select the OK button. (See Figure 6–70.) An exploded presentation view is constructed.

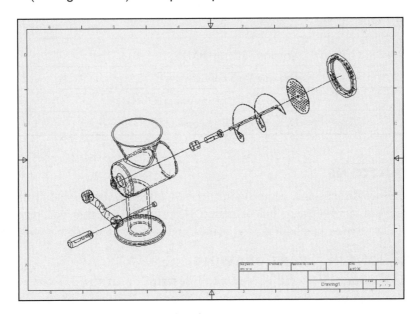

Figure 6–70 *Presentation view constructed*

12. The drawing views of the assembly drawing are complete. Save and close your file (file name: Foodgrinder.idw).

DWG OUTPUT

You can output a drawing file to various file formats: BMP, DWF, DWG, and DXF. By selecting Save Copy As from the File menu, you specify a file format. (See Figure 6–71 and Table 6–4)

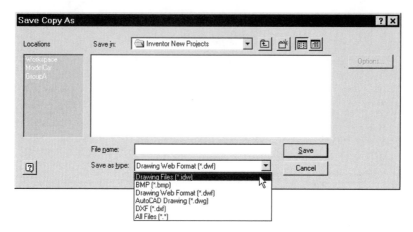

Figure 6–71 *Save Copy As dialog box*

Table 6–4 Output file formats

BMP	Windows Bitmap (BMP)
DWF	Drawing Web Format (DWF)
DWG	AutoCAD Drawing Format (DWG)
DXF	Drawing Exchange Format (DXF)

ANNOTATIONS

Annotations are dimensions, centerlines, surface texture symbols, weld symbol, geometric tolerance symbol, text, bill of materials, and balloons that you place in the drawing. They serve as supplements to the information provided by the drawing views.

ANNOTATIONS IN A PART DRAWING

A drawing file for a 3D solid part is a full description of the 3D object. Along with the orthographic views that show the shape and silhouettes of the object, you place dimensions, centerlines, surface texture symbols, geometric tolerance symbol, and text.

- You use dimensions to depict the size of the object.
- You use centerlines to illustrate axis and center locations.
- You use surface texture symbols to mandate surface finish requirement.
- You use geometric tolerance symbols to control the geometric shape of the object.
- You use text to provide textual description.

ANNOTATIONS IN AN ASSEMBLY DRAWING

An assembly drawing is an engineering document describing how various component parts of the assembly are put together. Because you use part drawings to depict the 3D object, you do not need to repeat the dimensions, surface finish requirement, and geometric tolerances of the individual component parts in an assembly drawing. Basically, you need to have a bill of materials and a set of balloons included in the assembly drawing.

- You use a bill of materials to outline the particulars of the individual component parts.
- You use a set of balloons in conjunction with the bill of materials to illustrate the locations of the individual parts in the assembly.

In addition to the bill of materials and the set of balloons, you place dimensions, centerlines, surface texture symbols, geometric tolerance symbol, text, and weld symbol.

- You use dimensions to depict the distances between the objects in the assembly.
- You use centerlines to illustrate axis and center locations and alternate locations of the component parts.
- You use surface texture symbols to describe the surface finish requirement of the set of objects after you assemble them together.
- You use geometric tolerance symbols to control the geometric relationships between the components in the assembly.
- You use text to provide description.
- You use weld symbols to illustrate how you will weld the component parts together in the assembly.

DIMENSIONS

Although dimensional information is already an integral part of the 3D part and assembly database, the dimensions of the objects are not readily perceivable if you do not display them explicitly on the drawing. To depict the actual size of a 3D solid in order to eliminate any possible errors that might arise in measuring the drawing, you have to add dimensions to your drawing.

Dimensioning Principles

There are two basic principles to follow when you add dimensions to a drawing:

- Each dimension required for the accurate definition of a feature should appear only once in the drawing. You should not assign more than one dimension to a feature.

- As far as possible, you should not require the reader of your drawing to do calculation in order to obtain the dimension of a feature.

Because of the second principle, you may find it essential to add more than one dimension to a feature. In that case, you should put the additional dimension within parentheses to indicate an auxiliary reference dimension.

Components of a Dimension

A dimension has four components:

- A dimension value
- A dimension line that is parallel to the direction of the described features
- A pair of arrowheads
- A pair of extension lines projecting from the feature to which the dimension refers

There should be a small gap between the end of the extension line and the feature. The extension line should project a short distance away from the intersection of the dimension line. (See Figure 6–72.)

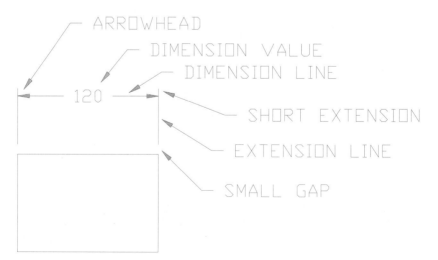

Figure 6–72 *Components of a dimension*

Dimension Types

You use dimensions to depict the size of an object. There are two kinds of dimensions: general dimensions and ordinate dimensions. Figure 6–72 above shows a general dimension. Ordinate dimensions are datum dimensions. They display the X and Y ordinate of a selected point.

> 1. Now open the file Base.idw. You will add dimensions to the drawing.

To add a dimension to your drawing, select the General Dimension and the Ordinate Dimension buttons of the Drawing Annotation toolbar. (See Figure 6.73.)

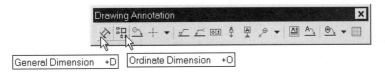

Figure 6–73 *General dimensions and ordinate dimensions buttons*

Set Dimension Style

Before you add a dimension to your drawing, you can set the shape of the arrowhead, the format and precision of the dimension value, and the geometry of the dimension. These parameters are set in the Drafting Standards dialog box.

> 2. Select Standards from the Format menu and then select the appropriate tab in the Drafting Standards dialog box (see Figures 6–74 through 6–76: Terminator tab, Dimension Value tab, and Dimension Geometry tab).
>
> 3. Select Apply to have any changes applied.

| Common | Sheet | Terminator | Dimension Value | Dimension Geometry | Centermark |

General: ⟶► Filled ▼

Size: 2.500

Aspect: 1.000

Datum: ⟶◄ Datum 45 Fille ▼

Figure 6–74 *Terminator tab*

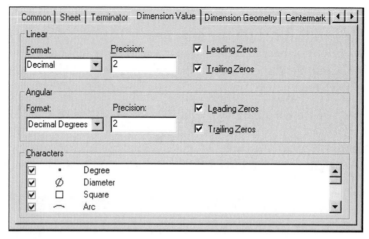

Figure 6–75 *Dimension Value tab*

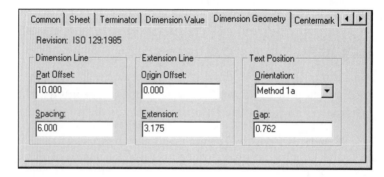

Figure 6–76 *Dimension Geometry tab*

4. Now add some dimensions to the drawing.

CENTERLINES

You use centerlines to indicate a center point and an axis of a cylindrical object. There are four ways to construct centerlines. You select the Center Mark, Center Line Bisector, Center Line, and Centered Pattern buttons from the Drawing Annotation toolbar.

Set Center Line Style

5. To set the display of centerlines, select the Centermark tab in the Drafting Standards dialog box. (See Figure 6–77.)

6. Select Apply to have any changes applied.

7. Add a centerline.

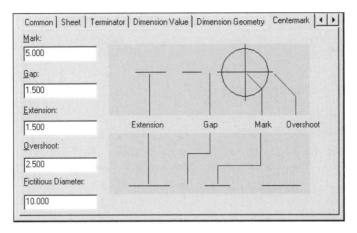

Figure 6–77 *Centermark tab*

SURFACE TEXTURE SYMBOL

You use a surface texture symbol to specify surface finish and the machining requirement for an object. (See Figure 6–78.)

Figure 6–78 *Surface texture symbols*

The basic surface texture symbol resembles the letter "V," with two variations: material removal prohibited and material removal required. You use the material removal prohibited symbol to prohibit any machining processes from being applied on the surface of the material. You use the material removal required symbol to mandate a machining requirement.

When you use the material removal required symbol, you specify the maximum surface roughness allowed. Optionally, you specify the machining allowance to be included and the direction of lay of machining.

You can specify the machining process and the sampling length. When you measure the surface texture, you take a sample; sampling length specifies the length of the sample.

8. To add a surface texture symbol, select Surface Texture Symbol from the Drawing Annotation toolbar. (See Figure 6–79.)

Figure 6–79 *Surface texture symbols button*

Set Surface Texture Symbol Style

9. To set the display of surface texture symbols, select the Surface Texture tab from the Drafting Standards dialog box. (See Figure 6–80.)

10. Select Apply to have any changes applied.

11. Add a surface texture symbol.

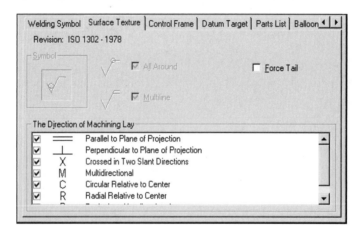

Figure 6–80 *Surface Texture tab*

WELD SYMBOL

You use a weld symbol to illustrate how parts are welded together. (See Figure 6–81.)

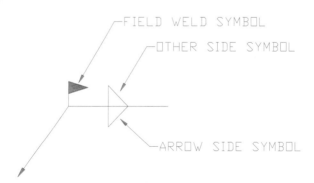

Figure 6–81 *Weld symbol*

The arrow side symbol specifies the kind of weld you will place at side of the joint indicated by the arrow. The other side symbol specifies the kind of weld you will construct at the other side of the joint indicated by the arrow. The field weld symbol is optional. It specifies that you will construct the weld joint on site.

12. To add a weld symbol, select Weld Symbol from the Drawing Annotation toolbar. (See Figure 6–82.)

Figure 6–82 *Weld Symbol button*

Set Weld Symbol Style

13. To set the display of the weld symbol, select the Welding Symbol tab in the Drafting Standards dialog box. (See Figure 6–83.)

14. Select Apply to have any changes applied.

15. Add a weld symbol.

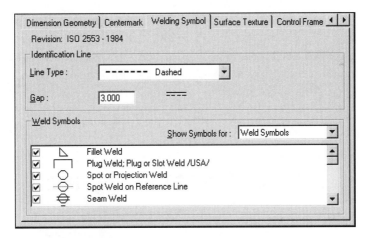

Figure 6–83 *Welding Symbol tab*

GEOMETRIC TOLERANCE SYMBOL

You use geometric tolerance symbols to detail the tolerance applied to the geometric shape of the 3D object. (See Figure 6–84.)

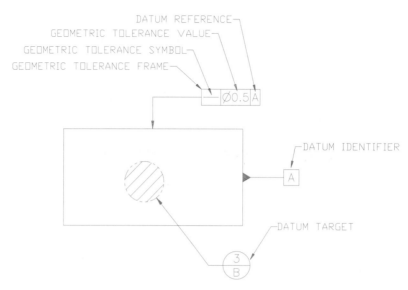

Figure 6–84 *Geometric tolerance symbols*

The main body of a geometric tolerance is a control frame. In the control frame, there are two or more compartments: In the first compartment, you put a geometric tolerance symbol depicting the kind of control you are imposing on the geometry. In the second compartment, you put the geometric tolerance value. The third compartment is optional; you can put the datum reference name. When you specify datum reference(s) in the geometric tolerance symbol, you specify a datum by using a datum identifier or a datum target. A datum identifier specifies the entire face of the indicated feature as datum reference. A datum target specifies a zone of the face of a feature as datum reference.

16. To add a geometric tolerance to your drawing, use the buttons shown in Figure 6–85.

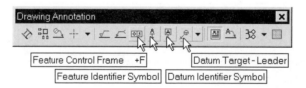

Figure 6–85 *Buttons for making geometric tolerance symbol*

Set Geometric Tolerance Symbol Style

17. To set the display of the geometric tolerance symbols, select the Control Frame tab, the Datum Target tab, and the Terminator tab in the Drafting Standards dialog box. (See Figures 6–86 through 6–88.)

18. Select OK to close the dialog box.

19. Add a geometric tolerance symbol.

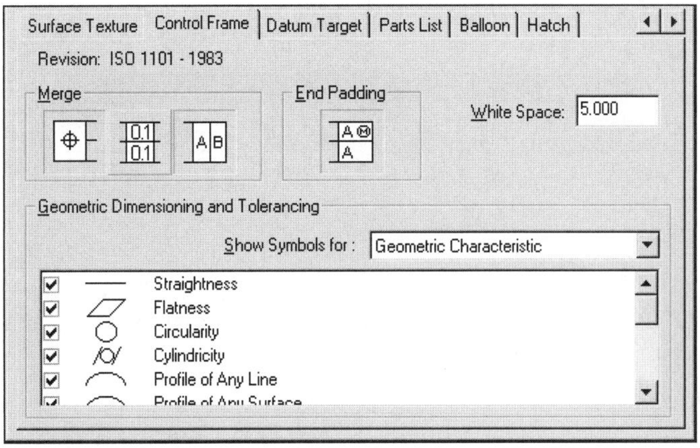

Figure 6–86 *Control Frame tab*

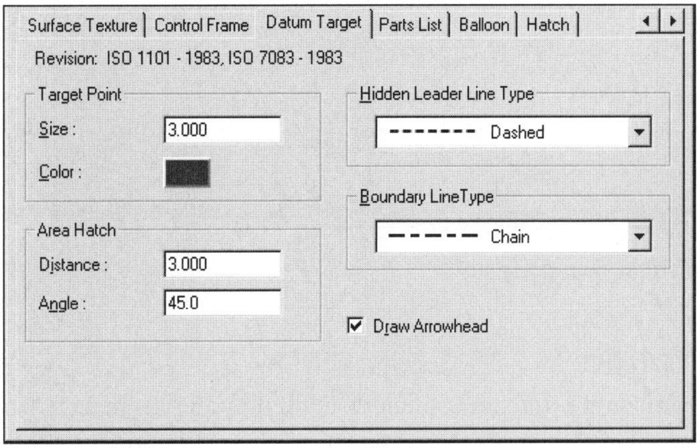

Figure 6–87 *Datum Target tab*

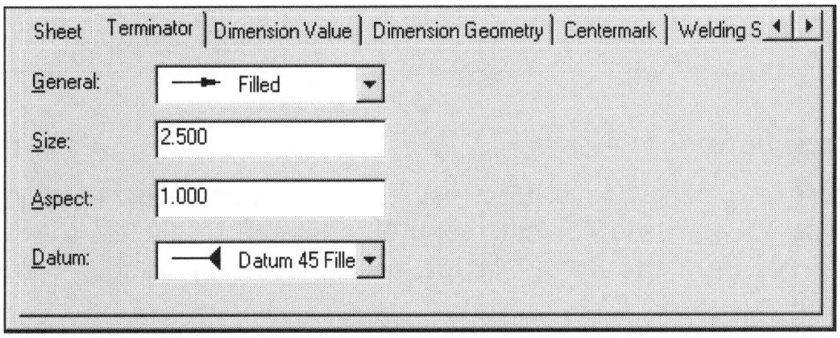

Figure 6–88 *Terminator tab*

TEXT

You add text to the drawing to depict textual information.

20. To place text in a drawing, select the Text and Leader Text buttons from the Drawing Annotation toolbar. (See Figure 6–89.)

Figure 6–89 *Text and Leader Text buttons*

21. After selecting the Text or Leader Text buttons, set the text style in the Format Text dialog box. (See Figure 6–90.)

22. Save and close your file.

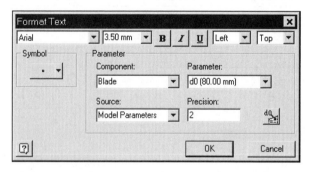

Figure 6–90 *Format Text dialog box*

BILL OF MATERIALS

It is standard engineering practice to include a bill of materials in an assembly. A bill of materials is a parts list that provides information about the quantity and references to the parts of the assembly. You can include all the components or select a set of components for the bill of materials. You can also place the parts list in the design notebook.

1. You will add a parts list to an assembly drawing. Open the drawing file Foodgrinder.idw.

Set Bill of Materials Style

2. Before you construct a parts list in your assembly drawing, select Standards from the Format menu and set the display of the bill of materials from the Parts List tab of the Drafting Standards dialog box. (See Figure 6–91.)

3. Select the OK button to close the dialog box.

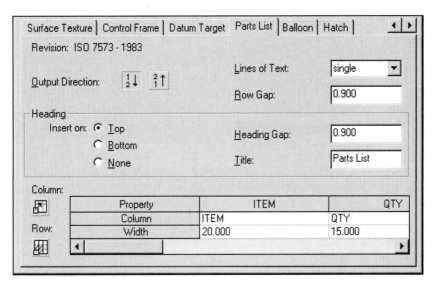

Figure 6–91 *Parts List tab*

Add Bill of Materials

4. To include a parts list, select Parts List from the Drawing Annotation toolbar.

5. Select a drawing view of the assembly. (See Figure 6–92.) In the Create Parts List dialog box, there are two areas: Level and Range. Level refers to the level in the hierarchy of an assembly.

If you simply put all the parts together in an assembly, you have one level of parts. If you put some parts in sub-assemblies and put the sub-assemblies in the assembly, you have two levels of components. Because you can have an assembly that consists of sub-assemblies and parts, you have to choose whether you want the parts list to display only the parts or all objects in the first level.

6. Choose First-Level Components and select OK.

If you specify Only Parts in the parts list, you can choose to display all the parts or a range of parts. Because you chose first level components in the Level box, there is only one option in the Range box: All.

7. Now select a point as indicated in Figure 6–93 to construct the parts list.

BALLOONS

In conjunction with the parts list, you need to include a set of leaders referencing the individual parts in the drawing view and the parts list. This special kind of leader is called balloon.

556

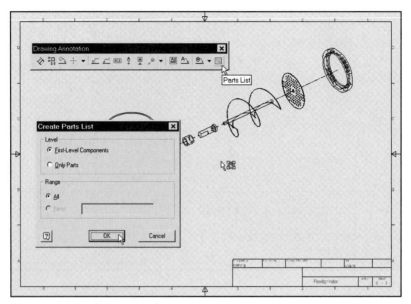

Figure 6–92 *Parts list being constructed*

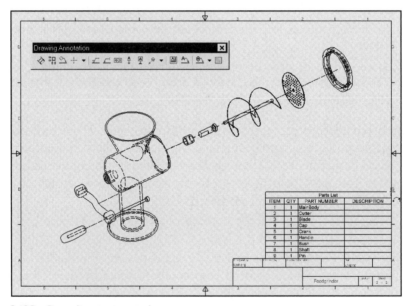

Figure 6–93 *Parts list constructed*

Set Balloon Style

8. Before you construct a set of balloons, select Standards from the Format menu and set the display of the bill of materials from the Balloon tab of the Drafting Standards dialog box. (See Figure 6–94.)

9. Select the OK button.

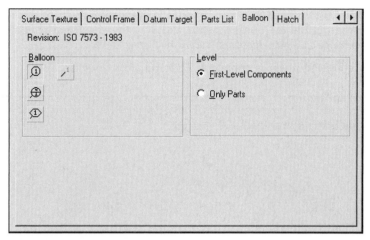

Figure 6–94 *Balloon tab*

Add Balloons

There are two ways to construct balloons in a drawing: You can construct individual balloons by selecting individual components of the drawing, and you can construct a set of balloons by selecting a drawing view. To construct individual balloons, select Balloon from the Drawing Annotation toolbar. To construct a set of balloons collectively, select Balloon All from the Drawing Annotation toolbar.

10. Now select Balloon All from the Annotation toolbar and select the isometric view. A set of balloons is constructed. (See Figure 6–95.)

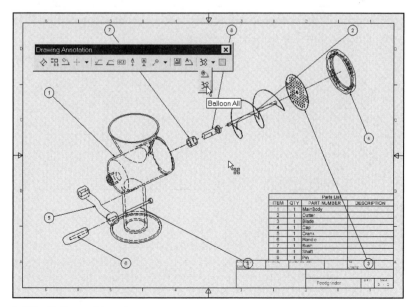

Figure 6–95 *Balloons placed*

11. Because construction of the balloons is automatic, some of them are placed outside the border. Select and drag the views and the balloons to rearrange the placement as shown in Figure 6–96.

12. The drawing sheet is complete. Save and close your file.

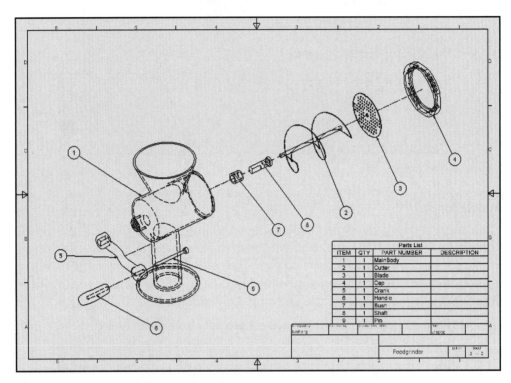

Figure 6–96 *Drawing view and balloons rearranged*

Exercises

I. DRAWING SHEET

Design and construct a drawing sheet. Then save the file as a template.

2. ENGINEERING DRAWINGS

Construct engineering drawings for the solid part and assembly files that you constructed in Chapters 2, 3, 4, and 5.

SUMMARY

An engineering drawing is a set of orthographic drawing views depicting a 3D object. To construct an engineering drawing, you use a drawing file and select a solid part, an assembly, or a presentation. Generation of orthographic views from the selected 3D objects is automatic. You need to specify only a drawing view type and a location.

A drawing file has two kinds of objects—drawing resources and drawing sheet. There are four kinds of drawing resources: standard drawing sheet templates, drawing borders, engineering title blocks, and sketched symbols. You specify a drawing sheet or select a drawing sheet from the resource. Then you add a border, a title block, and sketched objects. Apart from the standard drawing resources provided, you can construct your own borders, title blocks, and sketched objects for insertion in the drawing sheet.

After setting up a drawing sheet with borders, title blocks, and sketched objects, you select a 3D object (solid part, assembly, or presentation) and specify the drawing views. In a drawing file, you can have more than one drawing sheet. Therefore, you can select different solid part and assembly files for each drawing sheet in the drawing file. In practice, you should use one solid part or assembly for each drawing file to avoid confusion.

There are five kinds of drawing views—base view, projected view, auxiliary view, section view, and detail view. You construct these drawing views for part, sheet metal, and assembly drawings. In addition, you include a flat pattern view for a sheet metal drawing and include a presentation view for an assembly drawing.

To complete an engineering drawing, you add annotations such as dimensions, geometric tolerances, surface texture symbols, and welding symbols. Before you add annotations to a drawing, you need to select a drawing standard and make appropriate modifications to the standard if necessary.

REVIEW QUESTIONS

1. What are the four kinds of files? Which one will you use to construct an engineering drawing and how?

2. What are the four kinds of drawing resources in a drawing file? Briefly describe the ways to construct these resources. How can you make these resources available to other projects?

3. How many kinds of drawing views can you construct in a drawing file? Use simple sketches to illustrate your answer.

Imported Solids

You can import three kinds of solids by opening them: Mechanical Desktop solids (DWG), ACIS solids, and STEP solids. If you install Mechanical Desktop and Autodesk Inventor on the same computer and start both applications, you can open a Mechanical Desktop solid part file in Inventor and retain all the parametric information of the original Mechanical Desktop file. If you open an ACIS or a STEP file, you get a base solid feature. Basically, the base solid feature is static and is non-parametric. You can add sketched features, placed features, and work features to the base solid. The additional features are parametric and editable.

To learn how to manipulate a SAT or STEP solid, you will construct a SAT solid file, open the SAT file, and edit the base solid feature.

SAVING A SAT FILE

1. Open the file Box.ipt that you constructed in Exercise 9, Chapter 2.

2. Select Save Copy As from the File menu.

3. In the Save Copy As dialog box, select SAT (*.sat) in the Save as type box and specify a file name, Box. (See Figure A–1.)

4. Then close the file.

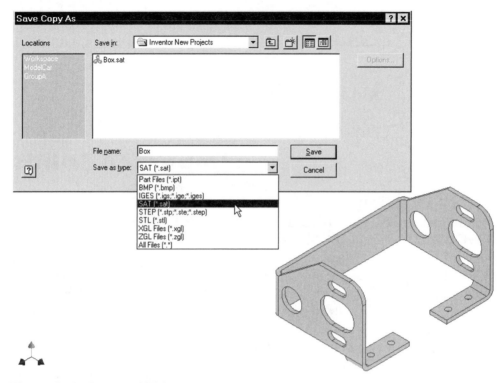

Figure A–1 *Save as a SAT file*

OPENING AND EDITING A SAT FILE

1. Now select Open from the File menu.

2. In the Open dialog box, select SAT (*.sat) in the Files of type box and select the file Box. The SAT file is opened.

3 In the browser bar, you will find an object Base1. This is the base solid feature. Select this feature, right-click, and select Edit Solid. (See Figure A–2.)

To modify the base solid, you use the Solids toolbar and panel, which provide tools to manipulate imported non-parametric solid parts. There are six button areas:

- **Move Face**   Enables you to move a selected face of the solid.

- **Extend or Contract Body**   Enables you to extend or contract the body of a solid.

- **Work Plane**   Enables you to construct a work plane.

- **Work Axis**   Enables you to construct a work axis.

- **Work Points**   Enables you to construct a work point.

- **Toggle Precise UI**   Toggles the display of the Precise Input dialog box.

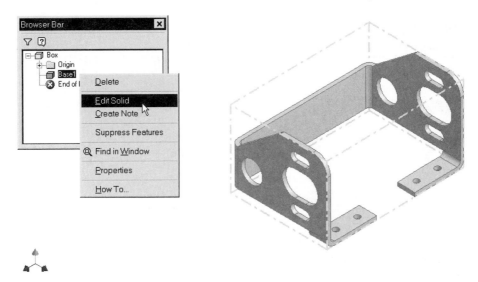

Figure A–2 *SAT file opened and base solid selected in the browser bar*

Move Face

4. Now you will move a face. Select Move Face from the Solids toolbar or panel.

5. Select the cylindrical face indicated in Figure A–3.

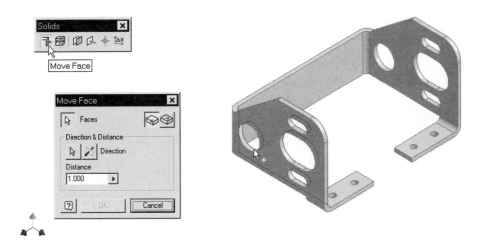

Figure A–3 *Cylindrical face selected*

6. Select the Direction button, select the lower edge, and flip the direction in accordance with Figure A–4.

7. Then set the distance to 3 units.

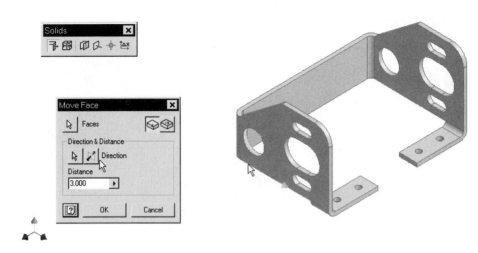

Figure A–4 *Lower edge selected*

8. Select the OK button. The face is moved.

Extend or Contract Body

9. Now you will extend the body of the solid. Select Extend or Contract Body from the Solids toolbar or panel.

10. Select the face indicated in Figure A–5.

11. Set the distance to 10 units and select the OK button.

12. The base solid is modified. Right-click and select Finish Solid Edit to exit edit mode.

13. Then save and close your file.

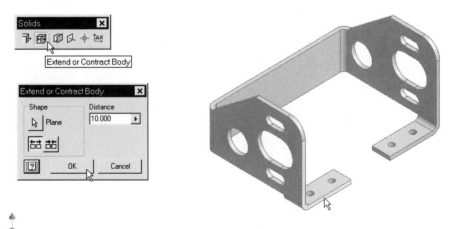

Figure A–5 *Face of a body selected*

The body is extended. See Figure A–6.

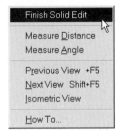

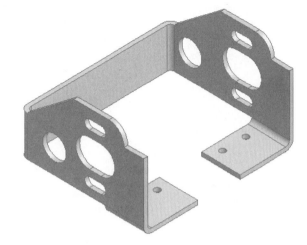

Figure A–6 *Body extended*

Inserted Objects

Along with geometric data of the 3D solids, you can insert various kinds of objects in your Inventor part files.

Select Object... from the Insert pull-down menu. (See Figure B–1.)

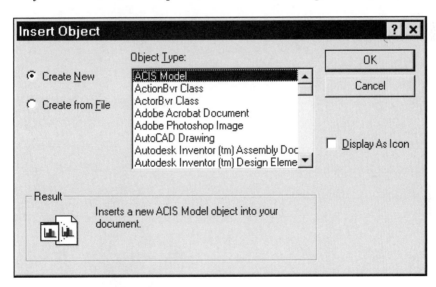

Figure B–1 *Insert Object dialog box*

In the dialog box, you select objects of various kinds that are available in your computer. You can create a new file or use an existing file. To construct a new file, select Create New. To use an old file, select Create from File.

INDEX

Note: Page numbers in **bold type** reference figures and tables.

LICENSE AGREEMENT FOR AUTODESK PRESS

THOMSON LEARNING™

Educational Software/Data

You the customer, and Autodesk Press incur certain benefits, rights, and obligations to each other when you open this package and use the software/data it contains. BE SURE YOU READ THE LICENSE AGREEMENT CAREFULLY, SINCE BY USING THE SOFTWARE/DATA YOU INDICATE YOU HAVE READ, UNDERSTOOD, AND ACCEPTED THE TERMS OF THIS AGREEMENT.

Your rights:

1. You enjoy a non-exclusive license to use the enclosed software/data on a single microcomputer that is not part of a network or multi-machine system in consideration for payment of the required license fee, (which may be included in the purchase price of an accompanying print component), or receipt of this software/data, and your acceptance of the terms and conditions of this agreement.

2. You own the media on which the software/data is recorded, but you acknowledge that you do not own the software/data recorded on them. You also acknowledge that the software/data is furnished "as is," and contains copyrighted and/or proprietary and confidential information of Autodesk Press or its licensors.

3. If you do not accept the terms of this license agreement you may return the media within 30 days. However, you may not use the software during this period.

There are limitations on your rights:

1. You may not copy or print the software/data for any reason whatsoever, except to install it on a hard drive on a single microcomputer and to make one archival copy, unless copying or printing is expressly permitted in writing or statements recorded on the diskette(s).

2. You may not revise, translate, convert, disassemble or otherwise reverse engineer the software/data except that you may add to or rearrange any data recorded on the media as part of the normal use of the software/data.

3. You may not sell, license, lease, rent, loan, or otherwise distribute or network the software/data except that you may give the software/data to a student or and instructor for use at school or, temporarily at home.

Should you fail to abide by the Copyright Law of the United States as it applies to this software/data your license to use it will become invalid. You agree to erase or otherwise destroy the software/data immediately after receiving note of Autodesk Press' termination of this agreement for violation of its provisions.

Autodesk Press gives you a LIMITED WARRANTY covering the enclosed software/data. The LIMITED WARRANTY can be found in this product and/or the instructor's manual that accompanies it.

This license is the entire agreement between you and Autodesk Press interpreted and enforced under New York law.

Limited Warranty

Autodesk Press warrants to the original licensee/ purchaser of this copy of microcomputer software/ data and the media on which it is recorded that the media will be free from defects in material and workmanship for ninety (90) days from the date of original purchase. All implied warranties are limited in duration to this ninety (90) day period. THEREAFTER, ANY IMPLIED WARRANTIES, INCLUDING IMPLIED WARRANTIES OF MERCHANTABILITY AND FITNESS FOR A PARTICULAR PURPOSE ARE EXCLUDED. THIS WARRANTY IS IN LIEU OF ALL OTHER WARRANTIES, WHETHER ORAL OR WRITTEN, EXPRESSED OR IMPLIED.

If you believe the media is defective, please return it during the ninety day period to the address shown below. A defective diskette will be replaced without charge provided that it has not been subjected to misuse or damage.

This warranty does not extend to the software or information recorded on the media. The software and information are provided "AS IS." Any statements made about the utility of the software or information are not to be considered as express or implied warranties. Autodesk Press will not be liable for incidental or consequential damages of any kind incurred by you, the consumer, or any other user.

Some states do not allow the exclusion or limitation of incidental or consequential damages, or limitations on the duration of implied warranties, so the above limitation or exclusion may not apply to you. This warranty gives you specific legal rights, and you may also have other rights which vary from state to state. Address all correspondence to:

Autodesk Press
3 Columbia Circle
P. O. Box 15015
Albany, NY 12212-5015